Graduate Texts in Mathematics 121

Graduate Texts in Mathematics

1 TAKEUTI/ZARING. Introduction to Axiomatic Set Theory. 2nd ed.
2 OXTOBY. Measure and Category. 2nd ed.
3 SCHAEFFER. Topological Vector Spaces.
4 HILTON/STAMMBACH. A Course in Homological Algebra.
5 MAC LANE. Categories for the Working Mathematician.
6 HUGHES/PIPER. Projective Planes.
7 SERRE. A Course in Arithmetic.
8 TAKEUTI/ZARING. Axiomatic Set Theory.
9 HUMPHREYS. Introduction to Lie Algebras and Representation Theory.
10 COHEN. A Course in Simple Homotopy Theory.
11 CONWAY. Functions of One Complex Variable. 2nd ed.
12 BEALS. Advanced Mathematical Analysis.
13 ANDERSON/FULLER. Rings and Categories of Modules.
14 GOLUBITSKY/GUILLEMIN. Stable Mappings and Their Singularities.
15 BERBERIAN. Lectures in Functional Analysis and Operator Theory.
16 WINTER. The Structure of Fields.
17 ROSENBLATT. Random Processes. 2nd ed.
18 HALMOS. Measure Theory.
19 HALMOS. A Hilbert Space Problem Book. 2nd ed., revised.
20 HUSEMOLLER. Fibre Bundles. 2nd ed.
21 HUMPHREYS. Linear Algebraic Groups.
22 BARNES/MACK. An Algebraic Introduction to Mathematical Logic.
23 GREUB. Linear Algebra. 4th ed.
24 HOLMES. Geometric Functional Analysis and its Applications.
25 HEWITT/STROMBERG. Real and Abstract Analysis.
26 MANES. Algebraic Theories.
27 KELLEY. General Topology.
28 ZARISKI/SAMUEL. Commutative Algebra. Vol. I.
29 ZARISKI/SAMUEL. Commutative Algebra. Vol. II.
30 JACOBSON. Lectures in Abstract Algebra I: Basic Concepts.
31 JACOBSON. Lectures in Abstract Algebra II: Linear Algebra.
32 JACOBSON. Lectures in Abstract Algebra III: Theory of Fields and Galois Theory.
33 HIRSCH. Differential Topology.
34 SPITZER. Principles of Random Walk. 2nd ed.
35 WERMER. Banach Algebras and Several Complex Variables. 2nd ed.
36 KELLEY/NAMIOKA et al. Linear Topological Spaces.
37 MONK. Mathematical Logic.
38 GRAUERT/FRITZSCHE. Several Complex Variables.
39 ARVESON. An Invitation to C^*-Algebras.
40 KEMENY/SNELL/KNAPP. Denumerable Markov Chains. 2nd ed.
41 APOSTOL. Modular Functions and Dirichlet Series in Number Theory. 2nd ed.
42 SERRE. Linear Representations of Finite Groups.
43 GILLMAN/JERISON. Rings of Continuous Functions.
44 KENDIG. Elementary Algebraic Geometry.
45 LOÈVE. Probability Theory I. 4th ed.
46 LOÈVE. Probability Theory II. 4th ed.
47 MOISE. Geometric Topology in Dimensions 2 and 3.

Serge Lang

Cyclotomic Fields I and II

Combined Second Edition

With an Appendix by Karl Rubin

Springer Science+Business Media, LLC

Serge Lang
Department of Mathematics
Yale University
New Haven, CT 06520
U.S.A.

Mathematical Subject Classifications (1980): 12A35, 12B30, 12C20, 14G20

Library of Congress Cataloging-in-Publication Data
Lang, Serge, 1927-
Cyclotomic fields I and II (Combined Second Edition)/Serge Lang
p. cm. -- (Graduate texts in mathematics; 121)
Bibliography: p.
Includes index.
ISBN 978-0-387-96671-7

1. Fields, Algebraic. 2. Cyclotomy. I. Title. II. Series
QA247.L33 1990
512'.3--dc19 87-35616

This book is a combined edition of the books previously published as *Cyclotomic Fields* and *Cyclotomic Fields II*, by Springer Science+Business Media, LLC, in 1978 and 1980, respectively. It contains an additional appendix by Karl Rubin.

9 8 7 6 5 4 3 2 1

ISBN 978-0-387-96671-7 ISBN 978-1-4612-0987-4 (eBook)
DOI 10.1007/978-1-4612-0987-4

Contents

APPENDIX BY KARL RUBIN

The Main Conjecture 397

Notation

$\mathbf{Z}(N)$ = integers mod $N = \mathbf{Z}/N\mathbf{Z}$.

If A is an abelian group, we usually denoted by A_N the elements $x \in A$ such that $Nx = 0$. Thus for a prime p, we denote by A_p the elements of order p. However, we also use p in this position for indexing purposes, so we rely to some extent on the context to make the intent clear. In his book, Shimura uses $A[p]$ for the kernel of p, and more generally, if A is a module over a ring, uses $A[\mathfrak{a}]$ for the kernel of an ideal $\mathfrak{a}$ in A. The brackets are used also in other contexts, like operators, as in Lubin–Tate theory. There is a dearth of symbols and positions, so some duplication is hard to avoid.

We let $A(N) = A/NA$. We let $A^{(p)}$ be the subgroup of A consisting of all elements annihilated by a power of p.

Introduction

Kummer's work on cyclotomic fields paved the way for the development of algebraic number theory in general by Dedekind, Weber, Hensel, Hilbert, Takagi, Artin and others. However, the success of this general theory has tended to obscure special facts proved by Kummer about cyclotomic fields which lie deeper than the general theory. For a long period in the 20th century this aspect of Kummer's work seems to have been largely forgotten, except for a few papers, among which are those by Pollaczek [Po], Artin–Hasse [A–H] and Vandiver [Va].

In the mid 1950's, the theory of cyclotomic fields was taken up again by Iwasawa and Leopoldt. Iwasawa viewed cyclotomic fields as being analogues for number fields of the constant field extensions of algebraic geometry, and wrote a great sequence of papers investigating towers of cyclotomic fields, and more generally, Galois extensions of number fields whose Galois group is isomorphic to the additive group of p-adic integers. Leopoldt concentrated on a fixed cyclotomic field, and established various p-adic analogues of the classical complex analytic class number formulas. In particular, this led him to introduce, with Kubota, p-adic analogues of the complex L-functions attached to cyclotomic extensions of the rationals. Finally, in the late 1960's, Iwasawa [Iw 11] made the fundamental discovery that there was a close connection between his work on towers of cyclotomic fields and these p-adic L-functions of Leopoldt–Kubota.

The classical results of Kummer, Stickelberger, and the Iwasawa–Leopoldt theories have been complemented by, and received new significance from the following directions:

1. The analogues for abelian extensions of imaginary quadratic fields in the context of complex multiplication by Novikov, Robert, and Coates–Wiles. Especially the latter, leading to a major result in the direction of the

Birch–Swinnerton-Dyer conjecture, new insight into the explicit reciprocity laws, and a refinement of the Kummer–Takagi theory of units to all levels.

2. The development by Coates, Coates–Sinnott and Lichtenbaum of an analogous theory in the context of *K*-theory.

3. The development by Kubert–Lang of an analogous theory for the units and cuspidal divisor class group of the modular function field.

4. The introduction of modular forms by Ribet in proving the converse of Herbrand's theorem. The connection between cyclotomic theory and modular forms reached a culmination in the work of Mazur–Wiles, who proved the "main conjecture". This is one of the greatest achievements of the modern period of mathematics.

5. The connection between values of zeta functions at negative integers and the constant terms of modular forms starting with Klingen and Siegel, and highly developed to congruence properties of these constant terms by Serre, for instance, leading to the existence of the p-adic L-function for arbitrary totally real fields.

6. The construction of p-adic zeta functions in various contexts of elliptic curves and modular forms by Katz, Manin, Mazur, Vishik.

7. The connection with rings of endomorphisms of abelian varieties or curves, involving complex multiplication (Shimura–Taniyama) and/or the Fermat curve (Davenport–Hasse–Weil and more recently Gross–Rohrlich).

My two volumes on Cyclotomic Fields provided a systematic introduction to the basic theory. No such introduction existed when they first came out. Since then, Washington's book has appeared, covering some of the material but emphasizing different things. As my books went out of print, Springer-Verlag and I decided to continue making them available in a single volume for the convenience of readers. No changes have been made except for some corrections, for which I am indebted to Larry Washington, Neal Koblitz, and others. Thus the book is kept essentially purely cyclotomic, and as elementary as possible, although in a couple of places we use class field theory. No connection is made with modular forms. This would require an entire book by itself. However, in a major development, a purely cyclotomic proof of the "main conjecture", the Mazur–Wiles theorem, has been found, and I am very much indebted to Karl Rubin for having given me an appendix containing a self-contained proof, based on work of Thaine, Kolyvagin and Rubin himself. For details of the history, see Rubin's own introduction to his appendix.

My survey article [L 5] provides another type of introduction to cyclotomic theory. First, at the beginning in §2 it gives a quick and efficient summary of main results, stripped of their proofs which necessarily add bulk. Second, this article is also useful to get a perspective on cyclotomic fields in connection with other topics, for instance having to do with modular curves and elliptic curves. In that survey, I emphasize questions about class groups and unit groups in a broader context than cyclotomic fields. Specifically, in Theorem 4.2 of [L 5] I state how Mazur–

Wiles construct certain class fields (abelian unramified extensions) of cyclotomic fields by means of torsion points on the Jacobians of modular curves. The existence of class fields of certain degrees is predicted abstractly by the pure cyclotomic theory, but the explicit description of the irrationalities generating such class fields provides an additional basic structure. In that sense, the purely cyclotomic proof of the "main conjecture", and even the "main conjecture" itself, do not supersede and are not substitutes for the Mazur–Wiles theory.

The first seven chapters of the present book, together with Chapters 10, 11, 12 and 13 and Rubin's appendix develop systematically the basic structure of units and ideal class groups in cyclotomic fields, or possibly Galois extensions whose Galois group is isomorphic to the group of p-adic integers. We look at the ideal class group in fields such as $\mathbf{Q}(\boldsymbol{\mu}_{p^n})$ where $\boldsymbol{\mu}_{p^n}$ is the group of p^n-th roots of unity. We decompose these groups, as well as their projective limits, into eigenspaces for characters of $(\mathbf{Z}/p\mathbf{Z})^*$, and we attempt to describe as precisely as possible the structure of these eigenspaces. For instance, let h_p denote the class number of $\mathbf{Q}(\boldsymbol{\mu}_p)$. There is already a natural decomposition $h_p = h_p^+ h_p^-$, where h_p^+ is the order of the $(+1)$-eigenspace, and h_p^- is the order of the (-1)-eigenspace for complex conjugation, and similarly for p^n instead of p. Part of the problem is to determine as accurately as possible the p-divisibility of h_p^+ and h_p^-, and also asymptotically for p^n instead of p when $n \to \infty$.

A number of chapters are logically independent of each other. For instance, readers might want to read Chapter 10 on measures and Iwasawa power series immediately after Chapter 4, since the ideas of Chapter 10 are continuations of those of Chapter 4. This leads naturally into the Ferrero–Washington theorems, proving Iwasawa's conjecture that the p-primary part of the ideal class group in the cyclotomic $\mathbf{Z}_p$-extension of a cyclotomic field grows linearly rather than exponentially. This is first done for the minus part (the minus referring, as usual, to the eigenspace for complex conjugation), and then it follows for the plus part because of results bounding the plus part in terms of the minus part. Kummer had already proved such results. Another proof for the Ferrero–Washington theorem was subsequently given by Sinnott [Sin 2].

The first seven chapters suffice for the proof of the "main conjecture" in Rubin's appendix, which does not use the Ferrero–Washington theorem. However, using that theorem in addition gives a clearer picture of the projective limit of the ideal class groups as module over the projective limit of the group rings $\mathbf{Z}_p[G_n]$, where G_n is the Galois group of $\mathbf{Q}(\boldsymbol{\mu}_{p^n})$ over $\mathbf{Q}(\boldsymbol{\mu}_p)$, and therefore also as module over $\mathbf{Z}_p$. This module plays a role analogous to the Jacobian in the theory of curves. The Ferrero–Washington theorem states that up to a finite torsion group, this module is free of finite rank over $\mathbf{Z}_p$. The "main conjecture" gives some description of the characteristic polynomial of a generator for the Galois group playing an analogous role to the Frobenius endomorphism in the theory of curves. Questions then arise whether these characteristic polynomials behave in ways similar to those in the theory of curves over finite fields. These questions pertain both to the nature of these polynomials, e.g. their coefficients and their roots (Riemann type hypotheses); and also concerning the behavior of these polynomials for varying p. Cf. [L 5], p. 274.

After dealing mostly with ideal class groups and units, we turn to a more systematic study of Gauss sums. We do what amounts to "Dwork theory", to derive the Gross–Koblitz formula expressing Gauss sums in terms of the p-adic gamma function. This lifts Stickelberger's theorem p-adically. Half of the proof relies on a course of Katz, who had first obtained Gauss sums as limits of certain factorials, and thought of using Washnitzer–Monsky cohomology to prove the Gross–Koblitz formula.

Finally, we apply these latter results to the Ferrero–Greenberg theorem, showing that $L_p'(0, \chi) \neq 0$ under the appropriate conditions. We take this opportunity to introduce a technique of Washington, who defined the p-adic analogues of the Hurwitz partial zeta functions, in a way making it possible to parallel the treatment from the complex case to the p-adic case, but in a much more efficient way.

Some basic conjectures remain open, notably the Kummer–Vandiver conjecture that h_p^+ is prime to p. The history of that conjecture is interesting. Kummer made it in no uncertain terms in a letter to Kronecker dated 28 December 1849. Kummer first tells Kronecker off for not understanding properly what he had previously written about cyclotomic fields and Fermat's equation, by stating "so liegt hierin ein grosser Irrthum deinerseits ..."; and then he goes on (Collected Works, Vol. 1, p. 84):

> Deine auf dieser falschen Ansicht berühenden Folgerungen fallen somit von selbst weg. Ich gedenke vielmehr den Beweis des Fermatschen Satzes auf folgendes zu grunden:
>
> 1. Auf den noch zu beweisenden Satz, dass es für die Ausnahmszahlen λ stets Einheiten giebt, welche ganzen Zahlen congruent sind für den Modul λ, ohne darum λte Potenzen anderer Einheiten zu sein, oder was dasselbe ist, dass hier niemals D/Δ durch λ theilbar wird.

In our notation: $\lambda = p$ and $D/\Delta = h_p^+$. Kummer wrote D/Δ as a quotient of regulators, expressing the index of the cyclotomic units in the group of all units. This index happens to coincide with h_p^+ (cf. Theorem 5.1 of Chapter 3). Thus Kummer rather expected to prove the conjecture. According to Barry Mazur, who reviewed Kummer's complete works when they were published by Springer-Verlag, Kummer never mentioned the conjecture in a published paper, but he mentioned it once more in another letter to Kronecker on 24 April 1853 (loc cit p. 93):

> Hierein hängt auch zusammen, dass eines meiner Hauperesultate auf welches ich seit einem Vierteljahre gebaut hatte, dass der zweite Faktor der Klassen-

> zahl D/Δ niemals durch λ theilbar ist, falsch ist oder wenigstens unbewiesen ... Ich werde also vorlaufig hauptsachlich meinen Fleiss nur auf die Weiterführung der Theorie der complexen Zahlen wenden, und dann sehen ob etwas daraus entsteht, was auch uber jene Aufgabe Licht verbreitet.

So the situation was less clear than Kummer thought at first. Much later, Vandiver made the same conjecture, and wrote [Va 1]:

> ... However, about twenty-five years ago I conjectured that this number was never divisible by l [referring to h^+]. Later on, when I discovered how closely the question was related to Fermat's Last Theorem, I began to have my doubts, recalling how often conjectures concerning the theorem turned out to be incorrect. When I visited Furtwängler in Vienna in 1928, he mentioned that he had conjectured the same thing before I had brought up any such topic with him. As he had probably more experience with algebraic numbers than any mathematician of his generation, I felt a little more confident ....

On the other hand, many years ago, Feit was unable to understand a step in Vandiver's "proof" that $p \nmid h^+$ implies the first case of Fermat's Last Theorem, and stimulated by this, Iwasawa found a precise gap which is such that there is no proof.

The Iwasawa–Leopoldt conjecture that the p-primary part of C^- is cyclic over the group ring, and is therefore isomorphic to the group ring modulo the Stickelberger ideal, also remains open. For prime level, Leopoldt and Iwasawa have shown that this is a consequence of the Kummer–Vandiver conjecture. Cf. Chapter IV, §4.

Much of the cyclotomic theory extends to totally real number fields, as theorems or conjecturally. We do not touch on this aspect of the question. Cf. Coates' survey paper [Co 3], and especially Shintani [Sh].

Coates, Ribet, and Rohrlich had read the original manuscript and had made a large number of suggestions for improvement. I thank them again, as well as Koblitz and Washington, for their suggestions and corrections.

New Haven, 1989 SERGE LANG

Character Sums 1

Character sums occur all over the place in many different roles. In this chapter they will be used at once to represent certain principal ideals, thus giving rise to annihilators in a group ring for ideal classes in cyclotomic fields.

They also occur as endomorphisms of abelian varieties, especially Jacobians, but we essentially do not consider this, except very briefly in §6. They occur in the computation of the cuspidal divisor class group on modular curves in [KL 6]. The interplay between the algebraic geometry and the theory of cyclotomic fields is one of the more fruitful activities at the moment in number theory.

§1. Character Sums Over Finite Fields

We shall use the following notation.

$F = F_q$ = finite field with q elements, $q = p^n$.
$\mathbf{Z}(N) = \mathbf{Z}/N\mathbf{Z}$.
ε = primitive pth root of unity in characteristic 0. Over the complex numbers, $\varepsilon = e^{2\pi i/p}$.
Tr = trace from F to F_p.
μ_N = group of Nth roots of unity.
$\lambda: F \to \mu_p$ the character of F given by

$$\lambda(x) = \varepsilon^{\operatorname{Tr}(x)}.$$

$\chi: F^* \to \mu_{q-1}$ denotes a character of the multiplicative group.
We extend χ to F by defining $\chi(0) = 0$.

The field $\mathbf{Q}(\mu_N)$ has an automorphism σ_{-1} such that

$$\sigma_{-1}: \zeta \mapsto \zeta^{-1}.$$

If $\alpha \in \mathbf{Q}(\mu_N)$ then the **conjugate** $\bar{\alpha}$ denotes $\sigma_{-1}\alpha$. Over the complex numbers, this is the **complex conjugate**.

The Galois group of $\mathbf{Q}(\mu_N)$ over $\mathbf{Q}$ is isomorphic to $\mathbf{Z}(N)^*$, under the map

$$c \mapsto \sigma_c$$

where

$$\sigma_c : \zeta \mapsto \zeta^c.$$

Let f, g be functions on F with values in a fixed algebraically closed field of characteristic 0. We define

$$S(f, g) = \sum_{x \in F} f(x)g(x).$$

We define the **Fourier transform** Tf by

$$Tf(y) = \sum_{x \in F} f(x)\lambda(-xy) = \sum f(x)\varepsilon^{-\mathrm{Tr}(xy)}.$$

Then Tf is again a function on F, identified with its character group by λ, and T is a linear map.

Theorem 1.1. *Let f^- be the function such that $f^-(x) = f(-x)$. Then $T^2 f = qf^-$, that is*

$$T^2 f(z) = qf(-z).$$

Proof. We have

$$\begin{aligned} T^2 f(z) &= \sum_y \sum_x f(x)\lambda(-yx)\lambda(-zy) \\ &= \sum_x f(x - z) \sum_y \lambda(-yx). \end{aligned}$$

If $x \neq 0$ then $y \mapsto \lambda(yx)$ is a non-trivial character, and the sum of the character over F is 0. Hence this last expression is

$$= qf(-z)$$

as desired.

We define the **convolution** $f * g$ between functions by the formula

$$(f * g)(y) = \sum_x f(x)g(y - x).$$

A change of variables shows that

$$f * g = g * f.$$

Theorem 1.2. *For functions f, g on F we have*

$$T(f * g) = (Tf)(Tg)$$

$$T(fg) = \frac{1}{q} Tf * Tg.$$

Proof. For the first formula we have

$$T(f * g)(z) = \sum_y (f * g)(y)\lambda(-zy) = \sum_y \sum_x f(x)g(y - x)\lambda(-zy).$$

We change the order of summation, let $t = y - x$, $y = x + t$, and find

$$= \sum_x f(x)\lambda(-zx) \sum_t g(t)\lambda(-zt)$$

$$= (Tf)(Tg)(z),$$

thereby proving the first formula.

The second formula follows from the first because T is an isomorphism on the space of functions on F, so that we can write $f = Tf_1$ and $g = Tg_1$ for some functions f_1, g_1. We then combine the first formula with Theorem 1.1 to get the second.

We shall be concerned with the **Gauss sums (Lagrange resolvant)**

$$S(\chi, \lambda) = S(\chi) = \sum_u \chi(u)\lambda(u)$$

where the sum is taken over $u \in F^*$. We could also take the sum over x in F since we defined $\chi(0) = 0$. Since λ is fixed, we usually omit the reference to λ in the notation. The Gauss sums have the following properties.

GS 0. *Let χ_1 be the trivial character* 1 *on F^*. Then*

$$S(\chi_1) = -1.$$

This is obvious from our conventions. It illustrates right at the beginning the pervasive fact, significant many times later, that the natural object to consider is $-S(\chi)$ rather than $S(\chi)$ itself. We shall also write

$$S(1) = S(1, \lambda),$$

but the convention remains in force that even for the trivial character, its value at 0 is 0.

GS 1. *For any character $\chi \neq 1$, we have $T\chi = \chi(-1)S(\chi)\chi^{-1}$.*

Proof. We have

$$T\chi(y) = \sum_x \chi(x)\lambda(-yx).$$

If $y = 0$ then $T\chi(y) = 0$ (summing the multiplicative character over the multiplicative group). If $y \neq 0$, we make a change of variables $x = -ty^{-1}$, and we find precisely the desired value

$$\chi(-1)S(\chi)\chi(y^{-1}).$$

GS 2. *We have* $S(\bar{\chi}) = \chi(-1)\overline{S(\chi)}$ *and for* $\chi \neq 1$, $S(\chi)S(\bar{\chi}) = \chi(-1)q$, *so*

$$S(\chi)\overline{S(\chi)} = q, \quad \textit{for } \chi \neq 1.$$

Proof. Note that $T^2\chi = T(\chi(-1)S(\chi)\chi^{-1}) = S(\chi)S(\chi^{-1})\chi$. But we also know that $T^2\chi = q\chi^{-}$. This proves **GS 2,** as the other statements are obvious.

Over the complex numbers, we obtain the absolute value

$$\boxed{|S(\chi)| = q^{1/2}.}$$

We define the **Jacobi sum**

$$J(\chi_1, \chi_2) = -\sum_x \chi_1(x)\chi_2(1 - x).$$

Observe the minus sign, a most useful convention. We have

$$J(1, 1) = -(q - 2).$$

GS 3. *If* $\chi_1\chi_2 \neq 1$ *then*

$$J(\chi_1, \chi_2) = -\frac{S(\chi_1)S(\chi_2)}{S(\chi_1\chi_2)}.$$

In particular, $J(1, \chi_2) = J(\chi_1, 1) = 1$. *If* $\chi_1\chi_2 = 1$ *but not both* χ_1, χ_2 *are trivial, then*

$$J(\chi_1, \chi_2) = \chi_1(-1).$$

Proof. We compute from the definitions:

$$\begin{aligned} S(\chi_1)S(\chi_2) &= \sum_x \sum_y \chi_1(x)\chi_2(y)\lambda(x + y) \\ &= \sum_x \sum_y \chi_1(x)\chi_2(y - x)\lambda(y) \\ &= \sum_x \sum_{u \neq 0} \chi_1(x)\chi_2(u - x)\lambda(u) + \sum_x \chi_1(x)\chi_2(-x). \end{aligned}$$

If $\chi_1\chi_2 \neq 1$, the last sum on the right is equal to 0. In the other sum, we interchange the order of summation, replace x by ux, and find

$$\sum_u \chi_1\chi_2(u)\lambda(u) \sum_x \chi_1(x)\chi_2(1 - x),$$

thus proving the first assertion of **GS 3.** If $\chi_1\chi_2 = 1$, then the last sum on the right is equal to $\chi_1(-1)(q - 1)$, and the second assertion follows from **GS 2.**

Next we give formulas showing how the Gauss sums transform under Galois automorphisms.

GS 4. $$S(\chi^p) = S(\chi).$$

Proof. Raising to the pth power is an automorphism of F, and therefore

$$\mathrm{Tr}(x^p) = \mathrm{Tr}(x).$$

Thus $S(\chi^p)$ is obtained from $S(\chi)$ by permuting the elements of F under $x \mapsto x^p$. The property is then obvious.

Let m be a positive integer dividing $q - 1$, and suppose that χ has order m, meaning that

$$\chi^m = 1.$$

Then the values of χ are in $\mathbf{Q}(\mu_m)$ and

$$S(\chi) = S(\chi, \lambda) \in \mathbf{Q}(\mu_m, \mu_p).$$

For any integer c prime to m we have an automorphism $\sigma_{c,1}$ of $\mathbf{Q}(\mu_m, \mu_p)$ such that

$$\sigma_{c,1}: \zeta \mapsto \zeta^c \quad \text{and} \quad \sigma_{c,1} \text{ is identity on } \mu_p.$$

For any integer ν prime to p, we have an automorphism $\sigma_{1,\nu}$ such that

$$\sigma_{1,\nu}: \varepsilon \mapsto \varepsilon^\nu \quad \text{and} \quad \sigma_{1,\nu} \text{ is identity on } \mu_m.$$

We can select ν in a given residue class mod p *such that ν is also prime to m. In the sequel we usually assume tacitly that ν has been so chosen, in particular in the next property.*

GS 5. $$\sigma_{c,1}S(\chi) = S(\chi^c) \quad \textit{and} \quad \sigma_{1,\nu}S(\chi) = \bar{\chi}(\nu)S(\chi)$$

Proof. The first is obvious from the definitions, and the second comes by making a change of variable in the Gauss sum,

$$x \mapsto \nu^{-1}x.$$

Observe that

$$\sigma_{1,\nu}\lambda(x) = \varepsilon^{\nu \operatorname{Tr}(x)} = \varepsilon^{\operatorname{Tr}(\nu x)} = \lambda(\nu x).$$

The second property then drops out.

The diagram of fields is as follows.

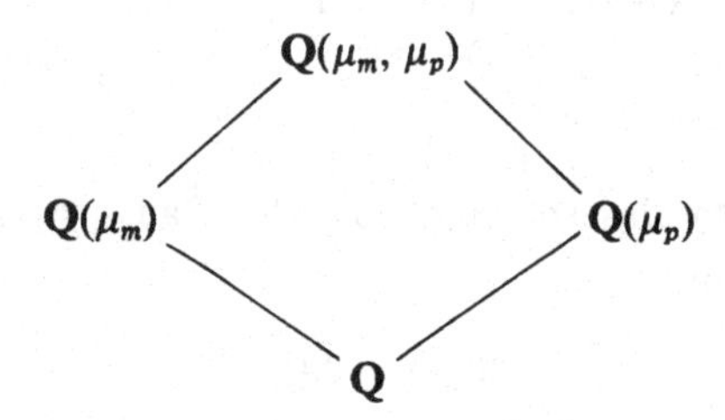

From the action of the Galois group, we can see that the Gauss sum (Lagrange resolvant) satisfies a Kummer equation.

Theorem 1.3. *Assume that χ has order m.*

(i) *$S(\chi)^m$ lies in $\mathbf{Q}(\mu_m)$.*
(ii) *Let b be an integer prime to m, and let $\sigma_b = \sigma_{b,1}$. Then $S(\chi)^{b-\sigma_b}$ lies in $\mathbf{Q}(\mu_m)$.*

Proof. In each case we operate on the given expression by an automorphism $\sigma_{1,\nu}$ with an integer ν prime to pm. Using **GS 5**, it is then obvious that the given expression is fixed under such an automorphism, and hence lies in $\mathbf{Q}(\mu_m)$.

§2. Stickelberger's Theorem

In the first section, we determined the absolute value of the Gauss sum. Here, we determine the prime factorization. We shall first express a character in terms of a canonical character determined by a prime.

Let $\mathfrak{p}$ be a prime ideal in $\mathbf{Q}(\mu_{q-1})$, lying above the prime number p. The residue class field of $\mathfrak{p}$ is identified with $F = F_q$. We keep the same notation as in §1. The equation $X^{q-1} - 1 = 0$ has distinct roots mod p, and hence reduction mod $\mathfrak{p}$ induces an isomorphism

$$\mu_{q-1} \xrightarrow{\approx} F^* = F_q^*.$$

Phrased another way, this means that there exists a unique character ω of F^* such that

$$\omega(u) \bmod \mathfrak{p} = u.$$

This character will be called the **Teichmuller character**. This last equation will also be written in the more usual form

$$\omega(u) \equiv u \pmod{\mathfrak{p}}.$$

The Teichmuller character generates the character group of F^*, so any character χ is an integral power of ω.

We let

$$\pi = \varepsilon - 1.$$

Let $\mathfrak{P}$ be a prime ideal lying above $\mathfrak{p}$ in $\mathbf{Q}(\mu_{q-1}, \mu_p)$. We use the symbol $A \sim B$ to mean that A/B is a unit, or the unit ideal, depending whether A, B are algebraic numbers or (fractional) ideals. We then have

$$\mathfrak{p} \sim \mathfrak{P}^{p-1}$$

because elementary algebraic number theory shows that p is totally ramified in $\mathbf{Q}(\varepsilon)$, and $\mathfrak{p}$ is totally ramified in $\mathbf{Q}(\mu_{q-1}, \mu_p)$.

Let k be an integer, and assume first that $0 \le k < q - 1$. Write the p-adic expansion

$$k = k_0 + k_1 p + \cdots + k_{n-1} p^{n-1}$$

with $0 \le k_i \le p - 1$. We define

$$\boxed{s(k) = k_0 + k_1 + \cdots + k_{n-1}.}$$

For an arbitrary integer k, we define $s(k)$ to be periodic mod $q - 1$, and defined by the above sum in the range first assumed. For convenience, we also define

$$\boxed{\gamma(k) = k_0!\, k_1! \cdots k_{n-1}!}$$

to be the product of the $k_i!$ in the first range, and then also define $\gamma(k)$ by $(q - 1)$-periodicity for arbitrary integers k. If the dependence on q is desired, one could write

$$s_q(k) \quad \text{and} \quad \gamma_q(k).$$

Theorem 2.1. *For any integer k, we have the congruence*

$$\frac{S(\omega^{-k}, \varepsilon^{\mathrm{Tr}})}{(\varepsilon - 1)^{s(k)}} \equiv \frac{-1}{\gamma(k)} \pmod{\mathfrak{P}}.$$

In particular,

$$\operatorname{ord}_{\mathfrak{P}} S(\omega^{-k}) = s(k).$$

Remark. Once more, we see how much more natural the negative of the Gauss sum turns out to be, for we have

$$\frac{-S(\omega^{-k}, \lambda)}{\pi^{s(k)}} \equiv \frac{1}{\gamma(k)} \pmod{\mathfrak{P}}$$

with 1 instead of -1 on the right-hand side.

Proof of Theorem 2.1. If $k = 0$ then the relation of Theorem 2.1 is clear because both sides of the congruence to be proved are equal to -1. We assume $1 \leq k < q - 1$, and prove the theorem by induction. Suppose first that $k = 1$. Then

$$\begin{aligned} S(\omega^{-k}) &= \sum_u \omega(u)^{-1}\varepsilon^{\mathrm{Tr}(u)} \\ &= \sum \omega(u)^{-1}(1 + \pi)^{\mathrm{Tr}(u)} \\ &= \sum \omega(u)^{-1}(1 + (\mathrm{Tr}\, u)\pi + O(\pi^2)) \end{aligned}$$

(interpreting $\mathrm{Tr}\, u$ as an integer in the given residue class mod p). But

$$\begin{aligned} \omega(u)^{-1}\,\mathrm{Tr}(u) &\equiv u^{-1}(u + u^p + \cdots + u^{p^{n-1}}) \bmod \mathfrak{P} \\ &\equiv 1 + u^{p-1} + \cdots + u^{p^{n-1}-1}. \end{aligned}$$

Each $u \mapsto u^{p^j-1}$ is a non-trivial character of F^*. Hence

$$\sum \omega(u)^{-1}\,\mathrm{Tr}(u) \equiv q - 1 \equiv -1 \pmod{\mathfrak{P}}$$

and therefore

$$\frac{S(\omega^{-1})}{\pi} \equiv -1 \pmod{\mathfrak{P}}$$

thus proving the theorem for $k = 1$.

Assume now the result proved for $k - 1$, and write

$$\omega^{-k} = \omega^{-1}\omega^{-(k-1)}$$

for $1 < k < q - 1$. We distinguish two cases.

Case 1. $p|k$, so we can write $k = pk'$ with $1 \leq k' < q - 1$. Then trivially

$$s(k) = s(k') \quad \text{and} \quad \gamma(k) = \gamma(k')$$

because k has the same coefficients k_i as k', shifted only by one index. Let $\sigma_p = \sigma_{p,1}$, so σ_p leaves ε fixed. Since

$$\sigma_p S(\omega^{-k'}) = S(\omega^{-pk'}) = S(\omega^{-k}),$$

we find that applying σ_p to the inductive congruence

$$\frac{S(\omega^{-k'})}{\pi^{s(k')}} \equiv \frac{-1}{\gamma(k')} \pmod{\mathfrak{P}}$$

yields a proof for the present case, because σ_p is in the decomposition group of $\mathfrak{P}$, whence $\sigma_p\mathfrak{P} = \mathfrak{P}$.

Case 2. $p \nmid k$. Then $1 \leq k_0$. Furthermore,

$$s(k) = s(k-1) + 1 \quad \text{and} \quad \gamma(k-1) = (k_0 - 1)!\, k_1! \cdots k_{n-1}!.$$

Then

$$\frac{S(\omega^{-k})}{\pi^{s(k)}} = \frac{S(\omega^{-1}\omega^{-(k-1)})}{\pi^{s(k)}} \equiv \frac{S(\omega^{-1})}{\pi} \frac{S(\omega^{-(k-1)})}{\pi^{s(k-1)}} \frac{-1}{J(\omega^{-1}, \omega^{-(k-1)})}$$

$$\equiv -1 \cdot \frac{-1}{\gamma(k-1)} \frac{-1}{J(\omega^{-1}, \omega^{-(k-1)})} \quad (\text{mod } \mathfrak{P}).$$

To conclude the proof, it will suffices to get the right congruence for J. We use **GS 3** from §1, to get:

$$-J(\omega^{-1}, \omega^{-(k-1)}) \equiv \sum u^{-1}(1-u)^{-(k-1)+q-1} \, (\text{mod } \mathfrak{P}),$$

and the sum is at first taken for $u \neq 0, 1$, but with the additional positive exponent $q - 1$ which does not change anything, we may then suppose that the sum is taken for $u \neq 0$ in F. Hence we get further

$$\equiv \sum_{u \neq 0} \sum_{j=0}^{q-k} (-1)^j \binom{q-k}{j} u^{j-1}.$$

If $j \neq 1$ then $\sum u^{j-1} = 0$, so we get the further congruence

$$-J(\omega^{-1}, \omega^{-(k-1)}) \equiv (-1)(q-k)(q- \,) \equiv -k_0 \quad (\text{mod } \mathfrak{P}),$$

thereby proving the theorem.

Having obtained the order of the Gauss sum at one prime above p, we also want the full factorization. Suppose that m is an integer > 1 and that $p \nmid m$. Let $\mathfrak{p}$ be a prime ideal above p in $\mathbf{Q}(\mu_m)$ and let

$$\mathbf{N}\mathfrak{p} = q = p^n.$$

Let k be an integer such that

$$\frac{k}{q-1} \text{ has order } m \text{ in } \mathbf{Q}/\mathbf{Z}.$$

Let $\langle t \rangle$ denote the smallest real number ≥ 0 in the residue class mod $\mathbf{Z}$ of a real number t. Let

$$G = \mathrm{Gal}(\mathbf{Q}(\mu_m)/\mathbf{Q}).$$

Define the **Stickelberger element** in the rational group ring

$$\boxed{\theta(k, \mathfrak{p}) = \sum_{c \in \mathbf{Z}(m)^*} \left\langle \frac{kc}{q-1} \right\rangle \sigma_c^{-1} \in \mathbf{Q}[G].}$$

Let $\mathfrak{P}$ be the prime ideal in $\mathbf{Q}(\mu_m, \mu_p)$ lying above $\mathfrak{p}$. Let ω as before be the Teichmuller character on F_q^*. We let $\sigma_c = \sigma_{c,1}$.

Theorem 2.2. *We have the factorization*

$$S(\omega^{-k}) \sim \mathfrak{P}^{(p-1)\theta(k,\mathfrak{p})} \sim \mathfrak{p}^{\theta(k,\mathfrak{p})}.$$

Proof. We have

$$\begin{aligned} \operatorname{ord}_{\sigma_c^{-1}\mathfrak{P}} S(\omega^{-k}) &= \operatorname{ord}_{\mathfrak{P}} \sigma_c S(\omega^{-k}) \\ &= \operatorname{ord}_{\mathfrak{P}} S(\omega^{-kc}) \\ &= s(kc) \end{aligned}$$

by Theorem 2.1. On the other hand, the isotropy group of $\mathfrak{p}$ in the Galois group G consists of the powers

$$\{\sigma_{p^i}\} \quad \text{for} \quad i = 0, \ldots, n-1.$$

Hence in the ideal $\mathfrak{p}^{\theta(k)}$ the prime $\sigma_c^{-1}\mathfrak{p}$ occurs with multiplicity

$$\sum_{i=0}^{n-1} \left\langle \frac{kcp^i}{q-1} \right\rangle.$$

Hence to prove Theorem 2.2 it will suffice to prove:

Lemma 1. *For any integer k we have*

$$s(k) = (p-1) \sum_{i=0}^{n-1} \left\langle \frac{kp^i}{q-1} \right\rangle.$$

Proof. We may assume that $1 \le k < q - 1$ since both sides are $(q-1)$-periodic in k, and the relation is obvious for $k = 0$. Since $p^n \equiv 1 \pmod{q-1}$ we find:

$$\begin{aligned} k &= k_0 + k_1 p + \cdots + k_{n-1}p^{n-1} \\ pk &\equiv k_{n-1} + k_0 p + \cdots + k_{n-2}p^{n-1} \pmod{q-1} \\ p^2 k &\equiv k_{n-2} + k_{n-1}p + \cdots + k_{n-3}p^{n-1} \pmod{q-1} \\ &\vdots \end{aligned}$$

Hence

$$\left\langle \frac{kp^i}{q-1} \right\rangle = \frac{\text{right-hand side of } i\text{th equation}}{q-1}.$$

Summing yields

$$\sum_{i=0}^{n-1} \left\langle \frac{kp^i}{q-1} \right\rangle = \frac{s(k)(1 + p + \cdots + p^{n-1})}{q-1} = s(k)\frac{1}{p-1},$$

thereby proving the lemma.

In Theorem 2.2 we note that the Gauss sum is not necessarily an element of $\mathbf{Q}(\mu_m)$, and the equivalence of ideals is true only in the appropriate extension field. Similarly, the Stickelberger element has rational coefficients. By the same procedure, we can both obtain an element in $\mathbf{Q}(\mu_m)$ and a corresponding element in the integral group ring, as follows.

For any integers $a, b \in \mathbf{Z}$ and any real number t, we have

$$b\langle t\rangle - \langle bt\rangle \in \mathbf{Z} \quad \text{and} \quad \langle at\rangle + \langle bt\rangle - \langle (a+b)t\rangle \in \mathbf{Z}.$$

The proof is obvious. Let us define $R = \mathbf{Z}[G]$, and

$$I = \text{ideal of } R \text{ generated by all elements } \sigma_b - b \text{ with } b \text{ prime to } m.$$

Then the above remark shows that

$$I\theta \subset R = \mathbf{Z}[G].$$

Although we won't need it, we may prove the converse for general insight. The matter is analyzed further in Chapter 2, §3.

Lemma 2. *We have $I\theta = R\theta \cap R$.*

Proof. Note that $m \in I$ because

$$m = -(\sigma_{1+m} - (1+m)).$$

Suppose that an element of $R\theta$ lies in R, that is

$$\sum z(b)\sigma_b\theta \in R$$

with $z(b) \in \mathbf{Z}$. Then

$$\sum z(b)\left\langle \frac{bc}{m} \right\rangle \in \mathbf{Z} \quad \text{for all } c$$

whence

$$\sum z(b)b \equiv 0 \pmod{m},$$

and $\sum z(b)b$ is in I. But then

$$\sum z(b)\sigma_b = \sum z(b)(\sigma_b - b) + \sum z(b)b$$

is in I, thus proving the lemma.

It will be convenient to formulate the results in terms of the powers of one character, depending on the integer m. Thus we let

$$\boxed{\chi_{\mathfrak{p}} = \omega_{\mathfrak{p}}^{-(\mathbf{N}\mathfrak{p}-1)/m}}$$

where $\omega_{\mathfrak{p}}$ is the Teichmuller character. We define the **Stickelberger element of level** m by

$$\theta(m) = \sum_{c \in \mathbf{Z}(m)^*} \left\langle \frac{c}{m} \right\rangle \sigma_c^{-1}.$$

As a special case of Theorem 2.2, we then obtain the factorization

FAC 1. $$S(\chi_{\mathfrak{p}}) \sim \mathfrak{p}^{\theta(m)}.$$

Therefore, if b is an integer prime to m, and $\sigma_b = \sigma_{b,1}$, then

FAC 2. $$S(\chi_{\mathfrak{p}})^{b-\sigma_b} \sim \mathfrak{p}^{\theta(m)(b-\sigma_b)}.$$

In **FAC 2** the algebraic number on the left lies in $\mathbf{Q}(\mu_m)$, and the group ring element $\theta(m)(b - \sigma_b)$ lies in $\mathbf{Z}[G]$, namely

$$(b - \sigma_b)\theta(m) = \sum_{c \in \mathbf{Z}(m)^*} \left(b\left\langle \frac{c}{m} \right\rangle - \left\langle \frac{bc}{m} \right\rangle \right) \sigma_c^{-1}.$$

Thus we have the ideal factorization of the $(b - \sigma_b)$-power of the Gauss sum in terms of powers of conjugates of the prime $\mathfrak{p}$ in $\mathbf{Q}(\mu_m)$.

We return later to the application of this factorization to the study of the ideal classes in the cyclotomic field, but it is worth while here to mention the simplest consequence. In every ideal class there exists an ideal prime to m. Since the ideal

$$\mathfrak{p}^{\theta(m)(b-\sigma_b)}$$

is principal for every prime $\mathfrak{p} \nmid m$, we find:

Theorem 2.3. *Let $\mathscr{C}$ be the ideal class group of $\mathbf{Q}(\mu_m)$. Then for all b prime to m,*

$$(b - \sigma_b)\theta(m)$$

annihilates $\mathscr{C}$.

For each integer r let

$$\theta_r(m) = \sum_c \left\langle \frac{rc}{m} \right\rangle \sigma_c^{-1}.$$

We are now allowing r to have common factors with m. Let:

$\mathscr{M}$ = module generated over $\mathbf{Z}$ by all elements θ_r with $r \in \mathbf{Z}$, called the **Stickelberger module**,

$\mathscr{S} = \mathscr{M} \cap R$, called the **Stickelberger ideal**.

Observe that $\mathscr{M}$ is also an R-module.

Theorem 2.4. *The Stickelberger ideal annihilates the ideal class group of* $\mathbf{Q}(\mu_m)$.

Proof. Let

$$\alpha = \sum_r z(r)\theta_r(m) \in R$$

be an element of the Stickelberger ideal, with $z(r) \in \mathbf{Z}$, and the sum taken with only a finite number of coefficients $\neq 0$. Then

$$\sum_r z(r)r \equiv 0 \bmod m.$$

By Theorem 2.2 we have the factorization

$$\prod S(\chi_{\mathfrak{p}}{}^r)^{z(r)} \sim \mathfrak{p}^\alpha,$$

and it is immediately verified that the left-hand side lies in $\mathbf{Q}(\mu_m)$ by using **GS 5** of the preceding section. This proves the theorem.

Next we look at the Jacobi sums. If d is an integer, then d operates in a natural way on $\mathbf{R}/\mathbf{Z}$ by multiplication. We denote this operation by $[d]$. Thus on representatives, we let

$$[d]\langle t\rangle = \langle dt\rangle, \quad t \in \mathbf{R}.$$

It is convenient to let

$$\Delta[a_1, a_2] = [a_1] + [a_2] - [a_1 + a_2].$$

Recall the **Jacobi sum** for $\chi_1\chi_2 \neq 1$:

$$J(\chi_1, \chi_2) = -\frac{S(\chi_1)S(\chi_2)}{S(\chi_1\chi_2)}.$$

Let a_1, a_2 be integers, $a_1 + a_2 \not\equiv 0 \bmod m$. Then from **FAC 1** we get:

FAC 3. $$J(\chi_{\mathfrak{p}}{}^{a_1} \chi_{\mathfrak{p}}{}^{a_2}) \sim \mathfrak{p}^{\Delta[a_1, a_2]\theta(m)},$$

where

$$\Delta[a_1, a_2]\theta(m) = \sum_c \left(\left\langle\frac{a_1c}{m}\right\rangle + \left\langle\frac{a_2c}{m}\right\rangle - \left\langle\frac{(a_1 + a_2)c}{m}\right\rangle\right)\sigma_c^{-1}.$$

and $\Delta[a_1, a_2]\theta(m) \in \mathbf{Z}[G]$ lies in the integral group ring. We know that the Jacobi sum lies in $\mathbf{Q}(\mu_m)$, so again we have an ideal factorization of an element of $\mathbf{Q}(\mu_m)$.

It will be convenient to introduce an abbreviation. Let

$$a = (a_1, a_2)$$

denote a pair of integers. We let

$$\Delta[a_1, a_2]\theta(m) = \theta(m)[a_1, a_2] = \theta(m)[a].$$

In several applications, e.g., in the next section, the level m is fixed, and consequently we omit m from the notation, and write simply

$$\theta(m)[a] = \theta[a].$$

If d is an integer prime to m then trivially

$$\sigma_d\theta[a] = \theta[da].$$

The next two sections are logically independent and can be read in any order. They pursue two different topics begun in §2.

§3. Relations in the Ideal Classes

Let $G = \mathrm{Gal}(\mathbf{Q}(\mu_m)/\mathbf{Q})$, so that elements of G can be written in the form σ_c, with $c \in \mathbf{Z}(m)^*$. We recall the **Stickelberger element**

$$\theta(m) = \sum_c \left\langle \frac{c}{m} \right\rangle \sigma_c^{-1}$$

from formulas **FAC 1** and **FAC 2**. Let

I = ideal of $\mathbf{Z}[G]$ generated by all elements $b - \sigma_b$, with integers b prime to m.

Let p be prime number prime to the Euler function $\phi(m)$. For instance, if $m = p$ itself, the prime p does not divide $p - 1$. The character group on G takes its values in $\phi(m)$th roots of unity. We let $q = p^n$ be a power of p such that $\phi(m)$ divides $q - 1$. We let $\mathfrak{o}_q$ be the ring of p-adic integers in the unramified extension of $\mathbf{Z}_p$ of degree n, so that $\mathfrak{o}_q/p\mathfrak{o}_q = \mathfrak{o}_q(p)$ is the finite field with $p^n = q$ elements. Then $\mathfrak{o}_q$ contains the $\phi(m)$th roots of unity. If $m = p$ then we take $q = p$ and $\mathfrak{o}_q = \mathbf{Z}_p$.

Let $\mathscr{C}$ be the ideal class group of $\mathbf{Q}(\mu_m)$, and $\mathscr{C}^{(p)}$ its p-primary component. We have an isomorphism

$$\mathbf{Z}_p \otimes \mathscr{C}^{(p)} \approx \mathscr{C}^{(p)}.$$

The elementary divisors of $\mathscr{C}^{(p)}$ over $\mathbf{Z}_p$ are the same as the elementary divisors of

$$\mathfrak{o}_q \otimes \mathscr{C}^{(p)} \quad \text{over } \mathfrak{o}_q.$$

If A is an $\mathfrak{o}_q$-ideal, on which G operates, we let $A(\chi)$ be the χ-eigenspace. We let

I_χ = $\mathfrak{o}_q$-ideal generated by all elements $b - \chi(b)$ with integers b prime to m.

By abuse of notation, we write often $\chi(b)$ instead of $\chi(\sigma_b)$. The important special case we shall consider is when $m = p$, in which case it is easy to determine I_χ. We assume $p \geq 3$.

Lemma 1. (i) *If $\chi = \omega$ is the Teichmuller character, then $I_\chi = (p)$.*
(ii) *If χ is non-trivial and not equal to the Teichmuller character, then $I_\chi = (1)$.*

Proof. For (i), we can take an integer b of the form

$$b = \zeta + pu$$

where u is a p-adic unit, and $\zeta = \omega(b)$ is a $(p - 1)$th root of unity. This makes (i) clear, and (ii) is obvious, from the definitions.

In the next sections we shall deal with Bernoulli numbers systematically. For the moment, we need only a special case, so we define *ad hoc* the first **Bernoulli polynomial**

$$\mathbf{B}_1(X) = X - \tfrac{1}{2}$$

and the first **Bernoulli number** $B_1 = -\frac{1}{2}$, its constant term. For any function f on $\mathbf{Z}(m)$ we define

$$B_{1,f} = \sum_{x \in \mathbf{Z}(m)} f(x)\mathbf{B}_1\left(\left\langle \frac{x}{m} \right\rangle\right).$$

In particular,

$$B_{1,\chi} = \sum_{c \in \mathbf{Z}(m)^*} \left(\left\langle \frac{c}{m} \right\rangle - \frac{1}{2}\right)\chi(c).$$

If χ is non-trivial, then $\sum \chi(c) = 0$, and hence in this case,

$$B_{1,\chi} = \sum_c \left\langle \frac{c}{m} \right\rangle \chi(c).$$

Then in the present terminology, Theorem 2.3 can be reformulated as follows.

Theorem 3.1. *For non-trivial χ, the ideal $B_{1,\bar{\chi}} I_\chi$ annihilates $\mathscr{C}^{(p)}(\chi)$.*

Corollary 1. *Assume that $m = p$ is prime ≥ 3. If χ is not equal to the Teichmuller character and is non-trivial, then*

$$\operatorname{ord} B_{1,\bar{\chi}} I_\chi = \operatorname{ord} B_{1,\bar{\chi}}.$$

Proof. Immediate from the lemma and the theorem.

Corollary 2. *If χ is equal to the Teichmuller character then $B_{1,\bar{\chi}} I_\chi = (1)$, and $\mathscr{C}^{(p)}(\chi) = 0$.*

Proof. Mod $\mathbf{Z}_p$, we have the congruence

$$B_{1,\omega^{-1}} = \frac{1}{p} \sum_{c=1}^{p-1} c\omega(c)^{-1} \equiv \frac{1}{p} \sum_{c=1}^{p-1} 1 \equiv \frac{p-1}{p} \quad (\operatorname{mod} \mathbf{Z}_p).$$

Hence $B_{1,\bar{\chi}}$ has a pole of order 1 at p. Lemma 1(i) concludes the proof.

Corollary 3 (Herbrand's theorem). *Assume again that $m = p$. Let $\chi = \omega^{1-k}$, with $2 \leq k \leq p - 2$. If $\mathscr{C}^{(p)}(\chi) \neq 0$, then $p | B_k$, where B_k is the* kth *Bernoulli number.*

Proof. In the next chapter Theorem 2.5, we shall prove the congruence

$$\frac{1}{n} B_{n,\omega^{k-n}} \equiv \frac{1}{k} B_k \pmod{p}$$

for k in the given range, and any positive integer n. By Corollary 1, we know that $B_{1,\bar{\chi}}$ annihilates $\mathscr{C}^{(p)}(\chi)$, and

$$B_{1,\bar{\chi}} = B_{1,\omega^{k-1}} \equiv \frac{1}{k} B_k \pmod{p}.$$

If p does not divide B_k, it follows that $B_{1,\bar{\chi}}$ is a p-unit, whence $\mathscr{C}^{(p)}(\chi) = 0$, thus proving Herbrand's theorem.

The converse of Herbrand's theorem has been proved by Ribet [Ri]. For analogues on the modular curves, see the [KL] series, especially [KL 6].

The reader interested in pursuing the ideas of this section may skip the rest of this chapter, read the first section of Chapter 2, and then go to Chapter 5.

§4. Jacobi Sums as Hecke Characters

Let ζ throughout this section be a fixed primitive mth root of unity. We consider the additive group

$$\mathbf{Z}(m)^{(2)} = \mathbf{Z}(m) \times \mathbf{Z}(m),$$

of order m^2. Its elements will be denoted by

$$a = (a_1, a_2), \qquad b = (b_1, b_2).$$

The dot product is the usual one, $a \cdot b = a_1 b_1 + a_2 b_2$. For any function f on $\mathbf{Z}(m)^{(2)}$ we have its **Fourier transform** $\hat{f}$, and the inversion formulas:

$$(*) \qquad f(a) = \sum_b \hat{f}(b)\zeta^{b \cdot a}$$

$$(**) \qquad \hat{f}(b) = \frac{1}{m^2} \sum_a f(a)\zeta^{-b \cdot a},$$

whose verifications are simple exercises.

For any prime ideal $\mathfrak{p}$ in $\mathbf{Q}(\mu_m)$ not dividing m, and $a \in \mathbf{Z}(m)^{(2)}$ we define

$$J(a, \mathfrak{p}) = J(\chi_{\mathfrak{p}}^{a_1}, \chi_{\mathfrak{p}}^{a_2}).$$

We extend the definition to fractional ideals of $\mathbf{Q}(\mu_m)$ prime to m by multiplicativity, thus defining $J(a, \mathfrak{a})$ for all $\mathfrak{a}$ prime to m. We have:

J 0. $$J(0, \mathfrak{p}) = -(\mathbf{N}\mathfrak{p} - 2).$$

We get $J(0, \mathfrak{a})$ by multiplicativity. We also need the congruence

J 1. $$J(0, \mathfrak{a})\mathbf{N}\mathfrak{a} \equiv 1 \bmod m^2.$$

By multiplicativity it suffices to prove it for prime ideals. In that case it is immediate, since m divides $\mathbf{N}\mathfrak{p} - 1$, and by **J 0**,

$$-(\mathbf{N}\mathfrak{p} - 2)\mathbf{N}\mathfrak{p} = 1 - (1 - \mathbf{N}\mathfrak{p})^2.$$

If a_1, or a_2, or $a_1 + a_2 \equiv 0 \bmod m$, then we shall say that a is **special**. Otherwise we say that a is **non-special**. The absolute value of the Gauss sum determined in **GS 2** immediately implies a corresponding result for the Jacobi sum, namely:

J 2. $$J(a, \mathfrak{a})\overline{J(a, \mathfrak{a})} = \mathbf{N}\mathfrak{a} \quad \text{if } a \text{ is non-special.}$$

If a is special, $a \neq 0$, note that $J(a, \mathfrak{a}) = 1$ or -1. In all cases, we have

J 3. $$J(a, \mathfrak{p}) = -\sum_u \chi_{\mathfrak{p}}^{a_1}(u)\chi_{\mathfrak{p}}^{a_2}(1 - u) = \sum_b \hat{J}(b, \mathfrak{p})\zeta^{b \cdot a}$$

where the Fourier coefficient $-\hat{J}(b, \mathfrak{p})$ is the number of solutions u of the equations

$$\chi_{\mathfrak{p}}(u) = \zeta^{b_1} \quad \text{and} \quad \chi_{\mathfrak{p}}(1 - u) = \zeta^{b_2}.$$

By multiplicativity, it follows that the Fourier coefficients $\hat{J}(b, \mathfrak{a})$ are integers for arbitrary $\mathfrak{a}$, that is

$$\hat{J}(b, \mathfrak{a}) \in \mathbf{Z}.$$

For the rest of this section, it will be convenient to assume that all number fields are contained in the complex numbers.

We have seen that $\theta[a]$ is in the integral group ring $\mathbf{Z}[G]$. For any non-zero element $\alpha \in \mathbf{Q}(\mu_m)$, we let

$$\begin{aligned} w(a, \alpha) &= J(a, (\alpha))\alpha^{-\theta[a]} && \text{if } a \text{ is non-special,} \\ w(a, \alpha) &= J(a, (\alpha)) && \text{if } a \text{ is special, } a \neq 0 \\ w(0, \alpha) &= 1. \end{aligned}$$

As usual, (α) is the principal (fractional) ideal generated by α.

If d is an integer prime to m, then trivially from **GS 5**,

$$\sigma_d J(a, \mathfrak{a}) = J(da, \mathfrak{a}) \quad \text{and} \quad \sigma_d w(a, \alpha) = w(da, \alpha).$$

Theorem 4.1. *The algebraic number $w(a, \alpha)$ is a root of unity.*

Proof. As (α) ranges over all principal fractional ideals, the numbers $w(a, \alpha)$ form a group. It will therefore suffice to prove that these numbers have absolute value 1, for then their conjugates also have absolute value 1, and these numbers form a finite group. In case a is special the theorem is true by definition. Otherwise we can use **J 2**, so that

$$J(a, (\alpha))\overline{J(a, (\alpha))} = \mathbf{N}\alpha.$$

On the other hand, the product of $\alpha^{\theta[a]}$ and its conjugate is equal to $\mathbf{N}\alpha$ under the hypothesis that $a_1 + a_2 \not\equiv 0 \bmod m$. Indeed, we have

$$\begin{aligned} \theta[a] + \theta[-a] = \sum \left(\left\langle \frac{a_1 c}{m} \right\rangle + \left\langle \frac{a_2 c}{m} \right\rangle - \left\langle \frac{(a_1 + a_2)c}{m} \right\rangle \right) \sigma_c^{-1} \\ + \sum \left\langle \left(\frac{-a_1 c}{m} \right\rangle + \left\langle \frac{-a_2 c}{m} \right\rangle - \left\langle \frac{-(a_1 + a_2)c}{m} \right\rangle \right) \sigma_c^{-1}. \end{aligned}$$

If t is a real number and not an integer, then

$$\langle t \rangle + \langle -t \rangle = 1,$$

and

$$\sum_{c \in \mathbf{Z}(m)^*} \sigma_c^{-1}$$

operates multiplicatively like the absolute norm. The desired relation for the product of $\alpha^{\theta[a]}$ and its conjugate follows at once. The theorem follows by using **J 2**, the analogous relation for the Jacobi sums.

The next theorem was proved originally by Eisenstein for prime level, and by Weil [We 2] in the general case, which we follow.

Theorem 4.2. *If α is an algebraic integer in $\mathbf{Q}(\mu_m)$, and $\alpha \equiv 1 \pmod{m^2}$ then for all a we have $w(a, \alpha) = 1$, so for a non-special,*

$$J(a, (\alpha)) = \alpha^{\theta[a]}.$$

Proof. We fix α and view J, w as functions of a, omitting α from the notation. In the Fourier inversion relation, we know that the Fourier coefficients $\hat{J}(b)$ are integers. But $\alpha \equiv 1 \pmod{m^2}$ implies that

$$w(a) \equiv J(a) \pmod{m^2}.$$

This is obvious from the definition if $a \neq 0$, and follows at once from **J 1** if $a = 0$. Hence $\hat{w}(b)$ is an algebraic integer for all b. Furthermore, for d prime to m,

$$\begin{aligned} \sigma_d \hat{w}(b) &= \frac{1}{m^2} \sum_a \sigma_d w(a) \zeta^{-da \cdot b} \\ &= \frac{1}{m^2} \sum_a w(da) \zeta^{-da \cdot b} \\ &= \hat{w}(b). \end{aligned}$$

It follows that $\hat{w}(b) \in \mathbf{Z}$ for all b. Now by the Plancherel formula,

$$\sum_b |\hat{w}(b)|^2 = \frac{1}{m^2} \sum_a |w(a)|^2.$$

Since we know that $|w(a)|^2 = 1$, and $\hat{w}(b)$ is an integer for all b, it follows that $\hat{w}(b) \neq 0$ for a single value of b, and is 0 for all other values of b. In particular, for this special b,

$$w(a) = \hat{w}(b) \zeta^{b \cdot a}.$$

But $w(0) = 1$, so $\hat{w}(b) = 1$. Putting $a = (1, 0)$ and $a = (0, 1)$ we get:

$$w(1, 0) = J(1, 0) = 1 \quad \text{and} \quad w(1, 0) = \zeta^{b_1}$$

$$w(0, 1) = J(0, 1) = 1 \quad \text{and} \quad w(0, 1) = \zeta^{b_2}.$$

It follows that

$$w(a) = 1$$

for all a, thus proving the theorem.

§5. Gauss Sums Over Extension Fields

We prove in this section a theorem of Davenport–Hasse [D–H].

Theorem 5.1. *Let $F = F_q$ be the finite field with q elements, and let E be a finite extension. Let*

$$T_{E/F} \quad \text{and} \quad N_{E/F}$$

be the trace and norm from E to F. Let

$$\chi_E = \chi \circ N_{E/F} \quad \text{and} \quad \lambda_E = \lambda \circ T_{E/F}.$$

Then

$$-S_E(\chi_E, \lambda_E) = (-S(\chi, \lambda))^{[E:F]}.$$

Proof. Let $m = [E : F]$. For any polynomial

$$f(X) = X^n + c_1 X^{n-1} + \cdots + c_0$$

with coefficients in F, define

$$\psi(f) = \lambda(c_1)\chi(c_0).$$

Then

$$\psi\colon \text{Monic polynomials of degree} \geq 1 \text{ over } F \to F$$

is a homomorphism, i.e., satisfies

$$\psi(fg) = \psi(f)\psi(g).$$

We write $n(f) = \deg f$. From unique factorization we have the formula

$$1 + \sum_f \psi(f) X^{n(f)} = \prod_P \frac{1}{1 - \psi(P) X^{n(P)}}$$

where the product is taken over all monic irreducible polynomials over F.
Suppose f is of degree 1, say $f(X) = X + c$. Then we see that

$$\sum_{n(f)=1} \psi(f) X^{n(f)} = S(\chi, \lambda) X.$$

On the other hand, if $n \geq 2$ we have

$$\sum_{n(f)=n} \psi(f) X^n = 0.$$

Indeed,

$$\sum_{n(f)=n} \psi(f) = q^{n-2} \sum_{c_1} \lambda(c_1) \sum_{c_0} \chi(c_0),$$

and the sum over c_1 in F on the right is 0, as desired.

Therefore we find

$$(1) \qquad 1 + S(\chi, \lambda)X = \prod_P \frac{1}{1 - \psi(P)X^{n(P)}}.$$

Mutatis mutandis, using the variable X^m instead of X, we get

$$(2) \qquad 1 + S_E(\chi_E, \lambda_E)X^m = \prod_Q \frac{1}{1 - \psi_E(Q)X^{mn(Q)}}.$$

where the product is taken over all monic irreducible polynomials Q over E, and

$$\psi_E(Q) = \chi_E(c_0(Q))\lambda_E(c_1(Q)).$$

We shall write the product over Q as

$$\prod_Q = \prod_P \prod_{Q|P}.$$

Each irreducible polynomial P splits in E into a product

$$P = Q_1 \cdots Q_r.$$

Let $n = n(P) = \deg P$. Then

$$\deg Q = n/r.$$

If α is any root of P, then $[F(\alpha):F] = n$ and the field $F(\alpha)$ is independent of the chosen root. We have the following lattice of fields.

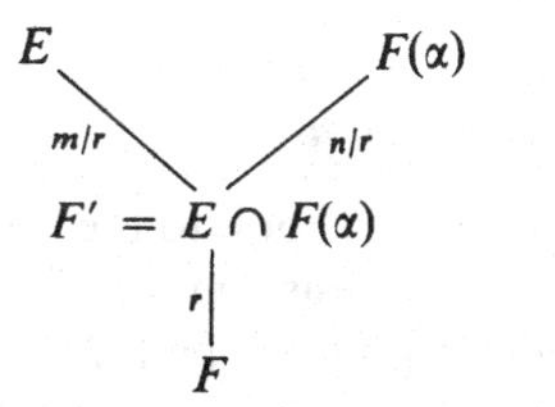

All the polynomials Q_i are conjugate over F, and their coefficients generate the field $F' = E \cap F(\alpha)$, of degree r over F. We have

$$r = (m, n).$$

These facts are all obvious from elementary field theory. Since

$$N_{E/F} = N_{F'/F} \circ N_{E/F'}, \qquad T_{E/F} = T_{F'/F} \circ T_{E/F'},$$

and

$$N_{F'/F} c_0(Q) = c_0(P), \qquad T_{F'/F} c_1(Q) = c_1(P),$$

we get

$$\begin{aligned}\psi_E(Q) &= (\chi(c_0(P))\lambda(c_1(P)))^{[E:F']} \\ &= \psi(P)^{m/r}.\end{aligned}$$

With a view towards (2), we conclude that

$$(3) \qquad \begin{aligned}\prod_{Q|P}(1 - \psi_E(Q)X^{mn(Q)}) &= (1 - \psi(P)^{m/r}X^{mn/r})^r \\ &= \prod_{\zeta^{m/r}=1}(1 - \psi(P)\zeta X^n)^r \\ &= \prod_{\zeta^m=1}(1 - \psi(P)(\xi X)^n).\end{aligned}$$

For this last step, we observe that the map

$$\xi \mapsto \xi^n$$

gives a surjection of $\mu_m \to \mu_{m/r}$, and the inverse image of any element of $\mu_{m/r}$ is a coset of μ_r since $r = (m, n)$. This makes the last step obvious.

Substituting (3) in (2), we now find

$$\begin{aligned}1 + S_E(\chi_E, \lambda_E)X^m &= \prod_{\xi^m=1}\prod_P \frac{1}{(1 - \psi(P)(\xi X)^{n(P)})} \\ &= \prod_{\xi^m=1}(1 + S(\chi, \lambda)\xi X) \\ &= 1 + (-1)^{m+1}S(\chi, \lambda)^m X^m.\end{aligned}$$

This proves the theorem.

§6. Application to the Fermat Curve

Although we do not return in this book to the applications of Gauss sums to algebraic geometry, we cannot resist giving the application of Davenport–Hasse [D–H], Hua–Vandiver [Hu–V], and Weil [We 1], [We 2], [We 3] to the computation of the zeta function of a Fermat curve.

We keep things to their simplest case, the method applies much more generally. We consider the Fermat curve $V = V(d)$ defined by

$$x^d + y^d + z^d = 0,$$

with $d \geq 2$, defined over a finite field F with q elements. Again for simplicity, we suppose that d divides $q - 1$, and therefore dth roots of unity are contained in F.

We let $\omega: F^* \to \mu_{q-1}$ be the Teichmuller character, and

$$\chi = \text{character such that } \chi(u) = \omega(u)^{(q-1)/d}.$$

If a is an integer mod d, we let $\chi^a(u)$ have the usual value if $u \neq 0$, and for $u = 0$ we let:

$$\chi^a(0) = 1 \quad \text{if } a = 0,$$
$$\chi^a(0) = 0 \quad \text{if } a \neq 0.$$

For u in F, we let:

$$N_d(u) = \text{number of solutions } x \in F \text{ such that } x^d = u.$$

Then

$$N_d(u) = \begin{cases} 1 & \text{if } u = 0 \\ 0 & \text{if } u \neq 0,\ u \text{ is not } d\text{th power in } F \\ d & \text{if } u \neq 0,\ u \text{ is } d\text{th power in } F. \end{cases}$$

Therefore

$$N_d(u) = \sum_{a \bmod d} \chi^a(u).$$

Theorem 6.1. *Let N be the number of points of $V(d)$ (in affine space) in the field F. Then*

$$N = q^2 - (q - 1) \sum \chi^{a+b}(-1)J(\chi^a, \chi^b).$$

The sum is taken over integers a, b satisfying $0 < a < d$ and $0 < b < d$, and $a + b \not\equiv 0 \pmod{d}$.

Proof. We have

$$N = \sum_{a,b,c} \sum_{L(u,v,w)=0} \chi^a(u)\chi^b(v)\chi^c(w)$$

where the sum over u, v, w is taken over triples of elements of F lying on the line

$$u + v + w = 0.$$

The sum over a, b, c is taken over elements in $\mathbf{Z}$ mod d.

The term for which $a = b = c = 0$ yields a contribution of q^2, that is the number of points on the line in F.

Next, suppose that in the remaining sum, one of a, b, c is 0 but not all are 0 in $\mathbf{Z}/d\mathbf{Z}$. Say $a = 0$ but $b \neq 0$. Then we may write the sum

$$\sum_{u+v+w=0} = \sum_{\text{certain } u,w} \chi^a(u)\chi^c(w) \sum_{\text{all } v \in F} \chi^b(v),$$

and the sum on the far right is 0. This shows that all the terms in the sum with one, but not all, of a, b, c equal to 0 give a contribution 0. Hence we get

$$N = q^2 + \sum_{0<a,b,c<d} \sum_{u+v+w=0} \chi^a(u)\chi^b(v)\chi^c(w)$$

where the sum over a, b, c is taken over positive integers satisfying the indicated inequality.

If $w = 0$ then $\chi^c(w) = 0$. We may therefore assume that in the inner sum, we have $w \neq 0$. We then put

$$u = u'w \quad \text{and} \quad v = v'w.$$

The inner sum then has the form

$$\sum_{w \neq 0} \chi^{a+b+c}(w) \sum_{u'+v'=-1} \chi^a(u')\chi^b(v').$$

If $a + b + c \not\equiv 0 \bmod d$, then the sum on the left is 0. Otherwise it is $q - 1$, which we assume from now on. Since $0 < a, b, c < d$, there is no such triple (a, b, c) with $a + b \equiv 0 \bmod d$, because any accompanying c would have to equal d. Hence the sum over a, b, c is for $a + b \not\equiv 0 \bmod d$, and then c is uniquely determined. Changing back the variables u', v' to $u'' = -u'$, $v'' = -v'$ and taking into account the value of the Jacobi sum yields the expression as stated in the theorem.

Let $\bar{N}$ be the number of points of $V(d)$ in projective space in the field F. Then

$$N = 1 + (q - 1)\bar{N}.$$

Therefore we obtain:

Corollary. $$\bar{N} = 1 + q - \sum \alpha_{a,b}$$

where $\alpha_{a,b} = \chi^{a+b}(-1)J(\chi^a, \chi^b)$, *and* (a, b) *are as in Theorem* 6.1.

Let $\bar{N}_\nu$ be the number of points of $V(d)$ in projective space over the field F_ν of degree ν over F. The theorem applied to F_ν instead of F yields an analogous expression, the character χ being replaced by χ_ν such that for $u \in F_\nu$,

$$\chi_\nu(u) = \omega(u)^{(q^\nu-1)/d} = \omega(u)^{(q^\nu-1)(q-1)/(q-1)d} = \omega(u^{1+q+\cdots+q^{\nu-1}})^{(q-1)/d}.$$

This last expression is nothing but χ composed with the norm map, in other words, it is precisely the character lifted to the extension as in the preceding section. The additive character is also lifted in a similar fashion. Therefore by Theorem 5.1 we find

$$\boxed{\bar{N}_\nu = 1 + q^\nu - \sum \alpha_{a,b}^\nu.}$$

Note that the power of $\chi(-1)$ also behaves in the same way as J when lifted to F_ν. Indeed, if q is odd then

$$1 + q + \cdots + q^{\nu-1} \equiv \nu \bmod 2,$$

and if q is even, then $1 = -1$ in F.

The **zeta function** $Z(V, T)$ is defined by the conditions

$$Z'/Z(T) = -\sum \bar{N}_\nu T^{\nu-1} \quad \text{and} \quad Z(0) = 1.$$

It is then immediate that

$$Z(V(d), T) = \frac{\prod (1 - \alpha_{a,b} T)}{(1 - T)(1 - qT)}.$$

This is best seen by taking the logarithmic derivative of the last expression on the right-hand side. The operator

$$f \mapsto f'/f$$

is a homomorphism, so we take the operator for each linear term. Inverting a geometric series we see that the logarithmic derivative of the last expression on the right-hand side has precisely the power series

$$\sum \bar{N}_\nu T^{\nu-1}.$$

Since it has the value 1 at $T = 0$, it is the unique function having the desired properties.

If finally one starts with the Fermat curve defined over the field of dth roots of unity, and one reduces mod primes $\mathfrak{p}$ not dividing d, one can take the product of the zeta functions for the reduced curve over the corresponding finite field. Then as Weil remarked, since the Jacobi sums are Hecke characters, it follows that the Hasse zeta function

$$\zeta(V(d), s) = \prod_{\mathfrak{p} \nmid d} Z(V(d), N\mathfrak{p}^{-s})$$

is equal to a Hecke L-series (up to the obvious factors of the zeta function of $\mathbf{Q}(\mu_d)$ at s and $s - 1$).

The computation of solutions in finite fields works in essentially the same way for diagonal equations

$$a_1 x_1^{d_1} + \cdots + a_r x_r^{d_r} = 0,$$

as in Hua–Vandiver [Hu–V] and Weil [We 1, 2, 3]. The additional connection with the Hasse zeta function for the curve over number fields was made by Weil.

2 Stickelberger Ideals and Bernoulli Distributions

The study of ideal classes or units in cyclotomic fields, or number fields (Iwasawa, Leopoldt), of divisor classes on modular curves (e.g., as in [KL]), of higher K-groups (Coates–Sinnott [Co 1], [Co 2], [C–S]) has led to purely algebraic theorems concerned with group rings and certain ideals, formed with Bernoulli numbers (somewhat generalized, as by Leopoldt). Such ideals happen to annihilate these groups, but in many cases it is still conjectural that the groups in question are isomorphic to the factor group of the group ring by such ideals.

However, it is possible to study these ideals, the structure of their factor group, and the orders of the factor groups in the group ring, without any allusion to the applications to ideal classes, divisors, or units. This chapter gives the foundations for such study, applicable to many contexts.

The first section gives Iwasawa's computation of the index of the Stickelberger ideal for $k = 1$, directly applicable to the ideal class group in cyclotomic fields. Next we deal with the basic theory of Bernoulli numbers and polynomials, and especially integrality theorems of Mazur and Coates–Sinnott. The sections concerning Stickelberger ideals for $k \geq 2$ are taken from Kubert–Lang [KL 8]. The last sections on distribution relations are from [KL 5] and Kubert [Ku].

For a discussion of conjectures in the case of totally real number fields, cf. Coates [Co 3], [Co 4], and the very general conjectures in Coates–Lichtenbaum [C–L].

The present chapter is organized so that a reader interested especially in the structure of the ideal class group in the cyclotomic tower (the basic substantial example of the theory) can read the first section, and then can go immediately to Chapter 3, followed by Chapter 5 without impairing the logical understanding of the material. I followed this pattern when I taught the course in 1977.

On the other hand, a reader especially eager to get into p-adic L-functions can concentrate on this chapter and then read Chapter 4 as a continuation omitting Chapter 3. Only the section on the p-adic regulator in Chapter 4 is related to Chapter 3. Chapter 2 may then be interpreted as giving the basic congruence properties of Bernoulli distributions, and Chapter 4 gives essentially more (p-adically) global measure theoretic properties.

A third alternative is to see Chapters 3 and 4 as forming a pair, describing side by side the complex and p-adic class number and regulator formulas originally conceived by Leopoldt.

§1. The Index of the First Stickelberger Ideal

Let $G \approx \mathbf{Z}(m)^*$ be the Galois group of $\mathbf{Q}(\mu_m)$, and assume that m is the conductor of that field, so that $m > 1$, m is odd, or m is divisible by 4. We let

$$M = \tfrac{1}{2} \text{ order of } G = \tfrac{1}{2}\phi(m).$$

We let

$$R = \mathbf{Z}[G], \qquad \varepsilon^- = \tfrac{1}{2}(1 - \sigma_{-1}), \qquad \varepsilon^+ = \tfrac{1}{2}(1 + \sigma_{-1}).$$

For any G-module, we let A^- be the (-1)-eigenspace for σ_{-1}. Then multiplication by ε^- is the projection operator on this eigenspace (provided 2 is invertible), and ε^- is the associated idempotent in the group algebra.

Lemma 1. *We have* $R^- = 2\varepsilon^- R = (1 - \sigma_{-1})R$ *and*

$$(\varepsilon^- R : R^-) = 2^M.$$

Proof. The inclusion $(1 - \sigma_{-1})R \subset R^-$ is clear. Conversely, let P be a set of representatives in $\mathbf{Z}(m)^*$ for $\mathbf{Z}(m)^*/\pm 1$. Let

$$\alpha = \sum z(c)\sigma_c^{-1} \in R^-$$

with coefficients $z(c) \in \mathbf{Z}$. Thus $\sigma_{-1}\alpha = -\alpha$. Then $z(-c) = -z(c)$. If we let

$$\beta = \sum_{c \in P} z(c)\sigma_c^{-1},$$

then $\alpha = (1 - \sigma_{-1})\beta$, thereby proving the lemma, because $\varepsilon^- R$ is a free abelian group of rank M.

We recall the **primitive Stickelberger element**

$$\theta' = \sum_{c \in \mathbf{Z}(m)^*} \left\langle \frac{c}{m} \right\rangle \sigma_c^{-1}.$$

We have written θ' instead of θ because we are now setting more permanent notation, and there is a more canonical element which has priority, namely

$$\theta = \sum \left(\left\langle \frac{c}{m} \right\rangle - \frac{1}{2} \right) \sigma_c^{-1} = \sum \mathbf{B}_1\left(\left\langle \frac{c}{m} \right\rangle \right) \sigma_c^{-1}.$$

It is immediately verified that

$$(*) \qquad \varepsilon^- \theta' = \theta, \quad \text{and so} \quad \theta = \theta^-.$$

We are interested in $R\theta \cap R$. The next lemma does away with a possible alternative definition of this ideal.

Lemma 2. $\quad R\theta \cap R = (R\theta' \cap R)^-$.

Proof. Let $T = R\theta' \cap R$. Clearly

$$T^- \subset \varepsilon^- R\theta = R\theta \quad \text{and} \quad T^- \subset R,$$

so the inclusion $\supset$ is obvious. Conversely, let $\alpha \in R\theta \cap R$. It will suffice to prove that $\alpha \in R\theta'$ (because $\alpha \in R$ and $\alpha = \alpha^-$). Write

$$\alpha = \sum z(b)\sigma_b\theta = \sum_c \sum_b z(b)\left(\left\langle \frac{cb}{m} \right\rangle - \frac{1}{2} \right) \sigma_c^{-1}.$$

From the hypothesis that α has integral coefficients, we conclude that

$$\sum_b z(b)\left(\frac{bc}{m} - \frac{1}{2} \right) \equiv 0 \pmod{\mathbf{Z}}$$

for all c prime to m, so that

$$\frac{1}{m} \sum_b z(b)b \equiv \frac{1}{2} \sum_b z(b) \pmod{\mathbf{Z}}.$$

We contend that

$$\sum z(b)b \equiv 0 \pmod{m} \quad \text{and} \quad \sum z(b) \equiv 0 \pmod{2}.$$

This is obvious if m is odd. Suppose m even, so m is divisible by 4. Write $m = 4m_0$. Each b is odd, and

$$\sum z(b)b \equiv 0 \pmod{2m_0}$$

so $\sum z(b)$ is even. Then

$$\sum z(b)b \equiv \frac{m}{2} \sum z(b) \pmod{m\mathbf{Z}},$$

thus proving also the first congruence. Only the second will be used.

Now let $s(G) = \sum \sigma$ be the sum of the elements of G in the group ring, and note that

$$\varepsilon^+\theta' = \tfrac{1}{2}s(G) \quad \text{and} \quad (1 + \sigma_{-1})\theta' = s(G).$$

Then

$$\begin{aligned}\alpha = \sum z(b)\sigma_b\varepsilon^-\theta' &= \sum z(b)\sigma_b(1 - \varepsilon^+)\theta' \\ &= \sum z(b)\sigma_b\theta' - \sum z(b)\sigma_b\varepsilon^+\theta' \\ &= \sum z(b)\sigma_b\theta' - \sum z(b)\tfrac{1}{2}s(G).\end{aligned}$$

Substituting $s(G) = (1 + \sigma_{-1})\theta'$ on the right and using $\sum z(b)$ even shows that α lies in $R\theta'$, and concludes the proof.

It is of interest to determine the index arising from Lemma 2. This is done in the next lemma. We let as usual:

$$w = \text{number of roots of unity in } \mathbf{Q}(\mu_m).$$

Lemma 3. $$(R\theta : R\theta \cap R) = w.$$

Proof. We define a homomorphism

$$T: R\theta \to \frac{1}{w}\mathbf{Z}/\mathbf{Z}$$

by mapping an element of the group algebra on its first coefficient mod $\mathbf{Z}$. In other words, if

$$\alpha = \sum a(c)\sigma_c,$$

we let $T\alpha = a(1)$. Note that

$$T(\theta) \equiv \frac{1}{m} - \frac{1}{2} \pmod{\mathbf{Z}},$$

and therefore that T is surjective. It now suffices to prove that its kernel is $R\theta \cap R$. But we have

$$\sigma_b\alpha\theta \equiv b\alpha\theta \pmod{R},$$

whence for odd b prime to m, and $\alpha \in R$, we get

$$T(\sigma_b\alpha\theta) \equiv bT(\alpha\theta) \pmod{\mathbf{Z}}.$$

If $\alpha\theta$ is in the kernel of T, it follows that $\alpha\theta$ also lies in R, thereby proving the lemma.

We now assume that $m = p^n$ is a prime power. Then

$$\mathscr{S} = R\theta \cap R$$

is called the **Stickelberger ideal**. We want to determine the index

$$(R^- : \mathscr{S}).$$

Define

$$B_{1,\chi} = \sum_{x \in \mathbf{Z}(m)} \chi(x)\mathbf{B}_1\left(\left\langle \frac{x}{m} \right\rangle\right)$$

for any character χ on $\mathbf{Z}(m)^*$. Let χ' be the primitive character associated with χ, and let m' be its conductor. Then it is easy to verify that if we replace m by m' and χ by χ' in the right-hand side, we obtain the same value, so $B_{1,\chi}$ is independent of whether we view χ as primitive character, or simply a character on $\mathbf{Z}(m)^*$. (The above fact is a special case of the distribution relation, discussed in the next section.)

Next, we shall use the fact that

$$\chi(\theta) = B_{1,\bar{\chi}} \neq 0$$

for odd characters χ. For primitive χ the non-vanishing of $B_{1,\chi}$ comes from its relation with the L-series, and will be briefly recalled in Chapter 3. Cf. also [L 3], Chapter 14, Corollary of Theorem 2.2.

Lemma 4. $$(R\theta : Rm\theta) = m^M.$$

Proof. This is obvious if one can show that $R\theta$ is a free abelian group of rank M. When m is a prime power, this results from the fact that for odd χ we have

$$\chi(\theta) = B_{1,\bar{\chi}} \neq 0.$$

We shall analyze $(R^- : \mathscr{S})$ by the sequence of groups and subgroups shown in the following diagram.

$$\begin{array}{ccccc}
\varepsilon^- R & \overset{2M}{\supset} & R^- & \overset{?}{\supset} & \mathscr{S} \\
m^M \prod_{\chi \text{ odd}} -B_{1,\chi} \; \cup & & & & \cup \; w \\
Rm\theta & \underset{m^M}{\subset} & & & R\theta
\end{array}$$

We have shown the inclusion relations, and we have also indicated the indices. All of them have been proved, except the one on the left-hand side. This will be the item in the final lemma, and we then find:

Theorem 1.1 (Iwasawa). *Assume that m is a prime power. Then*

$$(R^- : \mathscr{S}) = w \prod_{\chi \text{ odd}} -\frac{1}{2} B_{1,\chi}.$$

Remark. Even though some inclusions go opposite to each other in the diagram, to compute indices one still has multiplicativity, with opposite inclusions occurring with opposite exponents. Cf. §4 if you don't find this obvious.

Lemma 5. $$(\varepsilon^- R : \varepsilon^- Rm\theta) = \pm m^M \prod_{\chi \text{ odd}} B_{1,\chi}.$$

Proof. First observe that the sign is whatever is needed to make the right-hand side positive. Multiplication by $\varepsilon^- m\theta$ is an endomorphism of $\mathbf{Q}R^-$, which is a semisimple algebra, decomposing into a product of 1-dimensional algebras corresponding to the odd characters. Consequently we find

$$\det(\varepsilon^- m\theta) = \prod_{\chi \text{ odd}} \chi(m\theta) = m^M \prod_{\chi \text{ odd}} B_{1,\chi}.$$

On the other hand, $\varepsilon^- m\theta$ maps $\varepsilon^- R$ into itself, and by standard elementary linear algebra, the index is given by the absolute value of the determinant. This proves the lemma, and the theorem.

Remark. In Chapter 3 we shall prove that the index computed in Theorem 1.1 is the order of the (-1)-eigenspace of the ideal class group in the cyclotomic field, denoted by h^-. The analytic class number formula will show that the product of $-B_{1,\chi}$ yields the positive sign.

The theorem and its proof are due to Iwasawa [Iw 7]. It was generalized to composite levels m by Sinnott [Si]. In the composite case, one cannot deal any more with a single element θ, but one has to deal with the module generated by Stickelberger elements of all intermediate levels

$$\sum_{c \in \mathbf{Z}(m)^*} \mathbf{B}_1\left(\left\langle \frac{c}{d} \right\rangle\right) \sigma_c^{-1}$$

for all divisors d of m. A similar situation had already arisen in the analogous situation in dimension one higher, concerning the Stickelberger elements formed with $\mathbf{B}_2$ rather than $\mathbf{B}_1$, in the Kubert–Lang series [KL 2], [KL 3], [KL 5].

§2. Bernoulli Numbers

We recall first some general notions concerning distributions, defined by Mazur following the work of Iwasawa.

Let $\{X_n\}$ be a sequence of finite sets, and suppose given a sequence of surjective maps

$$\pi_{n+1}: X_{n+1} \to X_n,$$

so that we can consider the projective limit

$$X \to \cdots \to X_{n+1} \to X_n \to \cdots \to X_1.$$

For convenience, we took our family of sets indexed by the positive integers. In applications, it often occurs that the sets are ordered by the positive integers ordered by divisibility. For instance, the family of sets $\mathbf{Z}/N\mathbf{Z}$ arises in the sequel. We shall also consider the projective family

$$\{\mathbf{Z}/p^n\mathbf{Z}\},$$

with a fixed prime number p, and $n = 0, 1, 2, \ldots$. In each case, the connecting homomorphism

$$\mathbf{r}_M: \mathbf{Z}/N\mathbf{Z} \to \mathbf{Z}/M\mathbf{Z}$$

for $M | N$ is reduction mod M, denoted by $\mathbf{r}_M$.

This type of projective family will also arise in isomorphic form as follows. We have an isomorphism

$$\frac{1}{N}\mathbf{Z}/\mathbf{Z} \to \mathbf{Z}/N\mathbf{Z}$$

given by multiplication with N. We then have a commutative diagram

$$\begin{array}{ccc} \frac{1}{N}\mathbf{Z}/\mathbf{Z} & \to & \mathbf{Z}/N\mathbf{Z} \\ \downarrow{\scriptstyle N/M} & & \downarrow{\scriptstyle r_M} \\ \frac{1}{M}\mathbf{Z}/\mathbf{Z} & \to & \mathbf{Z}/M\mathbf{Z} \end{array}$$

where the left vertical arrow is multiplication with N/M, and the right arrow is reduction mod M. Thus the system

$$\left\{\frac{1}{N}\mathbf{Z}/\mathbf{Z}\right\}$$

is also a projective system, ordered by divisibility.

Let us now return to the general projective system $\{X_n\}$. For each n suppose given a function φ_n of X_n into an abelian group V. We say that the family $\{\varphi_n\}$ is **compatible** if for each n and $x \in X_n$ we have

$$\varphi_n(x) = \sum_{\pi_{n-1}y = x} \varphi_{n+1}(y).$$

The sum is taken over all the elements of X_{n+1} lying above x. In what follows, we often omit the subscripts, and write $\pi y = x$, for instance.

Let K be a ring of operators on V. Let f be a function on X_m for some integer m, with values in K. If $n \geq m$, then we view f as defined on X_n through the natural projection on X_m. We conclude at once from the compatibility relation that

$$\sum_{x \in X_n} f(x)\varphi_n(x) = \sum_{x \in X_m} f(x)\varphi_m(x).$$

Let X be the projective limit

$$X = \varprojlim X_n,$$

with the limit topology, so that X is a compact space. For each n we have a surjective map

$$\mathbf{r}_n \colon X \to X_n.$$

For each $x \in X_n$ the inverse image $\mathbf{r}_n^{-1}(x)$ is an open set in X, and the totality of such open sets for all n, x is a basis for the topology of X.

A function f on X is called **locally constant** if and only if there exists n such that f factors through X_n. Such functions are also called **step functions**, and their group is denoted by $St(X, K)$. For each such function, we can define its integral

$$\int f \, d\varphi = \sum_{x \in X_n} f(x)\varphi_n(x),$$

independent of the choice of n such that f factors through X_n. We then call the family $\{\varphi_n\}$, or the functional $d\varphi$, a **distribution** on X. It is an additive map

$$d\varphi \colon St(X, K) \to V.$$

Examples of such maps will be given later with Bernoulli numbers.

Let K be a complete field with respect to a non-Archimedean valuation, and suppose that V is a non-Archimedean Banach space over K, i.e., V is a complete vector space, with a norm

$$| \ | \colon V \to \mathbf{R}^+$$

satisfying

$$|v + w| \leq \max\{|v|, |w|\} \quad v, w \in V$$

$$|cv|_V = |c|_K |v|_V \qquad c \in K, v \in V.$$

If φ is bounded, i.e., $|\varphi_n(x)|$ is bounded for all n, $x \in X_n$, then we say that φ is **bounded**, or **quasi-integral** for the valuation. For any $f \in St(X, K)$ we have

$$\left|\int f\,d\varphi\right| \leq \|f\|\,\|\varphi\|,$$

where $\|f\|$ is the sup norm of f, and $\|\varphi\|$ is the sup norm of the values $|\varphi_n(x)|$. Indeed, if f factors through X_n, then

$$\left|\int f\,d\varphi\right| = \left|\sum_{x \in X_n} f(x)\varphi_n(x)\right| \leq \max_{x \in X_n} |f(x)|\,|\varphi_n(x)|$$

by the non-Archimedean property, so our assertion is clear.

In particular, if $f \in C(X)$ is a continuous function on X, then we can approximate f uniformly by a sequence $\{f_n\}$ of step functions, and since $\|f - f_n\| \to 0$, we get

$$\|f_n - f_m\| \to 0$$

for $m, n \to \infty$. Hence the integrals

$$\int f_n\,d\varphi$$

converge, and define the integral

$$\int f\,d\varphi$$

for such a continuous function, provided that φ is bounded. This will be the case in important examples, and bounded distributions are also called **measures**.

All this is preliminary to defining the distributions which are of importance to us, namely the Bernoulli distributions. If $x \in \mathbf{Z}(N)$ then x/N can be viewed as an element of $\mathbf{Q}/\mathbf{Z}$. For any $t \in \mathbf{R}/\mathbf{Z}$ we let $\langle t \rangle$ be the smallest real number ≥ 0 in the residue class of t mod $\mathbf{Z}$. What we want is for each positive integer k a polynomial P_k with rational coefficients, leading coefficient 1, such that the functions

$$x \mapsto N^{k-1}P_k\left(\left\langle \frac{x}{N} \right\rangle\right)$$

form a distribution on the projective system $\{\mathbf{Z}/N\mathbf{Z}\}$. Such polynomials will be given by the Bernoulli polynomials. Let the **Bernoulli numbers** B_k be defined by the power series

B 1.
$$F(t) = \frac{t}{e^t - 1} = \sum_{k=0}^{\infty} B_k \frac{t^k}{k!}.$$

Then for instance

$$B_0 = 1, \qquad B_1 = -\tfrac{1}{2}, \qquad B_2 = \tfrac{1}{6}.$$

Observe that

$$F(-t) - F(t) = t,$$

so that F is almost even, and in particular, we have

$$B_k = 0 \quad \text{if } k \text{ is odd, } k \neq 1.$$

We define the **Bernoulli polynomials** $\mathbf{B}_k(X)$ by the expansion

B 2. $$F(t, X) = \frac{te^{tX}}{e^t - 1} = \sum_{k=0}^{\infty} \mathbf{B}_k(X) \frac{t^k}{k!}.$$

Then it is clear that the Bernoulli numbers are the constant terms of the Bernoulli polynomials, that is

$$B_k = \mathbf{B}_k(0).$$

We find:

$$\mathbf{B}_0(X) = 1, \qquad \mathbf{B}_1(X) = X - \tfrac{1}{2}, \qquad \mathbf{B}_2(X) = X^2 - X + \tfrac{1}{6}.$$

The desired distribution relation is implied by the next formula.

B 3. $$\mathbf{B}_k(X) = N^{k-1} \sum_{a=0}^{N-1} \mathbf{B}_k\left(\frac{X + a}{N}\right).$$

Proof. On one hand, we have

$$\begin{aligned}
\sum_{a=0}^{N-1} \frac{te^{(X+a)t}}{e^{Nt} - 1} &= \frac{1}{N} \sum_{a=0}^{N-1} \frac{Nte^{[(X+a)/N]Nt}}{e^{Nt} - 1} \\
&= \frac{1}{N} \sum_{a=0}^{N-1} \sum_{k=0}^{\infty} \mathbf{B}_k\left(\frac{X + a}{N}\right) \frac{(Nt)^k}{k!} \\
&= \sum_{k=0}^{\infty} \left[\sum_{a=0}^{N-1} N^{k-1} \mathbf{B}_k\left(\frac{X + a}{N}\right)\right] \frac{t^k}{k!}.
\end{aligned}$$

On the other hand, summing the geometric series $\sum e^{at}$ directly from $a = 0$ to $a = N - 1$ and using the definition of the Bernoulli polynomials shows that the coefficient of $t^k/k!$ is precisely $\mathbf{B}_k(X)$, thereby proving the desired identity.

Relation **B 3** can also be written in the form

B 4.
$$N^{k-1}\sum_{t \bmod N}\mathbf{B}_k\left(\left\langle y+\frac{t}{N}\right\rangle\right) = \mathbf{B}_k(\langle Ny\rangle)$$

for $y \in \mathbf{R}/\mathbf{Z}$. This can be interpreted as follows.

On the projective system

$$\left\{\frac{1}{M}\mathbf{Z}/\mathbf{Z}\right\}$$

the association

$$x \mapsto M^{k-1}\mathbf{B}_k(\langle x\rangle) \quad \text{for } x \in \frac{1}{M}\mathbf{Z}/\mathbf{Z}$$

defines a distribution.

Proof. If $y \in (1/MN)\mathbf{Z}/\mathbf{Z}$ is one element such that $Ny = x$, then all elements in the inverse image of x by the mapping $(N \cdot id)^{-1}$ consist of

$$y + \frac{t}{N}, \quad \text{with } t \bmod N.$$

Multiplying **B 4** by M^{k-1} yields precisely the distribution relation.

Since the system $\{(1/M)\mathbf{Z}/\mathbf{Z}\}$ is isomorphic to the system $\{\mathbf{Z}/M\mathbf{Z}\}$, we can also express the distribution relation on the latter. It is convenient to normalize this distribution further and to give it a special symbol. For $x \in \mathbf{Z}/N\mathbf{Z}$ we define

$$E_k^{(N)}(x) = N^{k-1}\frac{1}{k}\mathbf{B}_k\left(\left\langle\frac{x}{N}\right\rangle\right).$$

Then the family $\{E_k^{(N)}\}$ *forms a distribution on* $\{\mathbf{Z}/N\mathbf{Z}\}$.

Remark. Historically, this distribution arose in the context of the partial zeta functions. Indeed, if $x \in (\mathbf{Z}/N\mathbf{Z})^*$, define

$$\zeta_N(x, s) = \sum_{\substack{n \in x \\ n > 0}} n^{-s}.$$

The Dirichlet series converges only for $\mathrm{Re}(s) > 1$, but it is classical and elementary that it can be analytically continued to the whole complex plane, and Hurwitz has shown that

$$\zeta_N(x, 1-k) = -E_k^{(N)}(x) \quad \text{for } k \geq 1.$$

Furthermore the partial zeta functions themselves satisfy the distribution relation. For a further discussion, cf. Example 4 at the end of the chapter. For distributions associated with zeta functions in connection with Cartan groups, see [KL 10].

For the applications, we shall use one more formula concerning the Bernoulli polynomials, namely

B 5. $$\mathbf{B}_k(X) = X^k - \tfrac{1}{2}kX^{k-1} + \text{lower terms}.$$

This is obvious by the direct multiplication of the series

$$\frac{t}{e^t - 1} = \sum B_k \frac{t^k}{k!} \quad \text{and} \quad e^{tX} = \sum X^k \frac{t^k}{k!}.$$

For what we have in mind, we don't care about the lower terms, which have rational coefficients.

Let N be a positive integer, and let f be a function on $\mathbf{Z}/N\mathbf{Z}$. We form the polynomial

$$F_f(t, X) = \sum_{a=0}^{N-1} f(a) \frac{te^{(a+X)t}}{e^{Nt} - 1}.$$

We define the generalized Bernoulli polynomials (relative to the function f) by

B 6. $$F_f(t, X) = \sum_{k=0}^{\infty} \mathbf{B}_{k,f}(X) \frac{t^k}{k!}.$$

In particular, the constant term of $\mathbf{B}_{k,f}(X)$ is the generalized Bernoulli number

$$B_{k,f} = \mathbf{B}_{k,f}(0).$$

For instance, f may be a Dirichlet character χ on $\mathbf{Z}(N)^*$, extended to $\mathbf{Z}/N\mathbf{Z}$ by the value 0 on integers not prime to N. Then $B_{k,\chi}$ is the generalized Bernoulli number of Leopoldt. Directly from the definition, we then find the expression

B 7. $$B_{k,f} = N^{k-1} \sum_{a=0}^{N-1} f(a) \mathbf{B}_k\left(\left\langle \frac{a}{N} \right\rangle\right).$$

In terms of the distribution relation, this can be written

$$\frac{1}{k} B_{k,f} = \int_{\mathbf{Z}_p} f \, dE_k.$$

The distribution $\{E_k^{(N)}\}$ is rational valued. We shall be interested in its p-adic integrality properties for a prime p. For this purpose, we describe a process which integralizes this distribution. For historical comments, see below, after Theorem 2.1.

Let c be a rational number. For N prime to c (i.e., prime to the numerator and denominator of c) we define

$$E_{k,c}^{(N)}(x) = E_k^{(N)}(x) - c^k E_k^{(N)}(c^{-1}x),$$

for $x \in \mathbf{Z}(N)$. Multiplication by c or c^{-1} is well defined on $\mathbf{Z}(N)$ so our expression makes sense. If N is a power of a prime p, then we could also take c to be a p-adic unit. We can write symbolically

$$\boxed{E_{k,c} = E_k - c^k E_k \circ c^{-1}.}$$

This distribution satisfies the following properties.

E 1. $$E_{1,c}^{(N)}(x) = \left\langle \frac{x}{N} \right\rangle - c\left\langle \frac{c^{-1}x}{N} \right\rangle + \frac{c-1}{2}.$$

Proof. We have

$$\begin{aligned} E_{1,c}^{(N)}(x) &= \mathbf{B}_1\left(\left\langle \frac{x}{N} \right\rangle\right) - c\mathbf{B}_1\left(\left\langle \frac{c^{-1}x}{N} \right\rangle\right) \\ &= \left\langle \frac{x}{N} \right\rangle - \frac{1}{2} - c\left(\left\langle \frac{c^{-1}x}{N} \right\rangle - \frac{1}{2}\right) \end{aligned}$$

whence the assertion is clear.

E 2. $$E_{k,c}^{(N)}(x) \equiv x^{k-1} E_{1,c}^{(N)}(x) \bmod \frac{N}{kD(k)} \mathbf{Z}[c, 1/c],$$

where $D(k)$ is a least common multiple of the denominators of the coefficients of the polynomial $\mathbf{B}_k(X)$.

Proof. We work with a representative integer x such that

$$0 \leq x \leq N - 1.$$

We write

$$c^{-1}x = b + yN$$

with an integer b satisfying $0 \leq b \leq N - 1$ and $y \in \mathbf{Z}[1/c]$. Then

$$\frac{c^{-1}x}{N} = \frac{b}{N} + y = \left\langle \frac{b}{N} + y \right\rangle + z$$

with some integer z. Since $\mathbf{B}_k(X) = X^k - \frac{1}{2}kX^{k-1} +$ lower terms, we find the following congruences mod $N/(D(k))\mathbf{Z}[c, 1/c]$:

$$N^{k-1}\left[\mathbf{B}_k\left(\left\langle\frac{x}{N}\right\rangle\right) - c^k\mathbf{B}_k\left(\left\langle\frac{c^{-1}x}{N}\right\rangle\right)\right]$$

$$\equiv N^{k-1}\left[\left(\frac{x}{N}\right)^k - \frac{k}{2}\left(\frac{x}{N}\right)^{k-1}\right]$$

$$- N^{k-1}c^k\left[\left(\frac{b}{N} + y - z\right)^k - \frac{k}{2}\left(\frac{b}{N} + y - z\right)^{k-1}\right]$$

$$\equiv \frac{x^k}{N} - \frac{k}{2}x^{k-1} - \left[N^{k-1}\left(\frac{x}{N} - cz\right)^k - c^k\frac{k}{2}(b + Ny - Nz)^{k-1}\right]$$

$$\equiv \frac{x^k}{N} - \frac{k}{2}x^{k-1} - \left[\frac{x^k}{N} - kx^{k-1}cz - c^k\frac{k}{2}b^{k-1}\right]$$

$$\equiv kx^{k-1}\left(\frac{x}{N} - c\left\langle\frac{c^{-1}x}{N}\right\rangle + \frac{c-1}{2}\right)$$

and Property **E 2** follows by using **E 1**.

The values of $E_{k,c}^{(N)}$ are in

$$\frac{1}{kD(k)}\mathbf{Z}[c, 1/c].$$

They will be called N-integral if they are p-integral for every prime dividing N.

Theorem 2.1. (i) *The values of $E_{k,c}^{(N)}$ are N-integral.*

(ii) *We have the congruence for every prime p dividing N:*

$$E_{k,c}^{(N)}(x) \equiv x^{k-1}E_{1,c}^{(N)}(x) \bmod N\mathbf{Z}_p.$$

(iii) *If c is an integer prime to $2kN$ and to the denominators of the Bernoulli polynomial $\mathbf{B}_k(X)$, then the values of $E_{k,c}^{(N)}$ lie in $\mathbf{Z}$.*

Proof. For large integer ν the values $N^\nu/kD(k)$ are N-integral. Let $M = N^\nu$. The distribution relation yields

$$E_{k,c}^{(N)}(x) = \sum_y E_{k,c}^{(M)}(y)$$

where the sum is taken over those y mod M which reduce to x mod N. The expression for $E_{1,c}^{(M)}$ is obviously N-integral except possibly for the term

$(c - 1)/2$. But if N is even then c is odd, so $(c - 1)/2$ is N-integral, and if N is odd, then $(c - 1)/2$ is N-integral. If we apply **E 2** to each term $E_{k,c}^{(M)}(y)$ then we see that the first two assertions are proved.

For case (iii), we take $M = (NkD(k))^{\nu}$ for large ν. The argument then proceeds as before, because the only denominators occurring in

$$\frac{1}{k}\mathbf{B}_k\left(\left\langle\frac{x}{N}\right\rangle\right) \quad \text{or} \quad \frac{1}{k}\mathbf{B}_k\left(\left\langle\frac{c^{-1}x}{N}\right\rangle\right)$$

contain only primes dividing $NkD(k)$.

For $k = 1$ the integralizing process already appears in the Stickelberger theorem, and was used extensively by Iwasawa. For $k > 1$, Coates–Sinnott obtained integral elements in group rings by this process [C–S 2], Theorem 1.3 and [C–S 3], Theorem 1. Mazur formulated this integralizing process in terms of measure theory and the distribution relation, which allows the jacking up argument used to prove Theorem 2.1.

For the rest of this section, we let $N = p^n$ with some fixed prime number p, so the distributions are defined on the projective limit of $\mathbf{Z}(p^n)$, which is none other than the p-adic integers $\mathbf{Z}_p$. We view the values of the distributions to be in $\mathbf{C}_p$, the completion of the algebraic closure of the p-adic numbers. We may express Theorem 2.1(ii) in the limit as follows.

Theorem 2.2. *Let c be a p-adic unit. Then*

$$E_{k,c}(x) = x^{k-1}E_{1,c}(x).$$

We shall now express Bernoulli numbers in terms of the integralized distributions.

Theorem 2.3. *Let $c \in \mathbf{Z}_p^*$ and let k be an integer ≥ 1 such that $c^k \neq 1$. Then*

$$\frac{1}{k}B_k = \frac{1}{1 - c^k}\int_{\mathbf{Z}_p} x^{k-1}\,dE_{1,c}(x).$$

Proof. By definition,

$$\frac{1}{k}B_k = \int_{\mathbf{Z}_p} dE_k = \int_{\mathbf{Z}_p} dE_{k,c} + \int_{\mathbf{Z}_p} c^k\,dE_k(c^{-1}x).$$

On the last integral to the right, we make the change of variable

$$x \mapsto cx,$$

which gives

$$\int_{\mathbf{Z}_p} dE_k(x) = \int_{\mathbf{Z}_p} dE_k(c^{-1}x).$$

The formula we want drops out by using Theorem 2.2.

Corollary 1 (Kummer Congruence). *Let α be a residue class* mod $p-1$ *and* $\alpha \neq 0$. *Then for even positive integers* $k \equiv \alpha \bmod p-1$, *the values* $(1/k)B_k$ *are all congruent* mod p, *and are* p*-integral.*

Proof. Select c to be a primitive root mod p so that

$$c^k \not\equiv 1 \bmod p.$$

Then $1 - c^k$ is a unit at p. The values $1 - c^k$ and $x^{k-1} \bmod p$ are independent of the choice of k in the residue class mod $p-1$, and the corollary then follows from the expression of $(1/k)B_k$ as the integral of the theorem.

Corollary 2 (Von Staudt Congruence). *Let* $k \equiv 0 \bmod p-1$, *and* k *even. Then*

$$B_k \equiv -\frac{1}{p} \bmod \mathbf{Z}_p.$$

Proof. Suppose p odd for simplicity. Let $c = 1 + p$. An easy induction shows that

$$c^k \equiv 1 + pk \bmod p^2k\mathbf{Z}_p.$$

Hence

$$\frac{1}{1-c^k} = -\frac{1}{pk}(1 + O(p)),$$

and so

$$B_k \equiv -\frac{1}{p}\int_{\mathbf{Z}_p^*} x^{k-1}\, dE_{1,c}(x),$$

because the integral over $p\mathbf{Z}_p$ is $\equiv 0 \bmod p$. An approximating sum mod p for the integral over $\mathbf{Z}_p^*$ is

$$\sum_{x=1}^{p-1} x^{k-1}\left(\left\langle\frac{x}{p}\right\rangle - c\left\langle\frac{c^{-1}x}{p}\right\rangle + \frac{c-1}{2}\right).$$

Since $c = 1 + p$ we have

$$\left\langle\frac{c^{-1}x}{p}\right\rangle = \frac{x}{p}.$$

The desired congruence follows from the fact that

$$\sum_{x=1}^{p-1} x^k \equiv -1 \bmod p.$$

We leave $p = 2$ as an exercise. We merely wanted to show how classical congruences can be handled systematically from integration theory.

Let f be any function on $\mathbf{Z}/N\mathbf{Z}$. We defined

$$B_{k,f} = N^{k-1} \sum_{a=0}^{N-1} f(a)\mathbf{B}_k\left(\left\langle \frac{a}{N} \right\rangle\right).$$

In terms of the distribution notation, this can be written

$$\frac{1}{k} B_{k,f} = \int_{\mathbf{Z}_p} f\, dE_k.$$

We shall apply this when f is a character of finite order on $\mathbf{Z}_p^*$, so that f is an ordinary Dirichlet character on $\mathbf{Z}(p^n)^*$ for some positive integer n. As usual, for such a character, we define its value to be 0 on elements of $\mathbf{Z}(p^n)$ which are not prime to p. Then by definition, for any character ψ of finite order on $\mathbf{Z}_p^*$ we have the formula for the Bernoulli–Leopoldt numbers

$$\frac{1}{n} B_{n,\psi} = \int_{\mathbf{Z}_p^*} \psi\, dE_n.$$

Note: When $\psi = 1$ we do *not* have $(1/n)B_{n,\psi} = (1/n)B_n$ because ψ is 0 on $p\mathbf{Z}_p$ by definition.

Theorem 2.4. *Let ψ be a character of finite order on $\mathbf{Z}_p^*$. Then*

$$\frac{1}{n} B_{n,\psi} = \frac{1}{1 - \psi(c)c^n} \int_{\mathbf{Z}_p^*} \psi(a)a^{n-1}\, dE_{1,c}(a).$$

Proof. We write $dE_n = dE_{n,c} + c^n\, dE_n \circ c^{-1}$, or in other words

$$\frac{1}{n} B_{n,\psi} = \int \psi\, dE_{n,c} + \int \psi(x)c^n\, dE_n(c^{-1}x).$$

Integrals are taken over $\mathbf{Z}_p^*$. We let $x \mapsto cx$ in the second integral. Then $\psi(c)$ comes out as a factor. Using Theorem 2.2 concludes the proof.

Theorem 2.5. *Let $2 \leq k \leq p - 2$. Let $\omega\colon \mathbf{Z}(p)^* \to \mathbf{Z}_p^*$ be the Teichmuller character such that*

$$\omega(a) \equiv a \pmod p.$$

For any integer $n \geq 1$ we have

$$\frac{1}{n} B_{n,\omega^{k-n}} \equiv \frac{1}{k} B_k \pmod{p}.$$

Proof. Let $\psi = \omega^{k-n}$. Choose c to be a primitive root mod p, so that $c^k \not\equiv 1 \bmod p$. By Theorem 2.3 we get

$$\frac{1}{k} B_k \equiv \frac{1}{1-c^k} \int_{\mathbf{Z}_p^*} x^{k-1}\, dE_{1,c}(x) \pmod{p}.$$

By Theorem 2.4 we have the congruence mod p:

$$\frac{1}{n} B_{n,\psi} - \frac{1}{k} B_k \equiv \int_{\mathbf{Z}_p^*} x^{k-1} \left[\frac{1}{1-\psi(c)c^n} - \frac{1}{1-c^k} \right] dE_{1,c}(x)$$

because $1 - \psi(c)c^n$ and $1 - c^k$ are p-units. Since the expression in brackets under the integral sign is $\equiv 0 \pmod{p}$, the theorem follows.

The next sections, §3 *through* §7, *taken from Kubert–Lang* [*KL 8*], *deal further with the integrality properties of Stickelberger ideals.*

§3. Integral Stickelberger Ideals

Let k be an integer ≥ 2. Let $N = p^n$ be a prime power with $p \geq 3$ until §6. We let:

$G = \mathbf{Z}(N)^*$ if k is odd
$G = \mathbf{Z}(N)^*/\pm 1$ if k is even.
$R = R_G = \mathbf{Z}[G]$ and $R_p = \mathbf{Z}_p[G]$.
deg: $R \to \mathbf{Z}$ is the augmentation homomorphism, such that

$$\deg\Big(\sum_{\sigma \in G} m_\sigma \sigma\Big) = \sum m_\sigma.$$

This augmentation homomorphism extends to the complex group algebra by linearity.

R_m = ideal of R consisting of those elements whose degree is $\equiv 0 \bmod m$.
If I is an ideal of R, we let $I_m = I \cap R_m$.
card $G = |G|$.
$s(G) = \sum_{\sigma \in G} \sigma$.
For any $\xi \in R$ we have

$$\xi s(G) = (\deg \xi) s(G).$$

If J is an ideal of R, we write $d = \deg J$ to mean that d is the smallest integer ≥ 0 which generates the $\mathbf{Z}$-ideal of elements $\deg \xi$ with ξ in J.

Let $\mathbf{B}_k(X)$ be the kth Bernoulli polynomial. We let

$$\theta_k(N) = N^{k-1} \sum_{a \in G} \frac{1}{k} \mathbf{B}_k\left(\left\langle \frac{a}{N} \right\rangle\right) \sigma_a^{-1}$$

$$\theta'_k(N) = N^{k-1} \sum_{a \in G} \frac{1}{k} \left(\mathbf{B}_k\left(\left\langle \frac{a}{N} \right\rangle\right) - \mathbf{B}_k(0)\right) \sigma_a^{-1}$$

$$= \theta'_k - \frac{N^{k-1}}{k} B_k s(G),$$

where $B_k = \mathbf{B}_k(0)$ is the kth Bernoulli number. We have:

$$\deg \theta \neq 0 \quad \text{and} \quad \deg \theta' \neq 0, \quad \text{for } k \text{ even}.$$

In fact, these degrees can be computed easily. We need only that they are $\neq 0$ for k even, but the computation is as follows. Suppose k is odd. We use the distribution relation. Summing over all primitive elements, i.e., elements of p^n yields the value of the distribution summed over all elements of level p^{n-1}. Continuing in this fashion reduces the computation to level 1. But

$$p^{k-1} \sum_{a \in \mathbf{Z}(p)} \frac{1}{k} \mathbf{B}_k\left(\left\langle \frac{a}{p} \right\rangle\right) = \frac{1}{k} \mathbf{B}_k(0) = \frac{1}{k} B_k.$$

The degree of θ arises from the same sum but with the term $a = 0$ omitted. Hence

$$\deg \theta = \frac{1 - p^{k-1}}{k} B_k$$

and

$$\deg \theta' = \deg \theta - \frac{N^{k-1}}{k} B_k |G|$$

or

$$\deg \theta' = \left(\frac{1 - p^{k-1}}{k} - \frac{N^{k-1}}{k} \phi(p^n)\right) B_k.$$

These formulas would also be valid for k even, except for our convention to take $G = \mathbf{Z}(N)^*/\pm 1$. This requires dividing the formulas by 2 to get $\deg \theta$ and similarly for θ'. The non-vanishing for k even comes from the functional equation of the zeta function.

Next we give the ideals used in integralizing the distribution.

$J^{(k)}(N)$ = ideal of elements $\sum m(b)\sigma_b$ such that

$$\sum m(b)b^k \equiv 0 \pmod{N}$$

$I^{(k)}(N)$ = ideal of elements $\sigma_c - c^k$ with integers c prime to N.

Since k and N remain fixed, we often write θ and θ' instead of $\theta_k(N)$ and $\theta'_k(N)$. Similarly, we write $J^{(k)}$ and $I^{(k)}$, or J and I. It is obvious that

$$J^{(k)} \supset I^{(k)}.$$

We shall determine the extent to which $J \neq I$ in Lemma 4.

We have:

$\deg I^{(k)}(N) = p^t$, where t is the maximum integer such that $k \equiv 0 \bmod \phi(p^t)$.

This is obvious, because $\deg I^{(k)}(N)$ is generated by the integers $1 - c^k$ with c prime to p.

Theorem 3.1. (i) *We have*

$$R\theta'_k \cap R = I^{(k)}\theta'_k.$$

In fact, if an element $\xi \in R$ is such that $\xi\theta' \in R$, then $\xi \in I^{(k)}$.

(ii) *On the other hand, letting $I_p^{(k)} = \mathbf{Z}_p I^{(k)}$, we have*

$$R_p\theta_k \cap R_p = I_p^{(k)}\theta_k.$$

If an element $\xi \in R_p$ is such that $\xi\theta \in R_p$ then $\xi \in I_p^{(k)}$.

Proof. First we prove that for any prime ≥ 2, we have

$$I\theta' \subset R, \quad \text{and} \quad I_p\theta \subset R_p.$$

A similar property is due to Mazur and Coates–Sinnott, as mentioned before. Indeed, we have

$$\sigma_c^{-1}(\sigma_c - c^k)\theta_k = \sum_{a \in G} E_{k,c}^{(N)}(a)\sigma_a^{-1}$$

where

$$E_{k,c}^{(N)}(x) = N^{k-1}\frac{1}{k}\left[\mathbf{B}_k\left(\left\langle\frac{x}{N}\right\rangle\right) - c^k\mathbf{B}_k\left(\left\langle\frac{c^{-1}x}{N}\right\rangle\right)\right].$$

The p-integrality then follows from Theorem 2.1(i). For other primes we need a lemma.

Lemma 1. *The polynomial* $(1/k)(\mathbf{B}_k(X) - \mathbf{B}_k(0))$ *maps* $\mathbf{Z}$ *into* $\mathbf{Z}$ *and maps* $\mathbf{Z}_l$ *into* $\mathbf{Z}_l$ *for every prime* l.

Proof. A standard property of Bernoulli polynomials states that

$$\frac{1}{k}(\mathbf{B}_k(X+1) - \mathbf{B}_k(X)) = X^{k-1}.$$

Hence for any integer m we see recursively that the first assertion of the lemma is true. The second, concerning l-adic integers, follows by continuity. The lemma is also valid for $p = 2$.

We may define $E'_{k,c}$ by using $\mathbf{B}_k(X) - \mathbf{B}_k(0)$ instead of $\mathbf{B}_k(X)$ in the definition of $E_{k,c}$. The lemma shows that $I\theta' \subset R$.

For convenience we let

$$\mathbf{B}'_k(X) = \mathbf{B}_k(X) - \mathbf{B}_k(0).$$

Lemma 2. (i) *Let* $\xi \in R$ *and suppose that* $\xi\theta' \in \mathbf{Z}_p[G] = R_p$. *Then* $\xi \in J$.

(ii) *Let* $\xi \in R_p$ *and suppose that* $\xi\theta \in R_p$. *Then* $\xi \in J_p = \mathbf{Z}_p J$.

Proof. Write $\xi = \sum z(b)\sigma_b$ with integral coefficients $z(b)$. Then

$$\xi\theta' = N^{k-1} \sum_c \sum_b z(b) \frac{1}{k} \mathbf{B}'_k\left(\left\langle \frac{bc}{N} \right\rangle\right) \sigma_c^{-1},$$

and therefore

$$\frac{N^{k-1}}{k} \sum_b z(b) \mathbf{B}'_k\left(\left\langle \frac{b}{N} \right\rangle\right) \text{ is } p\text{-integral.}$$

But an elementary formula for Bernoulli polynomials, obtained directly from the definition, gives for an integer b,

$$\frac{N^{k-1}}{k} \mathbf{B}_k\left(\frac{b}{N}\right) = \sum_{i=0}^{k} \frac{N^{k-1}}{k} \binom{k}{i} B_i \left(\frac{b}{N}\right)^{k-i}.$$

Comparing the leading term modulo all the lower order terms, and taking into account that $B_1 = -\frac{1}{2}$ is p-integral (here we use $p \neq 2$), and the Kummer theorem that B_i is p-integral for $i < p - 1$, we find

$$\frac{\sum z(b) b^k}{kN} \equiv 0 \bmod \frac{1}{k} \mathbf{Z}_p.$$

Multiplying both sides by kN proves the lemma.

Lemma 3. *Let p^s be the smallest power of p such that $p^s\theta'_k$ is p-integral. Then*

$$s = n + \operatorname{ord}_p k.$$

We have $I^{(k)} \cap \mathbf{Z} = (p^s)$.

Proof. The argument uses the same expression for the Bernoulli polynomial as in the previous lemma. We see that

$$p^s \sum \frac{N^{k-1}}{k}\binom{k}{i}B_i\left(\frac{1}{N}\right)^{k-i} \quad \text{is } p\text{-integral.}$$

The leading term is p^s/kN. The Bernoulli numbers B_i are p-integral for $i < p - 1$ by Kummer, and for $i \geq p - 1$ the power N^{k-1} in front integralizes $(1/N)^{k-i}$. It follows that

$$\frac{p^s}{kN} \text{ is } p\text{-integral,}$$

whence s has the stated value. Since we have already seen that $I\theta' \subset R$, it follows that the p-contribution of $I \cap \mathbf{Z}$ is exactly p^s. It is clear that $I \cap \mathbf{Z}$ is equal to (p^s), because we can always select

$$c \equiv 1 \bmod N \quad \text{and} \quad c \equiv 0 \bmod l$$

for any prime $l \neq p$ to see that $I \cap \mathbf{Z}$ contains elements prime to l. This proves the lemma.

Lemma 4. *We have $J = I + \mathbf{Z}N$, and $(J : I) = p^{s-n} = p^{\operatorname{ord} k}$.*

Proof. It is clear that $N \in J$. Conversely, write an element of J in the form

$$\sum m(c)(\sigma_c - c^k) + \sum m(c)c^k.$$

The first term is in I, and the second term is an integral multiple of N. This proves the lemma.

We may now conclude the proof of the theorem. We prove (i). Suppose $\xi \in R$ and $\xi\theta' \in R$. By Lemma 2, $\xi \in J$. By Lemma 4, we know that

$$\xi \equiv zN \bmod I \quad \text{for some } z \in \mathbf{Z}.$$

We know that $I\theta' \subset R$. Hence $zN\theta' \in R$. By Lemma 3, it follows that p^s divides zN, so $\xi \in I$, and the theorem (i) is proved. The part (ii) is proved the same way.

§4. General Comments on Indices

Let V be a finite dimensional vector space over the rationals, and let A, B be lattices in V, that is free **Z**-modules of the same rank as the dimension of V. Let C be a lattice containing both of them. We define the index

$$(A : B) = \frac{(C : B)}{(C : A)}.$$

It is an easy exercise to prove that this index is independent of the choice of C, and satisfies the usual multiplicativity property

$$(A : D)(D : B) = (A : B).$$

Furthermore, if E is a lattice contained in both A and B then

$$(A : B) = \frac{(A : E)}{(B : E)}.$$

We leave the proofs to the reader.

Suppose that A is not only a lattice, but is an algebra over $\mathbf{Z}$. Let θ be an element of $\mathbf{Q}A = V$ and let m be a positive integer such that $m\theta \in A$. Assume that θ is invertible in $\mathbf{Q}A$. Then

$$(A : A\theta) = \pm \det_{\mathbf{Q}A} \theta,$$

where the determinant is taken for the linear transformation of $\mathbf{Q}A$ equal to multiplication by θ. This is easily seen, because

$$(A : A\theta) = (A : Am\theta)(Am\theta : A\theta)$$

and

$$(Am\theta : A\theta) = (A\theta : Am\theta)^{-1}.$$

Since $m\theta$ lies in A, the index $(A : Am\theta)$ is given by the absolute value of the determinant of $m\theta$, which is $m^r \det \theta$, where r is the rank of A. This power m^r then cancels the other index.

Note that the determinant can be computed in the extension of scalars by the complex numbers. In particular, if A is a semisimple algebra, and is commutative, then

$$\det \theta = \prod \chi(\theta)$$

where χ ranges over all the characters of the algebra, counted with their multiplicities. In the applications, the algebra is essentially a group ring, so the multiplicities are 1, and the characters come from characters of the group.

This will be applied to the case when $\theta = \theta^{(k)}$. We recall the definition of generalized Bernoulli numbers according to Leopoldt:

$$B_{k,\chi} = N^{k-1} \sum_{a \in G} \chi(a) \mathbf{B}_k\left(\left\langle \frac{a}{N} \right\rangle\right).$$

Thus

$$\chi(\theta) = \frac{1}{k} B_{k,\bar{\chi}}.$$

Note that the Bernoulli number is defined with respect to G, so that for k even, we are summing over $\mathbf{Z}(N)^*/\pm 1$. This convention is the most useful for present applications in §5 and §6. (We revert to the other convention in §7.) For even k, it gives half the other values.

The classical theorem about the non-vanishing of $B_{k,\chi}$ when k and χ have the same parity gives the desired invertibility of the Stickelberger element θ_k in the corresponding part of the group algebra over $\mathbf{Q}$.

§5. The Index for k Even

We let $s = n + \mathrm{ord}_p\, k$, and t is defined as in §3, to be the maximum integer such that $k \equiv 0 \bmod \phi(p^t)$. We regard $R_0 \cap R\theta$ (for k even) as the Stickelberger ideal. We shall prove:

Theorem 5.1.

$$\boxed{(R_0 : R_0 \cap R\theta) = N p^{\mathrm{ord}\, k - t} \prod_{\chi \neq 1} \pm \frac{1}{k} B_{k,\chi}.}$$

First observe that since $\deg \theta$ and $\deg \theta' \neq 0$ we have

$$R_0 \cap R\theta = R_0 \cap R\theta'.$$

By Theorem 2.1, we conclude that

$$R\theta' \cap R = I\theta', \quad \text{and hence} \quad R\theta' \cap R_0 = I_0\theta'.$$

But $R_0 + I\theta' = R_d$ where

$$d = \deg I\theta' = (\deg I)(\deg \theta').$$

In §3 we had noted $\deg I = p^t$. The factor $\deg \theta'$ will cancel ultimately. In any case, we have:

$$\begin{aligned}(R_0 : R_0 \cap R\theta) &= (R_0 : R_0 \cap R\theta') \\ &= (R_0 : I_0\theta') \\ &= (R_d : I\theta') \\ &= \frac{(R : I\theta')}{(R : R_d)} \\ &= \frac{1}{d}(R : R\theta')(R\theta' : I\theta') \\ &= \frac{1}{d}\prod \chi(\theta')(R : I).\end{aligned}$$

The product is taken over all characters χ of G. We separate this product into a factor with the trivial character, giving $\deg \theta'$, canceling that same factor in d, and the product over the non-trivial characters. For χ non-trivial, we have $\chi(\theta) = \chi(\theta')$.

In the final step we also wrote $(R\theta' : I\theta') = (R : I)$. This is because θ' is invertible in the group algebra over $\mathbf{Q}$. Hence the map $\xi \mapsto \xi\theta'$ induces an isomorphism on R.

We are therefore reduced to proving a final lemma.

Lemma. *$(R : I) = p^s$ where $s = n + \operatorname{ord}_p k$.*

Proof. We have $(R : I) = (R : J)(J : I)$. Any element ξ in R can be written in the form

$$\xi = \sum m(c)\sigma_c = \sum m(c)(\sigma_c - c^k) + \sum m(c)c^k.$$

From this it is clear that $(R : J) = N$, and the index $(J : I)$ is obvious, thus concluding the proof.

Remark. Of course we have not determined the sign occurring in the product of the Bernoulli numbers. It is the sign which makes the product come out positive, and which one determines easily from the functional equation of the zeta function and the factorization in L-series. This is irrelevant for our purposes here.

§6. The Index for k Odd

Assume k is odd. Note that $\theta = \theta'$. Let

$$\varepsilon^- = \tfrac{1}{2}(1 - \sigma_{-1})$$

be the idempotent which projects on the (-1)-eigenspace. It is immediate from the definition that θ is odd, that is,

$$\varepsilon^-\theta = \theta.$$

The **Stickelberger ideal** in this case is $R\theta \cap R = I\theta$, and is odd. We shall prove:

Theorem 6.1.

$$(R^- : R\theta \cap R) = Np^{\operatorname{ord} k} \prod_{\chi \text{ odd}} \pm \frac{1}{2k} B_{k,\chi}.$$

The rest of the section is devoted to the proof.

Lemma 1. *We have* $R^- = 2\varepsilon^- R$ *and* $(\varepsilon^- R : R^-) = 2^{\phi(N)/2}$.

Proof. This is the same as Lemma 1 of §1.

We then proceed as in the even case. First we write

$$(R^- : I\theta) = \frac{(\varepsilon^- R : \varepsilon^- I\theta)}{(\varepsilon^- R : R^-)}.$$

and then

$$\begin{aligned}(\varepsilon^- R : \varepsilon^- I\theta) &= (\varepsilon^- R : \varepsilon^- R\theta)(\varepsilon^- R\theta : \varepsilon^- I\theta) \\ &= \prod_{\chi \text{ odd}} \chi(\theta)(\varepsilon^- R : \varepsilon^- I)\end{aligned}$$

because θ is invertible in $\varepsilon^- \mathbf{Q}[G]$. Furthermore,

$$\begin{aligned}(\varepsilon^- R : \varepsilon^- I) &= (\varepsilon^- R : R^-)(R^- : 2\varepsilon^- I)(2\varepsilon^- I : \varepsilon^- I) \\ &= (R^- : 2\varepsilon^- I)\end{aligned}$$

because $(2\varepsilon^- I : \varepsilon^- I) = 2^{-\phi(N)/2}$ since $\varepsilon^- I$ is free of rank $\phi(N)/2$. Finally,

Lemma 2. $(R^- : 2\varepsilon^- I) = p^s$ *where* $s = n + \operatorname{ord}_p k$.

Proof. The group $2\varepsilon^- I$ is generated by elements of the form

$$(\sigma_c - \sigma_{-c}) - c^k(\sigma_1 - \sigma_{-1}).$$

An element $\xi \in R^-$ lies in $\mathbf{Z}(\sigma_1 - \sigma_{-1}) \bmod I$. Hence the same argument as in the past case gives the desired index.

§7. Twistings and Stickelberger Ideals

The Stickelberger elements θ_k should really be indexed by the groups to which they correspond. We now want to compare factor groups of the group ring by various Stickelberger ideals, twisted in various ways. Consequently, it is not useful any more to have G different in the even or odd case. For this section, we let $N = p^n$ still, and *we allow* $p = 2$. We let

$$G_n = \mathbf{Z}(p^n)^*.$$

We define
$$\theta_{k,c}(p^n) = \sigma_c^{-1}(\sigma_c - c^k)\theta_k(p^n) \in \mathbf{Z}(p^n)[G_n].$$

This makes sense since we know from §1 that $\theta_{k,c}(p^n)$ is p-integral.

Let V be a $\mathbf{Z}(p^n)[G_n]$-module. We define its **twist** to be the tensor product with the roots of unity,
$$V(1) = V \otimes \mu_N.$$

Then σ in G operates diagonally,
$$\sigma(v \otimes \gamma) = \sigma v \otimes \sigma\gamma, \quad \text{and} \quad \sigma_a(v \otimes \gamma) = a(\sigma_a v \otimes \gamma).$$

We let γ be a basis for μ_N over $\mathbf{Z}(N)$. Note that the element a on the right makes sense as an element of $\mathbf{Z}(N)$ since $V \otimes \mu_N$ is a module over $\mathbf{Z}(N)$.

From the definitions we then get the formula

TW 1.
$$\boxed{\theta_{k,c}(v \otimes \gamma) = \theta_{k-1,c}v \otimes \gamma,}$$

resulting from Theorem 2.1(iii),
$$E_{k,c}(a) \equiv a^{k-1}E_{1,c}(a) \bmod N.$$

The distribution relation allows us in **E 2** to replace N by high powers of N at a higher level, and then return to level N to get this congruence.

In particular, if $\theta_{k-1,c}$ annihilates V, then $\theta_{k,c}$ annihilates $V(1)$. The argument simply extracts in a general context the argument given by Coates–Sinnott [C–S 2] in connection with the ideal class groups in cyclotomic fields, see their Theorem 2.1.

Take V to be $\mathbf{Z}(p^n)[G_n]$ itself, so that $V(1)$ is generated by a single element $\sigma_1 \otimes \gamma$. The map
$$\xi \mapsto \xi(\sigma_1 \otimes \gamma)$$
gives an isomorphism
$$\mathbf{Z}(p^n)[G_n] \to \mathbf{Z}(p^n)[G_n] \otimes \mu_{p^n}.$$

Let $\mathscr{S}_k(p^n) =$ ideal of $\mathbf{Z}(p^n)[G_n]$ generated by the elements $\theta_{k,c}(p^n)$. Then the isomorphism induces a bijection
$$\mathscr{S}_k(p^n) \to \mathscr{S}_{k-1}(p^n) \otimes \mu_{p^n}.$$

Hence we get an isomorphism

TW 2.
$$\boxed{\Lambda_n/\mathscr{S}_k(p^n) \xrightarrow{\approx} \Lambda_n \otimes \mu_{p^n}/\mathscr{S}_{k-1}(p^n) \otimes \mu_{p^n},}$$

where $\Lambda_n = \mathbf{Z}(p^n)[G_n]$ is the group ring.

We may then pass to the projective limit. The limit of Λ_n is the Iwasawa algebra. We let $\mathscr{S}_k$ be the ideal generated by the elements $\theta_{k,c}$ (projective limit of $\theta_{k,c}(p^n)$). We obtain an isomorphism with the twist,

$$\Lambda/\mathscr{S}_k \to \Lambda(1)/\mathscr{S}_{k-1}(1).$$

This isomorphism permutes the eigenspaces for the action of μ_{p-1}, and this can be interpreted in terms of congruence relations between Bernoulli–Leopoldt numbers (with characters) in the obvious manner.

We now make remarks concerning twistings, ideal classes, and modular curves. We assume that the reader is acquainted with the latter. Suppose $N = p$ is prime $\neq 2, 3$. The Iwasawa–Leopoldt conjecture predicts an isomorphism

$$C^- \approx (R^-/\mathscr{S}_1)^{(p)},$$

where C^- is the p-primary part of the (-1)-eigenspace of the ideal class group in $\mathbf{Q}(\mu_p)$. On the other hand, Kubert–Lang [KL 7] establish an isomorphism

$$\mathscr{C}^0(X_1(p)) \approx R_0/\mathscr{S}_2,$$

where $\mathscr{C}^0(X_1(p))$ is the cuspidal divisor class group on the modular curve $X_1(p)$, generated by the cusps lying above the rational cusp on $X_0(p)$. Consequently, we expect a commutative diagram:

$$\begin{array}{ccc}
R^+(p)/\mathscr{S}_2(p) & \xrightarrow{\approx} & R^-(p) \otimes \mu_p/\mathscr{S}_1(p) \otimes \mu_p \\
\downarrow & & \downarrow \\
\mathscr{C}^0(X_1(p))(p) & \xrightarrow[??]{\approx} & C^-(p) \otimes \mu_p
\end{array}$$

It remains a problem to give a direct isomorphism at the bottom, from some sort of geometric construction. This may in fact lead to a proof of the Iwasawa–Leopoldt conjecture.

§8. Stickelberger Elements as Distributions

In this section we follow Kubert–Lang [KL 5] to describe a "Stickelberger distribution" associated with a distribution on $\mathbf{Q}/\mathbf{Z}$, and to give its basic properties.

Let h be a function on $\mathbf{Q}/\mathbf{Z}$ (with values in some abelian group, but for the rest of this section, we shall take values in some algebraically closed field F of characteristic 0). We say that h is an **ordinary distribution** if it satisfies the relation

$$h(r) = \sum_{Dt=r} h(t)$$

for every element $r \in \mathbf{Q}/\mathbf{Z}$, and positive integer D. The sum is taken over those elements t such that $Dt = r$. In the application we have in mind, h

will be obtained from the first Bernoulli polynomial, and generalizations on $(\mathbf{Q}/\mathbf{Z})^{(k)}$ lead to the higher Bernoulli polynomials. See [KL 5] for $k > 1$.

We let $G(N) \simeq \mathbf{Z}(N)^*$, writing the isomorphism as $a \mapsto \sigma_a$. We let h be an ordinary distribution as above. We define

$$h_N(x) = h\left(\left\langle \frac{x}{N} \right\rangle\right) \quad \text{for } x \in \mathbf{Z}(N).$$

For any function f on $G(N)$ we define (as usual)

$$S_N(f, h_N) = \sum_a f(a) h_N(a),$$

with the sum taken over $a \in \mathbf{Z}(N)^*$. If we define f on $\mathbf{Z}(N)$ to be 0 outside $G(N)$ then we see that

$$S_N(f, h_N) = \int f\, dh.$$

By abuse of notation, we often write $a \in G(N)$ instead of $a \in \mathbf{Z}(N)^*$.

Let $Z_N = (1/N)\mathbf{Z}/\mathbf{Z}$ and let $r \in Z_N$. We define

$$g_N(r) = \frac{1}{|G(N)|} \sum_{a \in G(N)} h(ra)\sigma_a^{-1}.$$

If the values of h are in the field F, then the values of g_N are in the group algebra $F[G(N)]$. It is clear that if M is a denominator for r, i.e., $r \in Z_M$ and M divides N, then the image of $g_N(r)$ under the canonical homomorphism $G(N) \to G(M)$ is equal to $g_M(r)$. Thus we may define

$$g(r) = \lim g_N(r)$$

in the injective limit of the group algebras (as vector spaces over F), ordered by divisibility, with the injections from one level to a higher one given by sending one group element to the sum of all the group elements lying above it under the canonical homomorphism.

Theorem 8.1. *The function* $g\colon \mathbf{Q}/\mathbf{Z} \to \lim F[G(N)]$ *is an ordinary distribution.*

Proof. Immediate from the definitions.

We define g to be the **Stickelberger distribution associated with** h.

Let A_N be the vector space generated by the values $g(r)$ with $r \in Z_N$ (essentially the same as the vector space generated by the values $g_N(r)$). We observe that $g(0)$ is a constant multiple of the augmentation element, that is

$$g(0) = \frac{h(0)}{|G(N)|} \sum_{\sigma \in G(N)} \sigma.$$

Let χ be a character of $G(N)$ and let $m = m(\chi)$ be its conductor. We define

$$S(\chi, h) = S_m(\chi_m, h_m)$$

where χ_m is the character on $G(m)$ determined by χ. We let

$\hat{G}_h(N)$ = set of characters χ such that $S(\bar{\chi}, h) \neq 0$.

Theorem 8.2. *The dimension of A_N is equal to the cardinality of $\hat{G}_h(N)$.*

Proof. The space generated by the elements $g_N(r)$ with $r \in Z_N$ is clearly a $G(N)$-module since

$$\sigma_b g_N(r) = g_N(rb), \quad \text{for } b \in G(N).$$

We let the idempotent associated with χ be the usual

$$e_\chi = \frac{1}{|G(N)|} \sum_b \bar{\chi}(b)\sigma_b.$$

If M is the conductor of χ, then

$$g_N\left(\frac{1}{M}\right)e_\chi = S(\bar{\chi}, h)\frac{1}{|G(M)|}e_\chi.$$

as one sees at once from the fact that ra depends only on the residue class of a mod M, for $a \in G(N)$. Hence A_N has a non-trivial χ-component if $S(\bar{\chi}, h) \neq 0$. This shows that the dimension of A_N it at least that which we asserted.

On the other hand, let $r \in Z_N$ and suppose r has exact period M. Let χ be any character of $G(N)$. Then

$$\begin{aligned} g_N(r)e_\chi &= \frac{1}{|G(N)|} \sum_{a \in G(N)} h(ra)\bar{\chi}(a)e_\chi \\ &= \frac{1}{|G(N)|} \sum_{b \in G(M)} h(rb) \sum_{\text{red}_M a = b} \bar{\chi}(a)e_\chi. \end{aligned}$$

If the conductor of χ does not divide M, then χ is non-trivial on the kernel of the reduction map

$$\text{red}_M : G(N) \to G(M),$$

and the sum on the right is 0. If the conductor of χ divides M, then $\bar{\chi}(a) = \bar{\chi}(b)$ on the right, so

$$\begin{aligned} g_N(r)e_\chi &= \frac{1}{|G(N)|} \sum_{b \in G(M)} h(rb) \frac{|G(N)|}{|G(M)|} \bar{\chi}(b)e_\chi \\ &= \frac{1}{|G(M)|} \sum_{b \in G(M)} h(rb)\bar{\chi}(b)e_\chi. \end{aligned}$$

Since we can write $r = a/M$ with some a prime to M, a change of variables in the sums shows that up to a non-zero constant factor, $g_N(r)e_\chi$ is equal to

$$S_M(\bar{\chi}_M, h_M)e_\chi.$$

We now have to analyze this sum. The next lemma will show that this sum is equal to some factor times $S_m(\bar{\chi}_m, h_m)$.

Lemma. *Let χ be a character of $G(N)$ with conductor m.*

(i) *If every prime dividing N also divides m then*

$$S_N(\chi_N, h_N) = S_m(\chi_m, h_m).$$

(ii) *Let p be a prime dividing N but not dividing m. Write $N = p^n M$ with $p \nmid M$. Then*

$$S_N(\chi_N, h_N) = (1 - \chi_m(p))S_M(\chi_M, h_M).$$

Proof. The first statement is immediate from the distribution relation. Let us prove (ii). We have

$$\sum_{a \in \mathbf{Z}(N)^*} \chi(a)h_N(a) = \sum_{b \in \mathbf{Z}(M)^*} \chi(b) \sum_{\substack{a \equiv b(M) \\ a \in \mathbf{Z}(N)^*}} h\left(\frac{a}{N}\right).$$

By the distribution relation, we know that

$$h\left(\frac{b}{M}\right) = \sum_{\substack{x \in \mathbf{Z}(N) \\ x \equiv b(M)}} h\left(\frac{x}{N}\right) = \sum_{\substack{a \in \mathbf{Z}(N)^* \\ a \equiv b(M)}} h\left(\frac{a}{N}\right) + \sum_{\substack{a \notin \mathbf{Z}(N)^* \\ a \equiv b(M)}} h\left(\frac{a}{N}\right).$$

The elements a in $\mathbf{Z}(N)$ which are not primitive but are $\equiv b \bmod M$ are in bijection with the elements $c \in \mathbf{Z}(N/p)$ satisfying the conditions

$$a = pc \quad \text{and} \quad c \equiv p^{-1}a \bmod M,$$

under the map

$$c \mapsto pc$$

which sends $\mathbf{Z}(N/p)$ into $p\mathbf{Z}/N\mathbf{Z} \subset \mathbf{Z}/N\mathbf{Z}$. Therefore the sum over primitive elements lying above a given b can be expressed as a difference

$$\sum_{\substack{a \in \mathbf{Z}(N)^* \\ a \equiv b(M)}} h\left(\frac{a}{N}\right) = h\left(\frac{b}{M}\right) - \sum_{c} h\left(\frac{c}{N/p}\right)$$

(where the sum is taken over $c \in \mathbf{Z}(N/p)$, $c \equiv p^{-1}a \bmod M$)

$$= h\left(\frac{b}{M}\right) - h\left(\frac{p^{-1}b}{M}\right)$$

(by the distribution relation). Plugging this into the first relation, and making a change of variables $b \mapsto pb$, we find

$$\begin{aligned} S_N(\chi, h_N) &= S_M(\chi_M, h_M) - \sum_{b \in \mathbf{Z}(M)^*} \chi(b) h\left(\frac{p^{-1}b}{M}\right) \\ &= (1 - \chi_M(p)) S_M(\chi_M, h_M). \end{aligned}$$

This concludes the proof of the lemma.

In applying the lemma to the theorem, we note that the χ-component is at most one-dimensional, and has exactly dimension 1 under the stated condition $S(\bar{\chi}, h) \neq 0$. This concludes the proof of the theorem.

A distribution can be decomposed as a direct sum of an odd and an even distribution, provided that its image is contained in some module on which multiplication by 2 is invertible.

In the next section, we shall prove that the rank of the values on Z_N is at most $|Z_N^*|$, where Z_N^* is the set of primitive elements in Z_N.

If we take for h the distribution arising from the Bernoulli polynomial

$$h(r) = \mathbf{B}_1(\langle r \rangle) \quad \text{if } r \neq 0,\ h(0) = 0$$

then the non-vanishing of $B_{1,\chi}$ for odd characters χ shows that h has the maximal attainable rank for an odd distribution. Consequently, we find:

Thoerem 8.3. *The Stickelberger distribution g associated with $h(r)$ as above is the universal odd ordinary distribution into modules on which multiplication by 2 is invertible.*

So far, Theorem 8.3 has been proved only for distributions with values in a field of characteristic zero. However, the next section will give a result of Kubert showing that the universal distribution is generated on Z_N by free generators whose cardinality is $|Z_N^*|$. This will take care of the additional integrality possibilities allowed in the statement of Theorem 8.3.

Later in the book, we shall see that the cyclotomic units in the cyclotomic field form an even distribution, which has maximal rank by the class number-regulator formula, cf. Chapter 3, §3, and Chapter 6, §3.

The direct sum then yields a distribution of maximal attainable rank. This is one method to show that the universal distribution in Theorem 9.1(ii) has rank $|Z_N^*|$.

§9. Universal Distributions

In this section we give a theorem of Kubert [Ku 1], [Ku 2], constructing a free basis for the universal distribution on $(1/N)\mathbf{Z}/\mathbf{Z}$. In [Ku 2] Kubert gives a complete treatment of the ordinary universal distribution on $\mathbf{Q}^k/\mathbf{Z}^k$ for arbitrary k, as a $GL_k(\mathbf{A}_\mathbf{Z})$-module, where $\mathbf{A}_\mathbf{Z}$ is the ring of integral finite adeles. Here we limit ourselves to $k = 1$, and give only the abelian group structure.

For simplicity of notation we let

$$Z_N = \frac{1}{N}\mathbf{Z}/\mathbf{Z}, \quad \text{and} \quad e_N = \frac{1}{N} \bmod \mathbf{Z}.$$

We let

$$g: \mathbf{Q}/\mathbf{Z} \to \text{some abelian group}$$

be an ordinary distribution, in other words we suppose that for $r \in \mathbf{Q}/\mathbf{Z}$, and a positive integer D we have

$$\sum_{Dt=r} g(t) = g(r).$$

It is clear that such distributions form a category, and we wish to construct the universal distribution.

We let Z_N^* be the set of primitive elements in Z_N, i.e., elements having period exactly N in Z_N.

The prime power case

Let $N = p^n$ be a prime power and write $N = MD$, a factorization with $M > 1$. Let $r \in Z_M^*$. If $Dt = r$ then it is immediate that $t \in Z_N^*$ (N = prime power is used here). The distribution relation shows that $g(r)$ is an integral linear combination of the images of the primitive elements $g(t)$. Hence 0 and these primitive elements generate the universal distribution, at level N.

We have

$$\sum_{t \in Z_N} g(t) = g(0) \quad \text{and} \quad \sum_{t \in Z_{N/p}} g(t) = g(0).$$

Hence we get one relation among primitive elements,

$$\sum_{t \in Z_N^*} g(t) = 0.$$

Let

$$T_N^* = Z_N^* - \{e_N\} \quad \text{and} \quad T_1^* = \{0\}.$$

Let

$$T_N = T_N^* \cup T_1^*.$$

Theorem 9.1. (i) *The elements $g(T_N)$ generate the abelian group generated by $g(Z_N)$.*

(ii) *If g is the universal distribution then the elements $g(t)$ with $t \in T_N$ are free generators.*

(iii) *The cardinality of T_N is equal to that of Z_N^*.*

Proof. The first statement is obvious from the preceding remarks. The cardinality of T_N is clearly equal to that of Z_N^*. For (ii), we may consider the free abelian group generated by the elements of Z_N^* and $\{0\}$, modulo the single linear relation

$$\sum_{t \in Z_N^*} (t) = 0.$$

We can then define g on Z_N^* to be the canonical homomorphism in the factor group, and for $r \in Z_M^*$ with $M \neq N$ and $M | N$ we can define

$$g(r) = \sum_{Dt = r} g(t), \quad \text{with } D = N/M.$$

It is then clear that g defines a mapping on Z_N satisfying the distribution relation.

The proof of (ii) is in some sense natural, but in many ways it is better to exhibit mappings which are distributions and which have the appropriate rank to get the lower bound for the rank of the universal distribution. Cf. the end of §8, where we exhibit natural distributions in the theory of cyclotomic fields which have such rank.

The composite case

To state the theorem concerning the universal distribution in the composite case, we shall write elements of Z_N according to their partial fraction decomposition. Let

$$N = \prod_{i \geq 1} p_i^{n_i}.$$

Then

$$\frac{1}{N}\mathbf{Z}/\mathbf{Z} = \bigoplus \frac{1}{p_i^{n_i}}\mathbf{Z}/\mathbf{Z}$$

and

$$\frac{a}{N} = \sum \frac{a_i}{p_i^{n_i}} \bmod \mathbf{Z}$$

where a_i is well defined mod $p_i^{n_i}$, while a is well defined mod N. We let:

T_N = set of elements a/N as above, such that either a_i is prime to p_i and $a_i \neq 1$, or $a_i = 0$.

It is then clear that T_N has cardinality $\phi(N)$.

Theorem 9.2. *The preceding theorem holds with this definition of T_N, for composite N.*

Proof. The proof will be a simplification of Kubert's proof by Katz. Let A_N be the abelian group generated by $g(Z_N)$. A distribution having the lower bound $\phi(N)$ for its rank has been exhibited in §8. Since T_N has this cardinality, it will suffice to prove that $g(T_N)$ generates A_N. We first show that the elements $g(a/N)$ with a such that a_i is prime to p, or $a_i = 0$, generate A_N. We do this by induction.

Let

$$\sum \frac{b_i}{p_i^{n_i}}$$

be an arbitrary element of Z_N. Write $b_1 = p_1^r a_1$ where a_1 is 0 or prime to p. If $a_1 = 0$ then we are through by induction, so we can assume that a_1 is prime to p, and $1 \leq r < n_1$. Then:

$$g\left(\frac{p_1^r a_1}{p_1^{n_1}} + \sum_{i \geq 2} \frac{b_i}{p_i^{n_i}}\right) = g\left(p_1^r\left(\frac{a_1}{p_1^{n_1}} + \sum_{i \geq 2} \frac{c_i}{p_i^{n_i}}\right)\right)$$

$$= \sum_{j \bmod p^r} g\left(\frac{a_1}{p_1^{n_1}} + \frac{j}{p^r} + \sum_{i \geq 2} \frac{c_i}{p_i^{n_i}}\right)$$

by the distribution relation. Since $r < n_1$ it follows that

$$\frac{a_1}{p_1^{n_1}} + \frac{j}{p_1^r} = \frac{a_1'}{p_1^{n_1}},$$

where a_1' is prime to p.

Inductively, we may now repeat the same argument with respect to p_2, $p_3, \ldots$. It merely suffices to observe the following. In the first step of the argument, when we factored out p_1^r, thus changing b_i to c_i, if b_i is prime to p then c_i is prime to p. Thus performing the same argument inductively on the other primes does not destroy the desired property for those primes which have already been taken care of. This concludes the first part of the proof.

Secondly, we show that we can recover those elements a/N for which a_i may be equal to 1 from the prescribed set T_N. Let

$$N' = N/p_1^{n_1}, \qquad y \in \frac{1}{N'}\mathbf{Z}/\mathbf{Z}.$$

From the distribution relation, we find:

$$\sum_{j \bmod p_1^{n_1}} g\left(\frac{j}{p_1^{n_1}} + y\right) = g(p_1^{n_1} y)$$

$$\sum_{k \bmod p_1^{n_1 - 1}} g\left(\frac{k}{p_1^{n_1 - 1}} + y\right) = g(p_1^{n_1 - 1} y).$$

Subtracting yields

$$\sum_{\substack{j \bmod p_1^{n_1} \\ (j,p)=1}} g\left(\frac{j}{p_1^{n_1}} + y\right) \equiv 0 \bmod A_{N'},$$

where $A_{N'}$ is the group generated by $g(Z_{N'})$. This yields

$$-g\left(\frac{1}{p_1^{n_1}} + y\right) \equiv \sum_{\substack{a_1 \neq 1 \\ (a_1,p_1)=1}} g\left(\frac{a_1}{p_1^{n_1}} + y\right) \bmod A_{N'}.$$

Observe that the same quantity y occurs on both sides of this relation. We may now repeat the procedure inductively on the partial fraction decomposition of y. If we write

$$y_1 = y = \sum_{i \geq 2} \frac{a_i}{p_i^{n_i}},$$

and say $a_2 = 1$, we get a similar congruence

$$-g\left(\frac{a_1}{p_1^{n_1}} + y_1\right) \equiv \sum_{\substack{a_2 \neq 1 \\ (a_2,p_2)=1}} g\left(\frac{a_1}{p_1^{n_1}} + \frac{a_2}{p_2^{n_2}} + y_2\right) \bmod A_{N''},$$

where $N'' = N/p_2^{n_2}$. In this way we reduce the proof to the case when N contains fewer prime factors, and then can apply induction with respect to the number of prime factors to conclude the proof.

§10. The Davenport–Hasse Distribution

In this section we give a relation of Davenport–Hasse [D–H]. Let F_q be the field with $q = p^n$ elements, and let $q \equiv 1 \bmod m$. We follow the notation of Chapter 1, §1. We let $\mathfrak{p}$ be a prime in $\mathbf{Q}(\mu_{q-1})$ lying above p, and let $\mathfrak{P}$ be a prime in $\mathbf{Q}(\mu_{q-1}, \mu_p)$ lying above $\mathfrak{p}$. We write as usual

$$\alpha \equiv \beta \bmod^* \mathfrak{P}$$

to mean that $\alpha\beta^{-1} \equiv 1 \bmod \mathfrak{m}_{\mathfrak{P}}$, where $\mathfrak{m}_{\mathfrak{P}}$ is the maximal ideal in the local ring at $\mathfrak{P}$. We use similar notation mod* $\mathfrak{p}$ or mod* p. We let χ, ψ be characters on F_q^*, and put

$$\tau(\chi) = -S(\chi, \lambda).$$

Theorem 10.1. (Davenport–Hasse) *We have*

$$\prod_{\chi^m=1} \tau(\chi\psi) = \tau(\psi^m)C(\psi, m)$$

where $C(\psi, m) = \psi(m^{-m}) \prod_{\chi^m=1} \tau(\chi)$.

Proof. Let $u_m(\psi)$ be the quotient of the left-hand side by the right-hand side, that is

$$u_m(\psi) = \frac{\prod \tau(\chi\psi)}{\tau(\psi^m)C(\psi, m)}.$$

We have to show $u_m(\psi) = 1$. First note that $u_m(\psi)$ lies in $\mathbf{Q}(\mu_{q-1})$. This is immediate by looking at the action of $\sigma_{1,\nu}$, cf. **GS5** of Chapter 1, §1. From the fact that $|S(\psi)| = \sqrt{q}$ if $\psi \neq 1$ and $|S(\psi)| = 1$ if $\psi = 1$, we conclude that $|u_m(\psi)| = 1$. Similarly, all conjugates of $u_m(\psi)$ have absolute value 1. Since $S(\psi)S(\bar{\psi}) = \pm q$, we know that only primes dividing p occur in the factorization of $S(\psi)$. We shall prove that

$$(1) \qquad u_m(\psi) \equiv 1 \bmod \mathfrak{P}.$$

This will imply that $u_m(\psi)$ is a unit, and therefore a root of unity. If $p \neq 2$, this congruence (1) implies that $u_m(\psi) = 1$. If $p = 2$, we shall give the argument at the end of the proof.

To prove the congruence, we simplify the expression in Stickelberger's theorem. For any integer k we had defined $s(k) = s_q(k)$ and $\gamma(k) = \gamma_q(k)$ in Chapter 1, §2. We let $r(k) = r_q(k)$ be the unique integer such that

$$0 \leq r(k) < q - 1 \quad \text{and} \quad k \equiv r(k) \bmod q - 1.$$

Lemma 1. *Let* $0 \leq k < q - 1$. *Then*

$$k! \equiv (-p)^{\frac{k - s(k)}{p-1}} \gamma(k) \bmod^* p.$$

Proof. By induction. Suppose first that $p \nmid k$. Then $k_0 \geq 1$, and

$$s(k) = s(k - 1) + 1, \qquad \gamma(k) = \gamma(k - 1)k_0.$$

The assertion is then obvious from the inductive step for $k - 1$. Next suppose $p | k$, so $k = pk'$. Since

$$\operatorname{ord}_p k! = \left[\frac{k}{p}\right] + \cdots + \left[\frac{k}{p^{n-1}}\right]$$

and similarly for k', we see that

$$\operatorname{ord}_p k! - \operatorname{ord}_p k'! = k'.$$

In $k! = (k'p)!$, the factors not divisible by p give a contribution of

$$(p - 1)! \equiv -1 \bmod p,$$

taken k' times. The product of the factors divisible by p yields $k'!p^t$, where $t = \operatorname{ord}_p k'!$. The lemma is then immediate.

As in Chapter 1, we let $\varepsilon = e^{2\pi i/p}$. We let $\pi = \varepsilon - 1$. Then from

$$(\varepsilon - 1 + 1)^p - 1 = 0$$

we see at once that

$$\pi^{p-1} \equiv -p \bmod^* \pi.$$

From Stickelberger's theorem and Lemma 1, we conclude that

$$\tau(\omega^{-k}) \equiv \frac{\pi^{r(k)}}{r(k)!} \bmod^* \mathfrak{P}. \tag{2}$$

This reduces the proof of the congruence relation (1) to the proof of such a congruence for the expressions on the right-hand side of (2), corresponding to the way $u_m(\psi)$ is made up from expression $\tau(\omega^{-k})$ for appropriate values of k. We shall prove two relations for the residue function, namely:

$$\sum_{mx \equiv 0} r(x + y) = r(my) + \sum_{mx \equiv 0} r(x) \tag{3}$$

$$\prod_{mx \equiv 0} r(x + y)! \equiv r(my)!\, m^{-r(my)} \prod_{mx \equiv 0} r(x)! \tag{4}$$

In these relations, sums and products are taken over elements $x \bmod q - 1$ such that $mx \equiv 0 \bmod q - 1$. The theorem is immediate from these relations, taking into account

$$m^{r(my)} \equiv \omega(m)^{r(my)} \bmod \mathfrak{P},$$

applied to y such that $\psi = \omega^{-y}$.

We prove the two relations (3) and (4). To begin with, we note that the left-hand side and right-hand side of each relation is unchanged when we change y in a residue class mod $(q - 1)/m$. Consequently we may assume that

$$0 \leq y < \frac{q-1}{m}.$$

We choose the obvious representatives

$$x = \nu\frac{q-1}{m} \quad \text{with } \nu = 0, 1, \ldots, m - 1.$$

Then

$$r(x) = x, \qquad r(x + y) = x + y, \qquad r(my) = my.$$

This makes (3) obvious, and (4) takes the form:

$$m^{my}\frac{\prod\left(y + \nu\frac{q-1}{m}\right)!}{(my)!} = \prod\left(\nu\frac{q-1}{m}\right)! \bmod^* p. \tag{5}$$

The products are taken for $v = 0, 1, \ldots, m - 1$ and y is taken as above, with $0 \leq y < (q - 1)/m$. Let $F(y)$ be the left-hand side of (5). Then the right-hand side of (5) is equal to $F(0)$, and consequently, it suffices to prove that

$$\frac{F(y)}{F(y-1)} \equiv 1 \bmod^* p,$$

with $1 \leq y < (q - 1)/m$, or equivalently

$$m^m \frac{\prod\left(y + v\frac{q-1}{m}\right)}{\prod (my - v)} \equiv 1 \bmod^* p,$$

or also

$$\prod_{v=0}^{m-1} \frac{my + v(q-1)}{my - v} \equiv 1 \bmod^* p.$$

For this it will suffice to prove that each factor in the product is $\equiv 1 \bmod p$. But the power of p entering in $my - v$ is at most p^{n-1}. Dividing numerator and denominator of each factor by $my - v$ shows that

$$\frac{my - v + vq}{my - v} \equiv 1 \bmod^* p.$$

This proves the theorem except when $p = 2$, when we know only that $u_m(\psi) = \pm 1$. In this case we argue further as in [D–H].

Let l be a prime dividing $q - 1$. Let

$$\psi = \psi_l \psi_{l'}$$

be the decomposition of ψ into a product of a character of l-power order, and a character of order prime to l. Then

$$\psi^m = \psi_l^m \psi_{l'}^m$$

is the corresponding decomposition for ψ^m. Let l^w be the highest power of l dividing $q - 1$, and let ζ_{l^w} be a primitive l^wth root of unity. Let

$$\lambda = \zeta_{l^w} - 1.$$

Since $\psi_l \equiv 1 \bmod \lambda$, it follows that $\psi \equiv \psi_{l'} \bmod \lambda$. Therefore

$$\tau(\psi) \equiv (\psi_{l'}) \bmod \lambda \quad \text{and} \quad \tau(\psi_l) \equiv 1 \bmod \lambda.$$

In particular,

$$u_m(\psi) \equiv u_m(\psi_{l'}) \bmod \lambda \quad \text{and} \quad u_m(\psi_l) \equiv 1 \bmod \lambda.$$

Since $u_m(\psi) = \pm 1$, it follows that $u_m(\psi) = 1$, thereby proving the theorem.

Remark. In [Ya], Yamamoto shows that the Gauss sums form the universal odd distribution modulo 2-torsion.

Appendix

In this chapter we have looked at the distributions which are especially relevant to the cyclotomic theory discussed in the rest of the book. It is worthwhile to give here a number of examples of distributions occurring throughout mathematics, involving various classical objects. We make a list of a general nature, including those we have already met.

(1) **The Bernoulli distribution**, which is essentially given by a polynomial.

(2) **The Fourier–Bernoulli distribution**, giving rise to the Bernoulli distribution, as follows. For real θ we have the Fourier expansion

$$\mathbf{B}_k(\langle\theta\rangle) = -\frac{k!}{(2\pi i)^k} \sum_{n \neq 0} \frac{e^{2\pi i n\theta}}{n^k}.$$

Thus we may even define $\mathbf{B}_k$ on $\mathbf{R}/\mathbf{Z}$, and through this Fourier series, the function given at level N by

$$\theta \mapsto N^{k-1}\mathbf{B}_k(\langle\theta\rangle)$$

satisfies the distribution relation.

(3) **The holomorphic Bernoulli distribution.** Let

$$f_k(z) = \sum_{n=1}^{\infty} \frac{z^n}{n^k},$$

and restrict z to the unit circle, $z = e^{2\pi i\theta}$. Then $\{N^{k-1}f_k\}$ defines a distribution. The real part for k even and imaginary part for k odd are mere homomorphic images of this one, and give rise to the Bernoulli distribution of (2).

(4) **The partial zeta functions.** Let

$$\zeta(s, u) = \sum_{n=0}^{\infty} \frac{1}{(n + u)^s}$$

be the Hurwitz zeta function, for $0 < u \leq 1$. For each real number t, let $\{t\}$ be the unique number congruent to t mod $\mathbf{Z}$, and such that

$$0 < \{t\} \leq 1.$$

Then for $a \in \mathbf{Z}(M)$, the function

$$a \mapsto M^{-s}\zeta\left(s, \left\{\frac{a}{M}\right\}\right)$$

satisfies the distribution relation, namely

$$N^{-s} \sum_{b \equiv a(M)} \zeta\left(s, \left\{\frac{b}{N}\right\}\right) = M^{-s}\zeta\left(s, \left\{\frac{a}{M}\right\}\right).$$

The sum on the left is taken for b in $\mathbf{Z}(N)$ reducing to a mod M.

(5) **The gamma distribution.** Define

$$G(z) = \frac{1}{\sqrt{2\pi}} \Gamma(z).$$

We view G as defined on $\mathbf{Q}/\mathbf{Z}$ with the origin deleted, but then with values in the *factor group*

$$G\colon \mathbf{Q}/\mathbf{Z} - \{0\} \to \mathbf{C}^*/\mathbf{Q}_a^*$$

of the multiplicative group of complex numbers, modulo the multiplicative group of all algebraic numbers. The classical identity

$$\prod_{j=0}^{N-1} \frac{1}{\sqrt{2\pi}} \Gamma\left(z + \frac{j}{N}\right) = \frac{1}{\sqrt{2\pi}} \Gamma(Nz) N^{\frac{1}{2} - Nz}$$

shows that G defines a distribution.

Rohrlich has conjectured that G is then the universal odd distribution, with values in groups where multiplication by 2 is invertible. This is a conjecture in the theory of transcendental numbers. It also leads to the question (in algebraic independence) whether the distribution relations, the oddness relations and the functional equations generate an ideal of definition over the algebraic numbers for all *algebraic relations* among the values of the gamma function $(1/\sqrt{2\pi})\Gamma$, with rational arguments.

(6) **The cyclotomic units,** which we have discussed.

(7) **The modular units,** which may be defined by their q-expansions, namely

$$g(a) = -q^{\frac{1}{2}\mathbf{B}_2(a_1)} e^{2\pi i a_2(a_1 - 1)/2}(1 - q_z) \prod_{n=1}^{\infty} (1 - q^n q_z)(1 - q^n/q_z)$$

where $a = (a_1, a_2) \in \mathbf{Q}^2/\mathbf{Z}^2$ and $a \neq (0, 0)$, where

$$z = a_1\tau + a_2,$$

and where the value $g(a)$ is to be taken in the multiplicative group of the modular function field modulo roots of unity, cf. [KL 3], following the work or Ramachandra and Robert. The association

$$a \mapsto g(a)$$

is the universal even ordinary distribution on $\mathbf{Q}^2/\mathbf{Z}^2 - \{0\}$. The ordinary Bernoulli distribution (with $k = 2$) then appears as a homomorphic image of this one.

In the last three examples, the distribution is not defined at 0. In such cases, it is useful terminology to refer to the distribution as **punctured**.

Roughly speaking, I expect that in any classical situation where a distribution arises naturally, it is universal (odd, even, punctured, as the case may be), always subject to taking values in groups where 2 is invertible.

(8) **The Lobatchevski distribution.** I am indebted to Milnor for the following brief comments which might inspire the reader. Define the **Lobatchevski function**

$$\lambda(\theta) = -\int_0^\theta \log|2 \sin t|\, dt.$$

This is essentially the same as the integral

$$-\int_0^\theta \log|e^{2\pi i t} - 1|\, dt.$$

Since the function $t \mapsto |e^{2\pi i t} - 1|$ satisfies the distribution relation, one sees at once that $\lambda(\theta)$ satisfies the distribution relation in the sense that on $\{(1/N)\mathbf{Z}/\mathbf{Z}\}$ the family $\{N\lambda(\theta)\}$ is a distribution, which is odd.

Let H be hyperbolic 3-space. This is the set of points

$$(x_1, x_2, y) \in \mathbf{R} \times \mathbf{R} \times \mathbf{R}^+$$

so (x_1, x_2) is an ordinary point in the plane, and $y > 0$. We endow H with the metric

$$\frac{dx_1^2 + dx_2^2 + dy^2}{y^2}.$$

Select four distinct points in the plane, and let T be the tetrahedron in H whose vertices are at these points. Then it can be shown that opposite dihedral angles are equal. (The dihedral angles are the angles between the faces of the tetrahedron.) Let α, β, γ be the dihedral angles. Then

$$\alpha + \beta + \gamma = \pi,$$

and the volume of the tetrahedron is precisely given in terms of the Lobatchevski function by

$$\iiint_T \frac{dx_1\, dx_2\, dy}{y^3} = \operatorname{Vol} T = \lambda(\alpha) + \lambda(\beta) + \lambda(\gamma).$$

The search for relations among such volumes had led Milnor to consider the Lobatchevski function and its relations, now known as distribution relations, and to show that it had the maximum rank (its values being viewed as contained in a vector space over the rationals). Of course, Kubert's construction in fact gives free generators over $\mathbf{Z}$.

Finally, let $\mathfrak{o} = \mathbf{Z}[\zeta]$, where ζ is a primitive cube root of unity. Then

$$\mathrm{PSL}_2(\mathfrak{o}) \subset \mathrm{PSL}_2(\mathbf{C}) = \operatorname{Aut} H,$$

where $\operatorname{Aut} H$ is the group of automorphisms for the Riemannian structure, orientation preserving. The tetrahedron is essentially a fundamental domain for $\mathrm{PSL}_2(\mathfrak{o})$. This point of view leads into the problem of determining all relations for volumes of fundamental domains in the higher dimensional case.

Complex Analytic Class Number Formulas 3

The complex analytic class number formulas date back to the 19th century. They relate class numbers of cyclotomic fields and units. They arise by factoring the zeta function of a cyclotomic field in L-series, and looking at the factorization of the residue.

§1. Gauss Sums on $\mathbf{Z}/m\mathbf{Z}$

We have to redo the properties developed in Chapter 1, for the ring with divisors of zero $\mathbf{Z}(m) = \mathbf{Z}/m\mathbf{Z}$. The only additional feature arises from the presence of non-zero elements which are not units. We let $m = \prod p^{n(p)}$ be the prime power product. We then have product decompositions

$$\mathbf{Z}(m) = \prod \mathbf{Z}(p^{n(p)}) \quad \text{and} \quad \mathbf{Z}(m)^* = \prod \mathbf{Z}(p^{n(p)})^*.$$

From the product, for any character χ on $\mathbf{Z}(m)^*$ and any character λ on $\mathbf{Z}(m)$ we have a decomposition

$$\chi = \prod_p \chi_p \quad \text{and} \quad \lambda = \prod_p \lambda_p.$$

If $x \in \mathbf{Z}(m)$ and x is not prime to m, we define $\chi(x) = 0$. We let ζ be a primitive mth root of unity (chosen to be $e^{2\pi i/m}$ over the complex numbers), and

$$\lambda(x) = \zeta^x.$$

Observe that $\mathbf{Z}(m)$ is self dual under the pairing

$$(x, y) \mapsto \zeta^{xy}.$$

Let $d|m$. We have a natural surjective homomorphism

$$\mathbf{Z}(m) \to \mathbf{Z}(d)$$

and also a surjective homomorphism

$$\mathbf{Z}(m)^* \to \mathbf{Z}(d)^*.$$

If there does *not* exist $d|m$ and $d \neq m$ such that χ factors through $\mathbf{Z}(d)^*$, then we call χ **primitive**. Again to determine the smallest d such that a given character factors through $\mathbf{Z}(d)^*$, we may look at prime powers.

Suppose $m = p^n$ is a prime power, and χ is a character on $\mathbf{Z}(p^n)$. Let p^r be the smallest power of p such that χ is trivial on

$$1 + p^r\mathbf{Z}(p^n).$$

For convenience, let us abbreviate

$$A = \mathbf{Z}(p^n),$$

so $1 + p^\nu A$ is a group for any positive integer ν. The following criterion is immediate.

χ is primitive if and only if $r = n$.

The power $p^r = p^{r(p)}$ is called the **conductor** of χ.

In the composite case, we let the **conductor** be defined by the product

$$c(\chi) = \operatorname{cond}(\chi) = \prod_{p|m} p^{r(p)}.$$

It is then clear that $c(\chi)$ is the smallest d such that χ factors through $\mathbf{Z}(d)^*$. We define

$$S(\chi) = S(\chi, \lambda) = \sum_x \chi(x)\lambda(x),$$

and the sum could be taken only over those $x \in \mathbf{Z}(m)^*$. It is then obvious that we have a decomposition

$$\boxed{S(\chi, \lambda) = \prod_p S_p(\chi_p, \lambda_p)}$$

where the sum S_p is taken over $\mathbf{Z}(p^{n(p)})^*$.

If d is an integer prime to m, then, as with Gauss sums over finite fields, we have

$$S(\chi, \lambda\circ d) = \bar{\chi}(d)S(\chi, \lambda),$$

by making the change of variables $x \mapsto d^{-1}x$.

On the other hand, if d is not prime to m, we have one new significant feature.

Theorem 1.1. *If χ is primitive and d is not prime to m, then*

$$S(\chi, \lambda\circ d) = 0.$$

Proof. Using the prime power decomposition, we may assume without loss of generality that $m = p^n$ is a prime power. Abbreviate

$$A = \mathbf{Z}(p^n).$$

Also without loss of generality, we may assume $d = p^r$ for some integer $r \geq 1$, and $r < n$. Form a coset decomposition

$$A^* = \bigcup u_i(1 + p^{n-r}A).$$

Then

$$\begin{aligned} S(\chi, \lambda\circ p^r) &= \sum_i \sum_{x\in p^{n-r}A} \chi(u_i)\chi(1 + x)\lambda(p^r u_i) \\ &= \sum_i \chi(u_i)\lambda(p^r u_i) \sum_x \chi(1 + x). \end{aligned}$$

Since χ is assumed primitive, it is non-trivial on $1 + p^{n-r}A$, and the sum on the right is 0, thus proving the theorem.

From here on we have the same formalism as for Gauss sums over finite fields. For any function f on $\mathbf{Z}(m)$ we define its **Fourier transform**

$$Tf(y) = \sum_{x\in \mathbf{Z}(m)} f(x)\lambda(-xy).$$

Theorem 1.2. (i) *We have $T^2 f = mf^-$.*
(ii) *If χ is primitive, then*

$$T\chi = \chi(-1)S(\chi)\chi^{-1}.$$

(iii) *Again if χ is primitive, then*

$$S(\chi)\overline{S(\chi)} = m.$$

Proof. Part (i) is proved as for the finite field case. For (ii), if y is not prime to m, then $T\chi(y) = 0$ by Theorem 1.1. If y is prime to m then we can make the usual change of variables to get the right answer. Part (iii) is then proved as in the finite field case.

§2. Primitive L-series

Let χ be a character mod m. We consider the Dirichlet L-series for $\mathrm{Re}(s) > 1$:

$$L(s, \chi) = \sum_{n=1}^{\infty} \frac{\chi(n)}{n^s} = \sum_{a \in \mathbf{Z}(m)^*} \chi(a) \sum_{n \equiv a} \frac{1}{n^s}.$$

Let ζ again be a primitive mth root of unity. Then we have

$$\frac{1}{m} \sum_{x \in \mathbf{Z}(m)} \zeta^{(a-n)x} = \begin{cases} 0 & \text{if } n \equiv a \bmod m \\ 1 & \text{if } n \not\equiv a \bmod m. \end{cases}$$

Indeed, if $a \not\equiv n \pmod{m}$, then the character $x \mapsto \zeta^{(a-n)x}$ is non-trivial on $\mathbf{Z}(m)$. Consequently we can write the L-series in the form

$$L(s, \chi) = \sum_{a \in \mathbf{Z}(m)} \chi(a) \frac{1}{m} \sum_{n=1}^{\infty} \sum_{x} \zeta^{(a-n)x} \frac{1}{n^s}$$

whence also

$$L(s, \chi) = \frac{1}{m} \sum_{x \in \mathbf{Z}(m)} S(\chi, \lambda \circ x) \sum_{n=1}^{\infty} \frac{\zeta^{-nx}}{n^s}.$$

Theorem 2.1. *Assume that χ is a primitive character* mod m. *Then*

$$L(s, \chi) = \frac{1}{m} S(\chi) \sum_{b \in \mathbf{Z}(m)^*} \bar{\chi}(b) \sum_{n=1}^{\infty} \frac{\zeta^{-nb}}{n^s}.$$

Proof. If x is not prime to m then the Gauss sum is 0 by Theorem 1.1. If b is prime to m, we can make the change of variables which yields the desired expression.

So far we have worked with $\mathrm{Re}(s) > 1$. We now want to have the value of the L-series at $s = 1$. It is not difficult to prove that the L-series has an analytic continuation for $\mathrm{Re}(s) > 0$. Of course, it is also known (and a little more involved) how to prove the analytic continuation to the whole complex plane. For our purposes, to get the value at 1, we can work *ad hoc*, let s be real > 1, and take the limit as s approaches 1. Then we don't need anything else here.

We recall a lemma about series.

Lemma. *Let $\{a_n\}$ be a decreasing sequence of positive numbers, whose limit is 0 as $n \to \infty$. Let $\{b_n\}$ be a sequence of complex numbers, and assume that there is a number $C > 0$ such that for all n,*

$$\left| \sum_{k=1}^{n} b_k \right| \leq C,$$

i.e., the partial sums of the series $\{b_n\}$ *are bounded. Then the series* $\sum a_n b_n$ *converges, and in fact*

$$\left|\sum_{k=1}^{n} a_k b_k\right| \le C a_1.$$

The proof is immediate using summation by parts.

We apply the lemma to the series with $b_n = \zeta^{-nb}$ and $a_n = 1/n^s$ with s real >0. The partial sums of the b_n are clearly bounded (they are periodic).

Let

$$z_0 = \zeta^{-b} \neq 1.$$

For $|z| < 1$ we have

$$-\log(1 - z) = \sum_{n=1}^{\infty} \frac{z^n}{n}.$$

As $z \to z_0$, $-\log(1 - z)$ approaches $-\log(1 - z_0)$. On the other hand, let

$$z = rz_0 \quad \text{with } 0 < r \le 1.$$

Then the series $\sum z^n/n$ converges to $\sum z_0^n/n$ as z tends to z_0 along the ray (that is, r tends to 1). This is again obvious by estimating the tail end of the series using the lemma. Consequently, we find:

Theorem 2.2. *If* χ *is a primitive character, then*

$$L(1, \chi) = -\frac{S(\chi)}{m} \sum_{b \in \mathbf{Z}(m)^*} \bar{\chi}(b) \log(1 - \zeta^{-b}).$$

The picture of the roots of unity looks like in the figure.

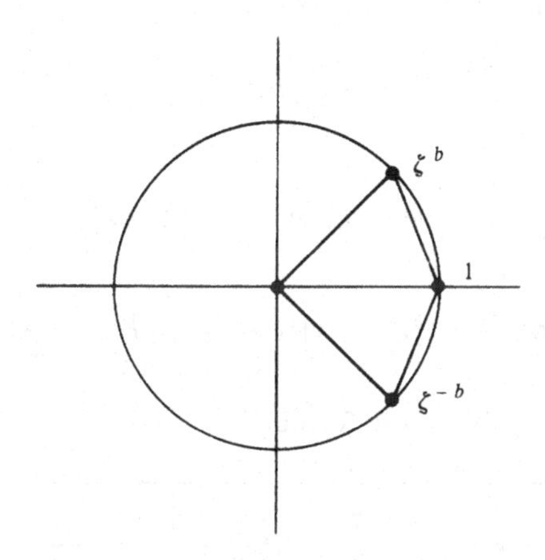

If

$$1 - \zeta^b = |1 - \zeta^b| e^{i\theta},$$

then the picture shows that

$$1 - \zeta^{-b} = |1 - \zeta^{-b}|e^{-i\theta}.$$

The branch of the logarithm is determined so that

$$-\frac{\pi}{2} < \theta < \frac{\pi}{2}.$$

Observe that we do not change the sum

$$\sum \bar{\chi}(b) \log(1 - \zeta^b)$$

if we replace b with $-b$. We shall distinguish two cases.

We say that χ is **even** if $\chi(-1) = 1$, and that χ is **odd** if $\chi(-1) = -1$. *We assume* $m > 2$, and $m = m(\chi)$ is the conductor of χ.

Case 1. χ **is even.**

In this case, adding the sum with b and $-b$ yields

$$2 \sum \bar{\chi}(b) \log(1 - \zeta^{-b}) = \sum \bar{\chi}(b)[\log(1 - \zeta^b) + \log(1 - \zeta^{-b})].$$

With χ even, we obtain the formula

$$\boxed{L(1, \chi) = -\frac{S(\chi)}{m} \sum_{b \in \mathbf{Z}(m)^*} \bar{\chi}(b) \log|1 - \zeta^b|.}$$

Case 2. χ **is odd.**

In this case, we let

$$\zeta = e^{2\pi i/m} \quad \text{and} \quad b = 1, \ldots, m - 1.$$

Then

$$\log(1 - \zeta^{-b}) = \log|1 - \zeta^{-b}| + i\left(\frac{\pi}{2} - \frac{\pi b}{m}\right)$$

$$\log(1 - \zeta^b) = \log|1 - \zeta^b| - i\left(\frac{\pi}{2} - \frac{\pi b}{m}\right).$$

Thus with χ odd, we obtain the formula

$$\boxed{L(1, \chi) = \frac{\pi i S(\chi)}{m} \sum_{b=1}^{m-1} \bar{\chi}(b)\left(\frac{b}{m} - \frac{1}{2}\right) = \frac{\pi i S(\chi)}{m} B_{1,\bar{\chi}}.}$$

Remark. Let m be an integer >1 and let χ be a non-trivial character on $\mathbf{Z}(m)^*$. Then either the conductor of χ is odd, or it is even, in which case it is divisible by 4. Hence for a primitive character, we cannot have $m = 2$.

This is in line with a field theoretic property. Consider the field

$$K = \mathbf{Q}(\mu_m).$$

Let m be the smallest positive integer for which we can write K in this fashion. Then either m is odd or m is divisible by 4. If m is odd, then the group of roots of unity μ_K in K consists of $\pm\mu_m$. If m is even, then $\mu_K = \mu_m$.

§3. Decomposition of L-series

For the applications we have in mind, we have to deal with two types of fields: The cyclotomic field $\mathbf{Q}(\mu_m)$ for some integer $m > 2$, and its maximal real subfield, over which it is of degree 2. We shall use a language which applies to the more general situation of an arbitrary abelian extension of the rationals (known to be contained in a cyclotomic field), but the reader may limit his attention to the two cases mentioned above. Certain proofs can be given *ad hoc* in these cases, while it is easiest to use general class field theory to deal with the general situation. I hope that the extent to which I recall certain proofs here will make the material readable to any reader not acquainted with class field theory.

Let K therefore be an abelian extension of $\mathbf{Q}$, and let K^+ be its real subfield. We let m be the smallest positive integer such that $K \subset \mathbf{Q}(\mu_m)$ (we call m the **conductor** of K). We assume $K \neq \mathbf{Q}$, and as said above, you may assume $K = \mathbf{Q}(\mu_m)$ or $K = \mathbf{Q}(\mu_m)^+$. We have a surjective homomorphism

$$\mathbf{Z}(m)^* \to \mathrm{Gal}(K/\mathbf{Q}) = G_{K/\mathbf{Q}}$$

Any character χ of $G_{K/\mathbf{Q}}$ gives rise to a character on $\mathbf{Z}(m)^*$, also denoted by χ. We let $m(\chi)$ be its conductor. We may view χ as factored through $\mathbf{Z}(m(\chi))^*$, in which case we speak of χ as the corresponding primitive character. If we need to make a distinction between χ as character on $\mathbf{Z}(m)^*$ or the corresponding primitive character on $\mathbf{Z}(m(\chi))^*$, then we denote this primitive character by χ_0. The context should always make clear which is meant.

Let

$$\zeta_K(s) = \prod \left(1 - \frac{1}{\mathrm{N}\mathfrak{p}^s}\right)^{-1}$$

be the zeta function associated with K. It is a fact that there is a decomposition

$$\zeta_K(s) = \prod_{\chi} L(s, \chi),$$

where the product is taken over all the *primitive* characters induced by the characters of $G_{K/\mathbf{Q}}$. We reproduce the proof in the case $K = \mathbf{Q}(\mu_m)$. In the last section we dealt with the L-series in its additive form. Here we use the multiplicative form

$$L(s, \chi) = \prod \left(1 - \frac{\chi(p)}{p^s}\right)^{-1}$$

where the product is taken over all primes p not dividing $m(\chi)$. All these series and products converge absolutely for $\mathrm{Re}(s) > 1$, and what is to be proved amounts to formal identities, localized at each prime p. Specifically, the decomposition is equivalent to proving for each prime number p:

$$\prod_{\mathfrak{p}|p} \left(1 - \frac{1}{N\mathfrak{p}^s}\right) = \prod_{\chi} \left(1 - \frac{\chi(p)}{p^s}\right).$$

It is therefore convenient to let $t = p^{-s}$. As usual, let

$$(p) = (\mathfrak{p}_1 \cdots \mathfrak{p}_r)^e, \qquad \mathrm{N}\mathfrak{p} = p^f$$

be the decomposition of p in prime ideals in K. Then

$$efr = [K : \mathbf{Q}].$$

The identity to be shown is then equivalent to

$$(1 - t^f)^r = \prod_{\chi} (1 - \chi(p)t).$$

Suppose first that $p \nmid m$. Then $e = 1$. The prime p generates a cyclic subgroup of order f in $\mathbf{Z}(m)^*$,

$$\mathbf{Z}(m)^* \supset \{p\} \supset \{1\}.$$

The value of a character χ viewed as character on $\mathbf{Z}(m)^*$ or as primitive character are the same on p. There are f distinct characters on the cyclic group $\{p\}$, corresponding to the fth roots of unity, each such character assigning one of these roots of unity to p. Each one of these characters then extends in r possible ways to $\mathbf{Z}(m)^*$. Since trivially we have the factorization

$$1 - t^f = \prod_{\zeta^f = 1} (1 - \zeta t),$$

we have proved our identity in the case $p \nmid m$. The argument is, by the way, entirely similar if $K \neq \mathbf{Q}(\mu_m)$.

Suppose secondly that $p|m$. Write $m = p^k m'$ with $(p, m') = 1$. If $p | m(\chi)$

then by definition $\chi(p) = 0$. If $p \nmid m(\chi)$ then χ factors through $\mathbf{Z}(m')^*$. We are therefore reduced to proving the identity

$$(1 - t^f)^r = \prod (1 - \chi(p)t)$$

where the product is taken over those χ whose conductor $m(\chi)$ is not divisible by p, and hence such that χ factors through $\mathbf{Z}(m')^*$. The arguments are then identical with the preceding arguments, replacing m by m'. This concludes the proof. (Cf. [L 1], Chapter XII, §1.)

As usual, we let r_1, r_2 be the number of real and complex conjugate embeddings of K.

If K is real then $r_1 = [K:\mathbf{Q}]$, $r_2 = 0$.

If K is not real, then $r_1 = 0$ and $r_2 = \frac{1}{2}[K:\mathbf{Q}]$.

We let

$$N = [K:\mathbf{Q}] = N_K.$$

We assume known the analytic continuation of the zeta function and L-series at 1 (cf. [L 1], Chapter VII, [L 3], Chapter XIV). By comparing residues, we have the **class number formula**:

CNF.
$$\boxed{\frac{2^{r_1}(2\pi)^{r_2}hR}{wd^{1/2}} = \prod_{\chi \neq 1} L(1, \chi).}$$

As usual:

$w = w_K =$ number of roots of unity in K.

$h = h_K =$ class number of K.

$R = R_K =$ regulator of K.

$d = d_K =$ absolute value of the discriminant.

If K is real, so $r_2 = 0$, then $w = 2$ and the formula reads:

$$\frac{hR}{\sqrt{d}} = \prod_{\chi \neq 1} \tfrac{1}{2}L(1, \chi).$$

Leopoldt's p-adic analogue will be given in the next chapter. If K is not real, then we let h^+, R^+ denote the class number and regulator of its real subfield and N^+ is the degree of the real subfield,

$$N^+ = N/2 = r_2.$$

We shall also need another fact whose proof is somewhat more delicate.

Theorem 3.1. *We have product expressions:*

(i)
$$\prod_{\chi \neq 1} m(\chi) = d$$

(ii)
$$\prod_{\chi \neq 1} S(\chi) = \begin{cases} d^{1/2} & \text{if } K \text{ is real} \\ i^{r_2} d^{1/2} & \text{if } K \text{ is not real.} \end{cases}$$

Proof. It is possible to give essentially algebraic proofs for these facts (although the sign of the Gauss sums is always a little delicate, involving something about the complex numbers). The best way to see the theorem, however, is probably as in Hasse [Ha 1], using the functional equations of the zeta function and L-series. Indeed, under the change $s \mapsto 1 - s$, the functions

$$d^{s/2}(\pi^{-s/2}\Gamma(s/2))^N \zeta_K(s) \qquad \text{if } K \text{ is real}$$

$$d^{s/2}(\pi^{-s/2}\Gamma(s/2))^{N/2}\left(\pi^{-s/2}\Gamma\left(\frac{1+s}{2}\right)\right)^{N/2} \zeta_K(s) \quad \text{if } K \text{ imaginary}$$

are invariant. On the other hand, under the transformation

$$s \mapsto 1 - s \quad \text{and} \quad \chi \mapsto \bar{\chi},$$

the following functions (for non-trivial χ)

$$m(\chi)^{s/2}(\pi^{-s/2}\Gamma(s/2))L(s, \chi) \qquad \text{if } \chi \text{ is even}$$

$$m(\chi)^{s/2}\left(\pi^{-s/2}\Gamma\left(\frac{1+s}{2}\right)\right)L(s, \chi) \quad \text{if } \chi \text{ is odd}$$

take on the factor

$$\frac{\sqrt{\chi(-1)m(\chi)}}{S(\chi)}.$$

Dividing the functional equation of the zeta function by the functional equation of the L-series, one sees that under $s \mapsto 1 - s$,

$$\left(\frac{m(\chi)}{d}\right)^{s/2} \quad \text{takes on the factor} \quad \frac{\sqrt{\chi(-1)m(\chi)}}{S(\chi)}.$$

The theorem then follows at once.

If we combine the residue formula, Theorem 3.1, and the expressions for the values $L(1, \chi)$ for primitive characters χ found in the last section, we then get the following factorizations for the product hR in the two cases.

K real.

$$2^{N-1} hR = \prod_{\chi \neq 1} \sum_{b \bmod m(\chi)} -\chi(b) \log|1 - \zeta_{m(\chi)}^b|.$$

Warning: In this case, $N = N^+$, $h = h^+$, $R = R^+$ and characters are even.

K imaginary.

$$\frac{2^{N/2}hR}{w} = \prod_{\substack{\chi \text{ even} \\ \chi \neq 1}} \sum_{b \bmod m(\chi)} -\chi(b) \log|1 - \zeta_{m(\chi)}^b| \cdot \prod_{\chi \text{ odd}} -B_{1,\chi}.$$

In the real case, we observe that all characters are even. Also the number of roots of unity in K when K is real is equal to 2. Otherwise, the formulas are just obtained by plugging in.

It will be convenient to reformulate them slightly, to make the connection between imaginary K and the maximal real subfield clearer. We let:

$$\begin{aligned} E &= E_K = \text{group of units in } K \\ E^+ &= E_{K^+} = \text{group of units in } K^+ \\ \mu_K &= \text{group of roots of unity in } K \\ C_K &= \text{group of ideal classes in } K. \end{aligned}$$

Lemma. *We have the index*

$$(E : \mu_K E^+) = \frac{2^{(N/2)-1}R^+}{R}.$$

Proof. This is obvious by computing the regulator of the units in K^+ with respect to K, where local factors of 2 occur in each row of the determinant expressing the regulator, whereas a local factor of 1 occurs in the corresponding determinant giving the regulator of the units in K^+.

Following Hasse, we give a symbol for the index in the lemma, calling it the **unit index:**

$$\boxed{Q_K = Q = (E : \mu_K E^+).}$$

Reading the class number formula in the real case applied to K^+, we find:

Theorem 3.2. *For imaginary K,*

$$h = h^+ Q w 2^{-N/2} \prod_{\chi \text{ odd}} -B_{1,\chi}.$$

In the next section, we shall analyze more closely the decomposition

$$h = h^+ h^-,$$

where h^- is defined as h/h^+, and we shall see that h^- is an integer. In any case, we have the class number formula:

CNF$^-$.

$$\boxed{h^- = Qw \prod_{\chi \text{ odd}} -\tfrac{1}{2}B_{1,\chi}.}$$

In the next section, we shall prove that $Q = 1$ if $K = \mathbf{Q}(\mu_m)$ and m is a prime power. In addition, h^- will be interpreted as the order of the (-1)-eigenspace of the ideal class group. From Theorem 1.1 of Chapter 2, we find:

Theorem 3.3. *If m is a prime power, $K = \mathbf{Q}(\mu_m)$, $G = \mathrm{Gal}(K/\mathbf{Q})$, and $\mathscr{S}$ is the Stickelberger ideal, then*

$$h^- = (\mathbf{Z}[G]^- : \mathscr{S}).$$

Let p be a prime number. If A is an abelian group, we denote by $A^{(p)}$ its p-primary part. As Iwasawa observed [Iw 7], knowing the index immediately shows that:

The group $C_K^{-(p)}$ is generated by one element over $\mathbf{Z}[G]$ if and only if there is a $\mathbf{Z}[G]$-isomorphism

$$C_K^{-(p)} \approx (\mathbf{Z}[G]^-/\mathscr{S})^{(p)}.$$

Indeed, we know that the Stickelberger ideal annihilates the ideal classes, so the isomorphism is obvious if there exists one generating element by Theorem 1.1 of Chapter 2.

Let $m = p$ itself. Iwasawa [Iw 7] and Leopoldt [Le 5], [Le 10] have shown that if the Kummer–Vandiver conjecture h^+ prime to p is true, then the cyclicity follows for the p-primary part of C^-. (See Chapter 6, §4.) Proving the Kummer–Vandiver conjecture, or the Iwasawa–Leopoldt conjecture that $C_K^{-(p)}$ is cyclic over the group ring is therefore one of the major problems of algebraic number theory today.

In the Iwasawa–Leopoldt conjecture it is necessary in general to restrict the conjecture to the p-primary component. For example, let F be an imaginary quadratic field $\mathbf{Q}(\sqrt{-p})$, and suppose p is such that F is contained in the cyclotomic field $\mathbf{Q}(\mu_p) = K$. Then K over F is totally ramified above p, and the Hilbert class field of F lifts to an unramified extension of K of the same degree, so the ideal class group C_F is a factor group of the ideal class group C_K, and $C_F = C_F^-$. It is known that there exist such fields, e.g., $\mathbf{Q}(\sqrt{-3299})$, for which C_F contains a group of type (3, 3), see Scholz–Taussky [S–T]. Furthermore, 3 does not divide $p - 1$, with $p = 3299$. Consequently all the non-trivial eigenspaces for characters of $\mathbf{Z}(p)^*$ of the local ring group $\mathbf{Z}_3[G]$ are cyclic over $\mathbf{Z}_3$. This shows that there cannot be an isomorphism

$$C_K^- \approx \mathbf{Z}[G]^-/\mathscr{S}_1,$$

and thus in general, the Iwasawa–Leopoldt conjecture has to be restricted to the p-primary component.

For h^+ we also get a formula, and it is convenient to introduce the group

$$G = \mathbf{Z}(m)^*/\pm 1,$$

and for each even character χ the group

$$G_\chi = \mathbf{Z}(m(\chi))^*/\pm 1.$$

Then there are exactly $N/2$ even characters and $(N/2) - 1$ non-trivial even characters. Therefore we obtain the other class number formula:

CNF$^+$.

$$h^+ = \frac{1}{R^+} \prod_{\chi \neq 1} \sum_{b \in G_\chi} -\chi(b) \log|1 - \zeta_{m(\chi)}^b|.$$

The product over $\chi \neq 1$ is taken over the non-trivial characters of G_χ, or equivalently the non-trivial even characters of $\mathbf{Z}(m)^*$. This product will be interpreted as a determinant of certain units in §5, and it will follow that h^+ is equal to the index of a certain subgroup of the units in the group of all units.

§4. The (± 1)-eigenspaces

In this section we analyze in greater detail the factors h^+ and h^- of the class number, and the corresponding unit index. We assume that m is odd or $m \equiv 0 \bmod 4$.

Theorem 4.1. *Let $K = \mathbf{Q}(\mu_m)$. Then $Q_K = 1$ if m is a prime power, and 2 if m is not a prime power.*

Proof. Let $E = E_K$ be the unit group in K. For each unit u in E, the quotient $\bar{u}/u$ is a unit, of absolute value 1, and for any automorphism σ of K over $\mathbf{Q}$, we have

$$\sigma(\bar{u}/u) = \overline{\sigma u}/\sigma u$$

because σ commutes with complex conjugation (abelian Galois group). Hence all conjugates of $\bar{u}/u$ have absolute value 1. Hence $\bar{u}/u$ is a root of unity. Let

$$\varphi : E \to \mu = \mu_K$$

be the homomorphism $\varphi(u) = \bar{u}/u$. Then

$$\mu^2 \subset \varphi(E) \subset \mu,$$

because if u is a root of unity, then $\varphi(u) = u^{-2}$ so the image of φ contains the squares. Hence the index of $\varphi(E)$ in μ is 1 or 2 because μ is cyclic. Furthermore, we see at once:

$$Q_K = 2 \textit{ if and only if } \varphi \textit{ is surjective, i.e., } \varphi(E) = \mu.$$

Assume that m is composite. Let ζ be a generator of μ if m is even, and a generator of the odd part of μ if m is odd. Then $1 - \zeta$ is a unit (elementary fact, and easy exercise), and $\varphi(1 - \zeta) = -\zeta^{-1}$, so $\varphi(E) = \mu$, in other words φ is surjective. On the other hand, $\varphi(E^+\mu) = \mu^2$, so the index is 2 in this case.

Suppose next that m is a prime power, $m = p^n$. We contend that $\varphi(E) \neq \mu$. It will follow that $\varphi(E) = \mu^2$, and since the kernel of φ is E^+ the theorem also follows in this case. Suppose $\varphi(E) = \mu$. Let ζ be a primitive mth root of unity, and let u be a unit such that

$$\bar{u}/u = -\zeta^{-1}.$$

Let

$$\alpha = \frac{1 - \zeta}{u}.$$

Then $\alpha/\bar{\alpha} = 1$ so $\alpha = \bar{\alpha}$ and α is real. But $1 - \zeta$ is a prime element above p in K and so α is also a prime element, which cannot lie in the real subfield. This proves the theorem.

Theorem 4.2. *Let $K = \mathbf{Q}(\mu_m)$. The natural map*

$$C_{K^+} \to C_K$$

of ideal classes in K^+ into the ideal class group of K is injective.

Proof. Let $\mathfrak{a}$ be an ideal of K^+ and suppose $\mathfrak{a} = (\alpha)$ with α in K. Then $\bar{\alpha}/\alpha$ is a unit, and in fact a root of unity as one sees by an argument similar to that in Theorem 4.1. Suppose that m is composite. By Theorem 4.1, we know that $Q_K = 2$ and φ is surjective, so there exists a unit u such that

$$u/\bar{u} = \bar{\alpha}/\alpha.$$

Then αu is real, and generates the same ideal as α, thus proving the theorem in this case.

Suppose that $m = p^n$ is a prime power. Let ζ be a primitive mth root of unity, and let $\lambda = 1 - \zeta$, so λ is a prime element above p in K. We can write $\bar{\alpha} = \alpha z$ with some root of unity, and $\lambda/\bar{\lambda} = -\zeta$ is a generator of μ. Hence

$$z = (\lambda/\bar{\lambda})^k$$

for some positive integer k. Then $\alpha\lambda^k$ is real. Since the ideal generated by α comes from K^+, and since p is totally ramified in K, it follows that k is even.

Hence z is a square in μ, and therefore in the image of φ, say $z = u/\bar{u}$ for some unit u. Then αu is real, and generates the same ideal as α, thus proving that $\mathfrak{a}$ is principal, and also proving the theorem.

Theorems 4.1 and 4.2 are classical, see for instance Hasse [Ha 1], Chapter 3, for more general results. The elegant proofs given here are due to Iwasawa.

Theorem 4.3. *Let K be an imaginary abelian extension of* $\mathbf{Q}$. *Then the norm map*

$$N_{K/K^+}: C_K \to C_{K^+}$$

on the ideal class group is surjective.

Proof. We have to use class field theory, which gives the more general statement:

Lemma. *Let K be an abelian extension of a number field F. Let H be the Hilbert class field of F (maximal abelian unramified extension of F). If $K \cap H = F$ then the norm map $N_{K/F}: C_K \to C_F$ is surjective.*

Proof. For any ideal class c in K, the properties of the Artin symbol show that

$$(c, KH/K) \text{ restricted to } H = (N_{K/F}c, H/F).$$

We have natural isomorphisms of Galois groups:

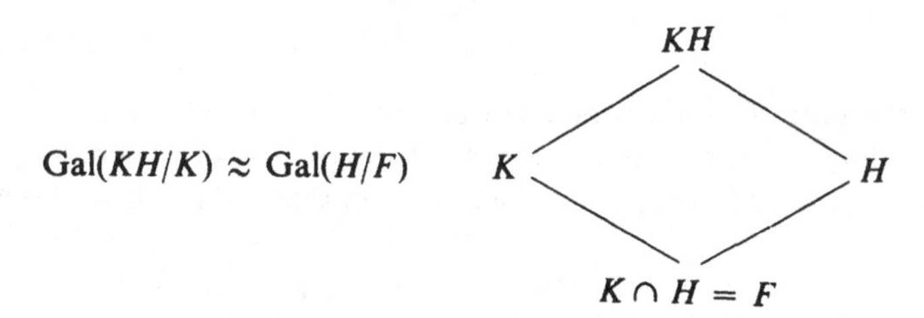

Hence the group $(N_{K/F}C_K, H/F)$ is the whole Galois group $\mathrm{Gal}(H/F)$, whence $N_{K/F}C_K = C_F$ since the Artin symbol gives an isomorphism of the ideal class group with the Galois group. This proves the lemma.

The theorem follows at once, because K over K^+ is ramified at the archimedean primes, and hence cannot intersect the Hilbert class field of F except in F.

Let τ denote complex conjugation. Let

$$\begin{aligned} C_K^- &= (-1)\text{-eigenspace of } C_K \\ &= \{c \in C_K \text{ such that } c^{1+\tau} = 1\}. \end{aligned}$$

Theorem 4.4. *Let* $K = \mathbf{Q}(\mu_m)$. *Then the sequence*

$$1 \to C_K^- \to C_K \xrightarrow{\text{norm}} C_{K^+} \to 1$$

is exact.

Proof. We consider the norm map followed by the injection,

$$C_K \xrightarrow{\text{norm}} C_{K^+} \xrightarrow{\text{inj}} C_K.$$

The kernel of this composite map is C_K^- by definition, so the theorem is obvious by what had already been proved.

Corollary. *The quotient* h/h^+ *is an integer, which is the order of the group* C_K^-.

Remark. The integer h^- is called the **first factor**, and h^+ is called the **second factor** of the class number, in older literature. This is poor terminology since the ordering seems arbitrary, and for several years this has been replaced by the plus and minus terminology.

§5. Cyclotomic Units

Let m again be the conductor of the cyclotomic field $\mathbf{Q}(\mu_m)$, so either m is odd > 1 or m is divisible by 4. Let ζ be a primitive mth root of unity. For b prime to m we let

$$g_b = \frac{\zeta^b - 1}{\zeta - 1}.$$

Then g_b is a unit called a **cyclotomic unit.** It is easy to see that g_b is equal to a real unit times a root of unity. Indeed, without loss of generality we may assume that b is odd, since ζ^b depends only on the residue class of b mod m. Then

$$\zeta^{-\nu} g_b \quad \text{for} \quad \nu = \frac{b-1}{2}$$

is real (i.e., fixed under σ_{-1}), as one sees immediately from the definitions. We let g_b^+ be this real unit, uniquely determined up to sign, and call it the **real cyclotomic unit.**

We let $\mathscr{E}$ be the group of units in $\mathbf{Q}(\mu_m)$ generated by the roots of unity and the cyclotomic units. We let $\mathscr{E}^+$ be the group of units in $\mathbf{Q}(\mu_m)^+$ generated by ± 1 and the real cyclotomic units. Then

$$E/\mathscr{E} \approx E^+/\mathscr{E}^+.$$

Observe that g_b and g_{-b} differ by a root of unity.

As before, let $N = [\mathbf{Q}(\mu_m) : \mathbf{Q}]$ and let

$$r = \frac{N}{2} - 1.$$

Then r is the rank of E, and also the rank of E^+. If $\varepsilon_1, \ldots, \varepsilon_r$ is a basis for E^+ (mod roots of unity), then the **regulator** R^+ is the absolute value of the determinant

$$\boxed{R(E) = R^+ = \pm \det_{a,j} \log|\sigma_a \varepsilon_j|}$$

where $j = 1, \ldots, r$ and $a \in \mathbf{Z}(m)^*/\pm 1$ and $a \not\equiv \pm 1 \pmod{m}$. It is convenient to let

$$G = \mathbf{Z}(m)^*/\pm 1$$

so we may view $a \in G$, $a \neq 1$ in G.

On the other hand, we may form the **cyclotomic regulator**

$$\boxed{R(\mathscr{E}) = R_{\text{cyc}} = \pm \det_{a,b \neq 1} \log|\sigma_a g_b|}$$

again with $a, b \in G$, and of course it does not matter if we write g_b or g_b^+ since the absolute value of a root of unity is 1.

For composite levels m the cyclotomic units are not necessarily independent, and so we now turn to prime power level,

$$m = p^n.$$

We shall prove in this case that the cyclotomic units are independent.

Interpreting the regulator as the volume of a fundamental domain for the lattice generated by the log vectors of units in $\mathbf{R}^r$, we see that

$$\boxed{(E : \mathscr{E}) = (E^+ : \mathscr{E}^+) = R_{\text{cyc}}/R^+.}$$

Remark. For composite m, as with the index of the Stickelberger ideal, it is necessary to consider the group generated by cyclotomic units of all intermediate levels to get a group of units of the right rank.

Theorem 5.1. *Let $K = \mathbf{Q}(\mu_m)$ and $h = h_K$. Assume $m = p^n$ is a prime power. Then*

$$h^+ = (E^+ : \mathscr{E}^+) = (E : \mathscr{E}).$$

Proof. Let G be any finite abelian group. Then we have the Frobenius determinant formula for any function f on G:

$$\prod_{\chi \neq 1} \sum_{a \in G} \chi(a) f(a^{-1}) = \det_{a,b \neq 1} [f(ab^{-1}) - f(a)].$$

The proof will be recalled later for the convenience of the reader. It is already clear that up to minor changes, this formula yields the theorem, taking into account the expression for h^+ obtained at the end of §3. We now make these changes explicit.

Lemma 1. *We have for* $G = \mathbf{Z}(m)^*/\pm 1$:

$$\begin{aligned} \pm \det_{a,b \neq 1} \log|\sigma_a g_b| &= \prod_{\chi \neq 1} \sum_{b \in G} \chi(b) \log|1 - \zeta^b| \\ &= \prod_{\chi \neq 1} \sum_{b \in G} \chi(b) \log|g_b|. \end{aligned}$$

Proof. The first expression comes from the Frobenius determinant formula (Theorem 6.2), and the second comes from the fact that for non-trivial χ,

$$\sum \chi(b) \log|1 - \zeta| = 0.$$

Lemma 2. *Let* $G_\chi = \mathbf{Z}(m(\chi))^*/\pm 1$. *For prime power* $m = p^n$, *we have*

$$\sum_{b \in G_\chi} \chi(b) \log|1 - \zeta_{m(\chi)}^b| = \sum_{b \in G} \chi(b) \log|1 - \zeta_m^b|.$$

Proof. Let $m(\chi) = p^s$. We write residue classes in $\mathbf{Z}(p^n)^*$ in the form

$$y = b + p^s c, \quad \text{with } 0 \leq c < p^{n-s},$$

and b ranges over a fixed set of representatives for residue classes of $\mathbf{Z}(p^s)^*$. Instead of the sums over G_χ and G respectively, it is easier now to work with sums over $\mathbf{Z}(p^s)^*$ and $\mathbf{Z}(p^n)^*$ respectively, and then divide by 2. The desired relation is then immediate from the identity

$$\prod_{\lambda^m = 1} (X - \lambda Y) = X^m - Y^m,$$

because we get

$$\sum_{y \bmod p^n} \chi(y) \log|1 - \zeta_{p^n}^y| = \sum_{b \bmod p^s} \chi(b) \log|1 - \zeta_{p^s}^b|.$$

This proves the lemma.

Theorem 5.1 is then immediate from the lemmas, and the class number formula for h^+ obtained from the L-series.

It is generally believed that the coincidence of group orders in Theorem 5.1 does not correspond to an isomorphism of the groups involved. Iwasawa has a counterexample at least that C^+ is not isomorphic to $E/\mathscr{E}$ as Galois module. Mazur has pointed out that the analogous statement for the case of elliptic curves with complex multiplication is definitely false.

We conclude this section by mentioning the most classical case of the quadratic subfield. For our purposes we are interested in the case of the *real* quadratic subfield. Thus for the end of this section, we let

$$m = p \quad \text{with } p \text{ prime} \neq 2, 3$$

and such that $K = \mathbf{Q}(\mu_m)$ contains a real subfield $F = \mathbf{Q}(\sqrt{D})$ with $D > 0$, so $D = p$, and D is the discriminant. Let $\varepsilon > 1$ be a fundamental unit of F, and h_F the class number. From

$$\zeta_F(s) = \zeta_{\mathbf{Q}}(s)L(s, \chi)$$

where χ has order 2, we get

$$\begin{aligned} \frac{2h_F \log \varepsilon}{\sqrt{D}} &= L(1, \chi) \\ &= -\frac{S(\chi)}{D} \sum_{a=1}^{D-1} \bar{\chi}(a) \log|1 - \zeta^a|. \end{aligned}$$

It is a simple matter of the theory of quadratic fields that the conductor $m(\chi)$ is exactly D (assumed > 0). The explicit value $S(\chi)$ can be determined in any number of ways (via functional equation, via Dirichlet's method as in my *Algebraic Number Theory*, Chapter IV, §3, etc.), and we have $S(\chi) = \sqrt{D}$. Thus we find:

Theorem 5.2. *For a real quadratic field $F = \mathbf{Q}(\sqrt{D})$ as above,*

$$2h_F \log \varepsilon = - \sum_{a \bmod D} \bar{\chi}(a) \log|1 - \zeta^a|.$$

We have the tower of fields:

$$\begin{array}{c} K \\ | \\ K^+ \\ | \\ F \\ | \\ \mathbf{Q} \end{array} \qquad K^+ / F: \ \mathbf{Z}(D)^*/\pm 1$$

Let $\mathscr{E}$ be the group of cyclotomic units in K^+ and let $\mathscr{E}_F$ be $\pm$ its norm group into F, so

$$\mathscr{E}_F = \pm N_{K^+/F}\mathscr{E}.$$

Then $\mathscr{E}_F \pmod{\pm 1}$ is infinite cyclic.

Theorem 5.3. $$h_F = (E_F : \mathscr{E}_F).$$

Proof. Let

$$\alpha = \prod_{\chi(a)=1} (1 - \zeta^a) \quad \text{and} \quad \alpha' = \prod_{\chi(a)=-1} (1 - \zeta^a),$$

where ζ is a fixed primitive Dth root of unity. Note that the character χ is even, so a and $-a$ occur simultaneously in each product. Therefore the norm from K^+ to F of any real cyclotomic unit

$$\zeta^b \frac{1 - \zeta^c}{1 - \zeta}$$

is a unit in F, and the group generated by these norms $\pmod{\pm 1}$ is infinite cyclic, generated by a unit $\eta > 0$ such that

$$\pm \eta^2 = \alpha'/\alpha.$$

From Theorem 5.2 we conclude that

$$h_F \log \varepsilon = \log \eta.$$

Thus $\eta = \varepsilon^h$, and since $\eta \pmod{\pm 1}$ generates the norms of cyclotomic units in K^+, this proves the index relation of Theorem 5.3.

This index relation is analogous to that of Theorem 5.1 for the full cyclotomic field. Since K^+ is totally ramified over F (at the prime p) it follows from class field theory that

$$h_F \text{ divides } h_K^+.$$

(*Proof*: Let H_F be the Hilbert class field of F. Then $H_F \cap K^+ = F$, so

$$[H_F K^+ : K^+] = [H_F : F] = h_F.$$

Since $H_F K^+/K^+$ is unramified, it follows by class field theory that h_F divides h_K^+.)

For tables of some h_F, see Borevich–Shafarevich, *Number Theory*, Academic Press, p. 424. It has been observed for a long time that h_F has very small values, and grows very slowly. It is unknown if there are infinitely many real quadratic fields of class number 1.

§6. The Dedekind Determinant

Let G be a finite abelian group and $\hat{G} = \{\chi\}$ its character group. We have the **Dedekind determinant** relation:

Theorem 6.1. *Let f be any (complex valued) function on G. Then*

$$\prod_{\chi \in \hat{G}} \sum_{a \in G} \chi(a) f(a^{-1}) = \det_{a,b} f(a^{-1}b).$$

Proof. Let F be the space of functions on G. It is a finite dimensional vector space whose dimension is the order of G. It has two natural bases. First, the characters $\{\chi\}$, and second the functions $\{\delta_b\}$, $b \in G$, where

$$\delta_b(x) = 1 \quad \text{if} \quad x = b$$
$$\delta_b(x) = 0 \quad \text{if} \quad x \neq b.$$

For each $a \in G$ let $T_a f$ be the function such that $T_a f(x) = f(ax)$. Then

$$(T_a \chi)(b) = \chi(ab) = \chi(a)\chi(b),$$

so that

$$T_a \chi = \chi(a)\chi.$$

So χ is an eigenvector of T_a. Let

$$T = \sum_{a \in G} f(a^{-1}) T_a.$$

Then T is a linear map on F, and for each character χ, we have

$$T\chi = \Big[\sum_{a \in G} \chi(a) f(a^{-1})\Big]\chi.$$

Therefore χ is an eigenvector of T, and consequently the determinant of T is equal to the product over all χ occurring on the left-hand side of the equality in Theorem 6.1.

On the other hand, we look at the effect of T on the other basis. We have

$$T_a \delta_b(x) = \delta_b(ax),$$

so that $T_a \delta_b$ is the characteristic function of $a^{-1}b$, and

$$T_a \delta_b = \delta_{a^{-1}b}.$$

Consequently

$$T\delta_b = \sum_{a \in G} f(a^{-1})\delta_{a^{-1}b}$$
$$= \sum_{a \in G} f(a^{-1}b)\delta_a.$$

From this we find an expression for the determinant of T which is precisely the right-hand side in Theorem 4.1. This proves our theorem.

Theorem 6.2. *The determinant in Theorem* 4.1 *splits into*

$$\det_{a,b} f(ab^{-1}) = \Big[\sum_{a \in G} f(a)\Big] \det_{a,b \neq 1} [f(ab^{-1}) - f(a)].$$

Therefore

$$\prod_{\chi \neq 1} \sum_{a \in G} \chi(a) f(a^{-1}) = \det_{a,b \neq 1} [f(ab^{-1}) - f(a)].$$

Proof. Let $a_1 = 1, \ldots, a_n$ be the elements of G. In the determinant

$$\det f(a_i a_j^{-1}) = \begin{vmatrix} f(a_1 a_1^{-1}) & f(a_1 a_2^{-1}) \cdots & f(a_1 a_n^{-1}) \\ \vdots & \vdots & \vdots \\ f(a_n a_1^{-1}) & f(a_n a_2^{-1}) \cdots & f(a_n a_n^{-1}) \end{vmatrix}$$

add the last $n - 1$ rows to the first. Then all elements of the new first row are equal to $\sum f(a^{-1}) = \sum f(a)$. Factoring this out yields

$$\Big[\sum_{a \in G} f(a)\Big] \begin{vmatrix} 1 & 1 & \cdots & 1 \\ f(a_2 a_1^{-1}) & f(a_2 a_2^{-1}) & \cdots & f(a_2 a_n^{-1}) \\ \vdots & \vdots & & \vdots \\ f(a_n a_1^{-1}) & f(a_n a_2^{-1}) & \cdots & f(a_n a_n^{-1}) \end{vmatrix}.$$

Recall that a_1 is chosen to be 1. Subtract the first column from each one of the other columns. You get the first statement.

On the other hand, the function f can be selected so that the elements $\{f(a)\}$, $a \in G$, are algebraically independent over $\mathbf{Q}$, and therefore the factorization given in this first statement for the determinant is applicable in the polynomial ring generated over $\mathbf{Z}$ by the variables $f(a)$. Combining the first statement with Theorem 6.1 yields the second relation where the product is taken only over $\chi \neq 1$.

Serre has pointed out to me that the determinant relation is due to Dedekind, February 1896, who communicated it to Frobenius in March. Cf.

T. Hawkins, "New light on Frobenius...", *Archive for History of Exact Sciences* **12** (1974), p. 223.

§7. Bounds for Class Numbers

In this book we have not emphasized questions having to do with the size of the class number. We shall here make some brief remarks concerning various possibilities to obtain bounds. We let h_m = class number of $\mathbf{Q}(\mu_m)$, and p is prime ≥ 3.

To begin we derive the expression of the class number h_p^- as a determinant following Carlitz–Olson [Ca–O]. We start with the expression

$$\begin{aligned} h_p^- &= 2p \prod_{\chi \text{ odd}} -\tfrac{1}{2}B_{1,\chi} \\ &= 2p \prod_{\chi \text{ odd}} -\tfrac{1}{2} \sum_{a \in \mathbf{Z}(p)^*} \chi(a)\left\langle \frac{a}{p} \right\rangle, \end{aligned}$$

because the characters are non-trivial, and the term with $\frac{1}{2}$ drops out. We try to rewrite this as a Dedekind determinant over the group

$$G = \mathbf{Z}(p)^*/\pm 1.$$

We have

$$\left\langle \frac{a}{p} \right\rangle + \left\langle \frac{-a}{p} \right\rangle = 1.$$

Let ω be the Teichmuller character such that $\omega(a) \equiv a \bmod p$. We write odd characters as products

$$\chi = \omega\psi$$

where ψ is even. Then we find

$$h_p^- = 2p \prod_{\psi} \frac{1}{2} \sum_{a \in G} \psi(a)\omega(a)\left(\left\langle \frac{a}{p} \right\rangle - \left\langle \frac{-a}{p} \right\rangle\right),$$

and this makes sense because the function

$$f(a) = \omega(a)\left(\left\langle \frac{a}{p} \right\rangle - \left\langle \frac{-a}{p} \right\rangle\right)$$

on $\mathbf{Z}(p)^*$ is actually well defined mod ± 1, so is defined on G. This expression is now in the form where we can apply the Dedekind determinant, thus getting

$$h_p^- = \frac{2p}{2^{(p-1)/2}} \det\left[\omega(ab)\left(\left\langle \frac{ab}{p} \right\rangle - \left\langle \frac{-ab}{p} \right\rangle\right)\right].$$

The size of the determinant is

$$\frac{p-1}{2}.$$

Let ζ be a primitive $(p-1)$-root of unity. Representatives for elements of G are given by the powers ζ^i with $1 \le i \le (p-1)/2$. The determinant may then be taken over indices

$$i, j = 1, \ldots, \frac{p-1}{2} \quad \text{with } a = \zeta^i, b = \zeta^j,$$

and $\omega(ab) = \zeta^{i+j}$. In the expansion of the determinant, every term contains a factor arising from these ζ^{i+j}, whose product is obviously 1. Consequently the determinant is the same as the determinant obtained by omitting these ζ^{i+j} from each term.

Let $R(a)$ be the smallest positive integer in the residue class of $a \bmod p$. Then $R(a)$ is an integer $\le p-1$, and

$$\left\langle \frac{a}{p} \right\rangle = \frac{R(a)}{p}.$$

We use the notation $R(\zeta^{i+j})$ and $R(-\zeta^{i+j})$ to denote similarly the smallest positive integers in the residue class of ζ^{i+j} and $-\zeta^{i+j}$ respectively. Then we have proved the following theorem.

Theorem 7.1. $\qquad \pm D_p = (2p)^{(p-3)/2} h_p^-$

where

$$D_p = \det[R(\zeta^{i+j}) - R(-\zeta^{i+j})].$$

Observe that each entry in the determinant D_p is an integer of absolute value $\le p-1$.

The absolute value of the determinant is the volume of the fundamental domain of its row vectors, say. This volume is bounded by the product of the Euclidean lengths of these vectors (Hadamard inequality). Carlitz [Ca] observed that this gives the bound

$$h_p^- < 2^{-(3p-7)/4}(p-1)^{(p+3)/4}.$$

As Carlitz–Olson relate it, the history of the determinant in Theorem 7.1 is amusing. The determinant

$$\det R(ab^{-1}), \quad a, b = 1, \ldots, \frac{p-1}{2}$$

was known classically as the Maillet determinant, conjectured to be $\neq 0$ by Maillet. Malo computed it for $p \le 13$, and found it equal to the appropriate power of p. He conjectured that it was always so equal, but Carlitz–Olson computed a bit further, and found extra factors. They derived that the Maillet determinant is equal to the determinant of Theorem 7.1 (up to the obvious power of 2), and then also to the class number times that power of p by using the expression of the class number as a product of generalized Bernoulli numbers (not called that at the time). Thus Malo had missed out the class number factor.

Metsänkylä [Me] gives growth estimates for h^-. Masley–Montgomery [M–M] also prove the inequalities

$$(2\pi)^{-p/2}p^{(p-25)/4} \leq h_p^- \leq (2\pi)^{-p/2}p^{(p+31)/4},$$

for primes $p > 200$. Thus the Carlitz bound is reasonably sharp. For applications of this see Ribet [Ri].

For primes p, it has been proved by Uchida [Uch] that $h_p = 1$ if and only if $p \leq 19$. More generally, Masley and Montgomery [M–M] subsequently proved that $h_m = 1$ for precisely 29 distinct values of m (always assumed $\not\equiv 2 \bmod 4$), the largest of which is $m = 84$. Masley [Mas 2] shows that $h_m = 2$ if and only if $m = 39, 56$. For Euclidean cyclotomic fields, see also [Mas 1].

4 The p-adic L-function

In this chapter we return to p-adic integration theory, and give Mazur's formulation of the p-adic L-function as Mellin transform. It turns out to be more convenient as a basic definition, than Iwasawa's previous formulation in terms of power series. The connection is made via Example 2 of §1. We derive further analytic properties, which allow us to make explicit its value at $s = 1$, thereby obtaining Leopoldt's formula in the p-adic case, analogous to that of the complex case. We also give Leopoldt's version of the p-adic class number formula and regulator.

The basic arguments are due to Leopoldt [Le 11]. However, we shall follow in §1 and §2 a course of Katz, which developed systematically operations on measures and their corresponding formulation on power series in the Iwasawa algebra. In this manner, constructions which appear slightly tricky in Leopoldt's paper here become completely natural, and even forced from these measure theoretic operations.

The Leopoldt transform then appears as an extension of an integral transform to a somewhat wider class of power series than those with p-adic integral coefficients. No use will be made of this, since only integral valued measures occur in the analysis of the p-adic L-function, but we include Leopoldt's results for completeness, for convenience of reference if the need ever arises for them.

The p-adic L-function in the case of elliptic curves is discussed in Robert [Ro], and especially Coates–Wiles [C–W 2], [C–W 3]. See also Lichtenbaum [Li 3], and Katz [Ka] for general comments concerning its connection with formal groups. For the case of totally real fields, Shintani's evaluation of the zeta function [Sh] presumably allows a development of the L-function similar to that of the cyclotomic case.

This chapter is used only in Chapter 7, and it can therefore be omitted without loss of the logical connections. On the other hand, if one leaves out the section on the p-adic regulator, then the chapter appears as a natural continuation of Chapter 2, and is essentially measure theoretic, independent of Chapter 3.

Throughout, we need the fact that if $\mathfrak{o}$ is the ring of integers in a p-adic field, then there is a natural isomorphism

$$\lim_{\leftarrow} \mathfrak{o}[X]/((1 + X)^{p^n} - 1) \approx \mathfrak{o}[[X]].$$

The limit is the projective limit, and is called the **Iwasawa algebra.** This is a basic fact of algebra. In the next chapter, we need further facts about this algebra and modules over it. For the convenience of the reader, all these facts and their proofs will be placed in the next chapter.

§1. Measures and Power Series

Let $\mathbf{C}_p$ be the completion of the algebraic closure of $\mathbf{Q}_p$, and let $\mathfrak{o} = \mathfrak{o}_{\mathbf{C}_p}$ be the ring of p-integers in $\mathbf{C}_p$. By a **measure** μ we shall mean an $\mathfrak{o}$-valued distribution on $\mathbf{Z}_p$. This means that for each integer $n \geq 0$ we have a function

$$\mu_n : \mathbf{Z}(p^n) \to \mathfrak{o}$$

such that the family $\{\mu_n\}$ is a distribution on the projective system $\mathbf{Z}(p^n)$.

Let $\mathrm{Cont}(\mathbf{Z}_p, \mathfrak{o})$ or $C(\mathbf{Z}_p, \mathfrak{o})$ be the space of continuous functions on $\mathbf{Z}_p$ into $\mathfrak{o}$, with sup norm. As usual, there is a bijection between measures and bounded functionals

$$\lambda : \mathrm{Cont}(\mathbf{Z}_p, \mathfrak{o}) \to \mathfrak{o}.$$

[A $\mathbf{Z}_p$-linear map λ is called **bounded** if there exists $C > 0$ such that

$$|\lambda(\varphi)| \leq C\|\varphi\| \quad \text{for all } \varphi \in \mathrm{Cont}(\mathbf{Z}_p, \mathfrak{o}).$$

The inf of such C is called the **norm** of λ, and denoted by $\|\lambda\|$. The bounded functionals form a p-adic space.] Indeed, it is clear that any measure μ gives rise to a functional

$$d\mu : \varphi \mapsto \int \varphi \, d\mu.$$

On the other hand, suppose λ is a bounded functional. If $x \in \mathbf{Z}(p^n)$, let φ_x be the characteristic function of the set of elements $y \in \mathbf{Z}_p$ such that

$$y \equiv x \bmod p^n.$$

Define

$$\mu_n(x) = \lambda(\varphi_x).$$

It is then clear that $\{\mu_n\}$ defines a measure. Since any continuous function on $\mathbf{Z}_p$ can be uniformly approximated by step functions, it follows easily that the correspondence

$$\mu \mapsto d\mu$$

is a bijection from $\mathfrak{o}$-valued measures on $\mathbf{Z}_p$ to bounded functionals.

Furthermore, define the **norm**

$$\|\mu\| = \sup_{n,x} |\mu_n(x)|,$$

taken for $x \in \mathbf{Z}(p^n)$ and all n. Then the map $\mu \mapsto d\mu$ is easily verified to be norm preserving.

The **Iwasawa algebra** is obtained as the projective limit

$$\Lambda_{\mathfrak{o}} = \lim \mathfrak{o}[X]/((1 + X)^{p^n} - 1) \approx \mathfrak{o}[[X]],$$

and

$$\mathfrak{o}[X]/((1 + X)^{p^n} - 1) = \mathfrak{o}[T]/(T^{p^n} - 1)$$

where $T = 1 + X$. Let $\gamma_n = T \bmod (T^{p^n} - 1)$, so $\gamma_n^{p^n} = 1$. Let as usual

$$\binom{r}{k} = \frac{r(r-1)\cdots(r-k+1)}{k!}.$$

The function μ_n on $\mathbf{Z}(p^n)$ can be viewed as an element of the group algebra $\mathfrak{o}[\gamma_n]$, namely

$$\begin{aligned}\sum_{r=0}^{p^n-1} \mu_n(r)\gamma_n^r &= \sum_{r=0}^{p^n-1} \mu_n(r) \sum_{k=0}^{p^n-1} \binom{r}{k} X^k \\ &= \sum_{k=0}^{p^n-1} \left(\sum_{r=0}^{p^n-1} \mu_n(r)\binom{r}{k}\right) X^k\end{aligned}$$

where the right-hand side is read mod $(1 + X)^{p^n} - 1$. Thus

$$\sum_{r=0}^{p^n-1} \mu_n(r)\gamma_n^r = \sum_{k=0}^{p^n-1} c_{n,k}X^k = P_n(X),$$

where the coefficients $c_{n,k}$ are given by

$$c_{n,k} = \sum_{r=0}^{p^n=1} \mu_n(r)\binom{r}{k}.$$

The canonical homomorphism $\mathbf{Z}(p^{n+1}) \to \mathbf{Z}(p^n)$ maps

$$P_{n+1}(X) \mapsto P_n(X),$$

and we let

$$P(X) = \lim P_n(X)$$

be the projective limit of these elements in the Iwasawa algebra. We call $P(X)$ the **power series associated** to μ, and also denote it by $(P\mu)(X)$ or $P\mu(X)$. Thus

$$P\colon \mathfrak{o}\text{-valued measures on } \mathbf{Z}_p \to \mathfrak{o}[[X]]$$

is an $\mathfrak{o}$-linear map. Conversely, any power series $f \in \mathfrak{o}[[X]]$ defines a compatible system of elements in the group algebras $\mathfrak{o}[\gamma_n]$, so the map P is bijective. We write

$$f = P\mu \quad \text{or} \quad \mu = \mu_f$$

to mean that f is the power series associated to μ as above. We call P the **Iwasawa isomorphism.**

For any $x \in \mathbf{Z}_p$ let

$$C_k(x) = \frac{x(x-1)\cdots(x-k+1)}{k!} = \binom{x}{k}.$$

Since $C_k(r)$ is an integer for any positive integer r, and since $\mathbf{Z}^+$ is dense in $\mathbf{Z}_p$, it follows that

$$C_k\colon \mathbf{Z}_p \to \mathbf{Z}_p$$

is a polynomial map of $\mathbf{Z}_p$ into itself, and in particular is continuous.

For fixed n, define

$$C_k^{(n)}(x) = \frac{r(r-1)\cdots(r-k+1)}{k!}$$

where $0 \leq r \leq p^n - 1$, and $r \equiv x \bmod p^n$. Then $C_k^{(n)}$ is a step function, defined at level n, and

$$\lim_{n\to\infty} C_k^{(n)} = C_k \text{ uniformly.}$$

Since the coefficients $c_{n,k}$ in the polynomial $P_n(X)$ are given by the sum of products of μ_n and the binomial coefficient, we obtain:

Theorem 1.1. *Let* $f(X) = \sum c_k X^k \in \mathfrak{o}[[X]]$. *Then*

$$c_k = \int_{\mathbf{Z}_p} \binom{x}{k}\, d\mu_f(x).$$

Theorem 1.2. *The power series $P\mu$ is the unique power series f such that for z in the maximal ideal of $\mathfrak{o}$, we have*

$$\int_{\mathbf{Z}_p} (1 + z)^x \, d\mu(x) = f(z).$$

Proof. We have

$$\int_{\mathbf{Z}_p} (1 + z)^x \, d\mu_f(x) = \int_{\mathbf{Z}_p} \sum_{k=0}^{\infty} \binom{x}{k} z^k \, d\mu_f(x).$$

We can interchange the sum and integral, apply Theorem 1.1, and we see that $P\mu$ has the desired property. Uniqueness is obvious since any power series is determined by its values.

Example 1. Let μ be the Dirac measure at a point $s \in \mathbf{Z}_p$, that is

$$\int_{\mathbf{Z}_p} \varphi \, d\mu = \varphi(s).$$

Then the associated power series f is

$$f(X) = \sum_{k=0}^{\infty} \binom{s}{k} X^k = (1 + X)^s.$$

Example 2. Let ν be a measure on $\mathbf{Z}_p$ whose support lies in the open closed subset $1 + p\mathbf{Z}_p$. Let γ be a topological generator of $1 + p\mathbf{Z}_p$, for instance $\gamma = 1 + p$. There is an isomorphism

$$\mathbf{Z}_p \to 1 + p\mathbf{Z}_p$$

such that

$$x \mapsto \gamma^x.$$

By pull back, there exists a unique measure $\mu = \mu_f$ on $\mathbf{Z}_p$ such that

$$\int_{1+p\mathbf{Z}_p} u^s \, d\nu(u) = \int_{\mathbf{Z}_p} \gamma^{sx} \, d\mu_f(x).$$

By Theorem 1.2, writing $\gamma^s = 1 + z$, we get

$$\int_{1+p\mathbf{Z}_p} u^s \, d\nu(u) = f(\gamma^s - 1).$$

The power series f is not easily determined in terms of ν. Iwasawa expressed his results on p-adic L-functions in terms of the power series f. Mazur gave the formulation in terms of the integral, see §3 below.

Theorem 1.3. (Mahler) *A function φ from $\mathbf{Z}_p$ into $\mathfrak{o}$ is continuous if and only if there exist elements $a_n \in \mathfrak{o}$ such that $|a_n| \to 0$ and*

$$\varphi(x) = \sum_{n=0}^{\infty} a_n \binom{x}{n}.$$

The sequence $\{a_n\}$ is uniquely determined by φ.

Proof. Given a sequence $\{a_n\}$ as above, it is clear that the function

$$\varphi(x) = \sum a_n \binom{x}{n}$$

is continuous. For uniqueness, let

$$\Delta\varphi(x) = \varphi(x+1) - \varphi(x).$$

Then $\varphi(0) = a_0$, and furthermore

$$\Delta\binom{x}{n} = \binom{x}{n-1}, \qquad \Delta^k\varphi(x) = \sum a_{n+k}\binom{x}{n}$$

and

$$\Delta^k\varphi(0) = a_k.$$

This proves uniqueness.

We now prove existence. In the applications, the measures will take values in the ring of p-adic integers in a finite extension of $\mathbf{Q}_p$. An argument using tensor products reduces the general case to this case, and we omit it since we have no use for it. The case of a finite extension is then reduced to the case when the measure is $\mathbf{Z}_p$-valued by taking a basis for the ring of values over $\mathbf{Z}_p$ and projecting on the coordinates. We now handle this case.

Let B be the Banach space of sequences (a_n) with $a_n \in \mathbf{Z}_p$, and $|a_n| \to 0$, under the sup norm. We have a $\mathbf{Z}_p$-linear map

$$B \to C(\mathbf{Z}_p, \mathbf{Z}_p) \quad \text{by} \quad (a_n) \mapsto \sum a_n \binom{x}{n}.$$

We have to show it is surjective. By completeness of $C(\mathbf{Z}_p, \mathbf{Z}_p)$ it suffices to prove that a given $f \in C(\mathbf{Z}_p, \mathbf{Z}_p)$ is congruent to the image of an element of B mod p^n for each n, and by a simple recursion, it suffices to do this mod p. In other words, it suffices to prove that the map

$$\{(a_n),\ a_n \in \mathbf{F}_p, \text{ almost all } a_n = 0\} \to C(\mathbf{Z}_p, \mathbf{F}_p)$$

given by the same formula as above, is surjective. But

$$C(\mathbf{Z}_p, \mathbf{F}_p) = \bigcup_N \operatorname{Maps}(\mathbf{Z}(p^N), \mathbf{F}_p)$$

because $\mathbf{F}_p$ is discrete and finite.

Lemma. *Let* $0 \leq k < p^N$. *Then the function*

$$x \mapsto \binom{x}{k} \bmod p$$

of $\mathbf{Z}_p$ *into* $\mathbf{F}_p$ *is periodic of period* p^N.

Proof. We have to show

$$\binom{x + p^N}{k} \equiv \binom{x}{k} \bmod p \quad \text{if } k < p^N.$$

Since

$$(1 + T)^{x+p^N} = (1 + T)^x(1 + T)^{p^N} \equiv (1 + T)^x(1 + T^{p^N}) \bmod p,$$

we prove the lemma by comparing the coefficients of T^k.

Now we are reduced to showing that

$$\{(a_n),\ a_n \in \mathbf{F}_p,\ a_n = 0 \text{ if } n > p^N\} \to \operatorname{Maps}(\mathbf{Z}(p^N), \mathbf{F}_p)$$

is bijective. Since both spaces have $\mathbf{F}_p$-dimension p^N, the surjectivity follows from injectivity, which is proved the same way we proved that the function $\varphi(x)$ has uniquely determined coefficients a_n. This proves Mahler's theorem.

Corollary. *If* $f(X) = \sum c_n X^n$ *and*

$$\varphi(x) = \sum a_n \binom{x}{n},$$

then

$$\int \varphi \, d\mu_f = \sum a_n c_n,$$

so

$$\left| \int \varphi \, d\mu_f \right| \leq (\sup|a_n|)\|f\| \leq \|f\|.$$

We define the **norms**:

$$\|f\| = \sup_n |c_n|$$

$$\|\mu\| = \sup_{n,x} |\mu_n(x)| \text{ as before.}$$

Theorem 1.4. *We have* $\|f\| = \|\mu_f\|$.

Proof. Since

$$c_n = \int \binom{x}{n} d\mu_f(x),$$

we get trivially $\|f\| \leq \|\mu_f\|$. Conversely, given a level p^n, let $x_0 \in \mathbf{Z}(p^n)$ and let φ be the locally constant function such that

$$\varphi(x_0) = 1, \quad \text{and} \quad \varphi(x) = 0 \quad \text{if } x \neq x_0,\, x \in \mathbf{Z}(p^n).$$

Then

$$\int \varphi d\mu_f = \mu_n(x_0),$$

and on the other hand, from the corollary of Theorem 1.3, we get

$$\left| \int \varphi \, d\mu_f \right| \leq \|f\|,$$

so $\|\mu_f\| \leq \|f\|$ as desired.

§2. Operations on Measures and Power Series

We shall give a list of integration formulas, or better, a list of operations on measures and their corresponding operations on power series.

Meas 0. $$\int_{\mathbf{Z}_p} d\mu_f = f(0).$$

Proof. Special case of Theorem 1.2 with $z = 0$.

For the next property, we let

$$\psi_z(x) = (1 + z)^x$$

if $z \in \mathfrak{m}$ = maximal ideal of $\mathfrak{o}$. Also (with formal groups in mind) we write

$$X[+]z = X + z + zX = (1 + z)(1 + X) - 1.$$

Meas 1. $\psi_z \mu_f = \mu_g$, *where* $g(X) = f(X[+]z)$.

Proof. For $w \in \mathfrak{m}$ we have

$$\int_{\mathbf{Z}_p} \psi_w \, d(\psi_z \mu_f) = \int_{\mathbf{Z}_p} \psi_w \psi_z \, d\mu_f$$

$$= \int_{\mathbf{Z}_p} (1 + w)^x (1 + z)^x \, d\mu_f(x)$$

$$= \int_{\mathbf{Z}_p} (1 + w + z + wz)^x \, d\mu_f(x).$$

The property is then clear from the definitions.

In particular, let ζ be a p^nth root of unity, and let $z = \zeta - 1$. Then

$$\psi_z(x) = \zeta^x$$

and we find:

Meas 2. $$\psi_{\zeta-1}\mu_f = \mu_g,$$

where $g(X) = f(\zeta(1 + X) - 1) = f(X[+](\zeta - 1))$.

As before, putting $T = 1 + X$, and $f(X) = f_{G_m}(T)$ if f is a rational function, we can write the power series $g(X)$ in **Meas 2** in the form

$$g(X) = f_{G_m}(\zeta T).$$

Moreover, let φ be a step function, constant on cosets mod p^n. Write the Fourier expansion

$$\varphi(x) = \sum_{\zeta^{p^n}=1} \hat{\varphi}(\zeta)\zeta^x$$

$$\hat{\varphi}(\zeta) = \frac{1}{p^n} \sum_{x \in \mathbf{Z}(p^n)} \varphi(x)\zeta^{-x}.$$

We find:

Meas 3. $$\varphi\mu_f = \mu_g$$

where $g(X) = \sum_{\zeta^{p^n}=1} \hat{\varphi}(\zeta) f(\zeta(1 + X) - 1).$

If $f(X) = f_{G_m}(T)$ *is a rational function, then*

$$g_{G_m}(T) = \sum_{\zeta^{p^n}=1} \hat{\varphi}(\zeta) f_{G_m}(\zeta T).$$

Let $\mathbf{U}_p = \mathbf{U}$ be the operator

$$\mathbf{U}f(X) = f(X) - \frac{1}{p} \sum_{\zeta^p=1} f(\zeta(1 + X) - 1).$$

We call $\mathbf{U}$ the **unitization operator** because of the next property.

Meas 4. *If* φ = *characteristic function of* $\mathbf{Z}_p^*$, *then*

$$\varphi\mu_f = \mu_{\mathbf{U}f}.$$

Proof. We compute trivially the Fourier expansion of φ:

$$\hat{\varphi}(\zeta) = \frac{1}{p}\sum_{x=1}^{p-1}\zeta^{-x} = \begin{cases} -1/p & \text{if } \zeta \neq 1 \\ \dfrac{p-1}{p} & \text{if } \zeta = 1. \end{cases}$$

Then **Meas 3** gives

$$g(X) = \sum_{\zeta^p=1}\hat{\varphi}(\zeta)f(\zeta(1+X)-1) = \mathbf{U}f(X),$$

as was to be shown.

Remark. Let A be the formal multiplicative group (cf. Chapter 8). In the notation of such groups, we can write the unitization operator in the form

$$\mathbf{U}f(X) = f(X) - \frac{1}{p}\sum_{z\in A_p} f(X\,[+]\,z).$$

Meas 5. *Let* χ *be a character on* $\mathbf{Z}_p^*$, *of finite order with conductor* N = *power of* p. *Let* ζ *be a primitive* Nth *root of unity, and let*

$$S(\chi, \zeta) = \sum_{a\in\mathbf{Z}(N)^*}\chi(a)\zeta^a.$$

Then

$$\chi\mu_f = \mu_g$$

where

$$g(X) = \frac{S(\chi,\zeta)}{N}\sum_{a\in\mathbf{Z}(N)^*}\bar{\chi}(a)f(\zeta^{-a}(X+1)-1).$$

If f *is a rational function, then*

$$g_{G_m}(T) = \frac{S(\chi,\zeta)}{N}\sum_a \bar{\chi}(a)f_{G_m}(\zeta^{-a}T).$$

Proof. It suffices to apply **Meas 3** and to compute the Fourier transform of χ. This is trivial, and we have

$$\sum_{y\in\mathbf{Z}(N)}\chi(y)\zeta^{xy} = \begin{cases} 0 & \text{if } x \equiv 0 \bmod p \\ \bar{\chi}(a)S(\chi,\zeta) & \text{if } x \equiv a \not\equiv 0 \bmod p. \end{cases}$$

Meas 6. $$x\mu_f(x) = \mu_{Df}(x)$$

where $D = (1 + X)D_X$. *In particular*

$$\int_{\mathbf{Z}_p} x^k \, d\mu_f(x) = \int_{\mathbf{Z}_p} d\mu_{D^k f} = D^k f(0).$$

Proof. Note that

$$x = \lim_{z \to 0} \frac{(1 + z)^x - 1}{z} = \lim_{z \to 0} \frac{\psi_z(x) - 1}{z}.$$

Hence for any step function φ we get

$$\begin{aligned} \int x\varphi(x) \, d\mu_f(x) &= \lim_{z \to 0} \int \frac{\psi_z(x) - 1}{z} \varphi(x) \, d\mu_f(x) \\ &= \lim_{z \to 0} \int \varphi(x) \, d\mu_{g_z}(x) \qquad \text{(by **Meas 1**)} \end{aligned}$$

where

$$g_z(X) = \frac{f(X + z + zX) - f(X)}{z} = (1 + X)f'(X) \bmod z$$

by Taylor's formula. The desired result follows by taking the limit as $z \to 0$ and using the non-trivial part of Theorem 1.4, that is:

$$\|\mu_{g_z} - \mu_{Df}\| \le \|g_z - Df\| \le |z|.$$

Remark. We shall deal throughout with three variables. Let T be the variable on the "multiplicative group." We put

$$T = e^Z, \qquad X = T - 1, \qquad T = 1 + X.$$

Then Z is the corresponding variable on the additive group. For any power series $f(X)$ (with coefficients in a field of characteristic 0) there is a corresponding power series denoted by $f^*(Z)$ or $f_{G_a}(Z)$ such that

$$f(X) = f(e^Z - 1) = f_{G_a}(Z) = f_{G_m}(T).$$

This last equality makes sense only when f is a rational function.

The differential operator D then can be expressed in terms of the three variables,

$$(1 + X)D_X = D_Z = TD_T.$$

The expression in terms of T applies only to rational functions of T (rational functions of X). The first two expressions in terms of X and Z apply to arbitrary power series, and for any positive integer k, the expression $D^k f$ makes sense whether we view f as power series in X or Z. Furthermore,

$$D^k f(0) = D_Z^k f_{G_a}(0).$$

If f is a rational function, this is also equal to $(TD_T)^k f_{G_m}(1)$.

Meas 7. *Let $g = \mathbf{U}g$ so μ_g is a measure on $\mathbf{Z}_p^*$. Then*

$$a^{-1}\mu_g(a) = \mu_{\mathbf{U}h}(a)$$

where h is any power series such that $Dh = g$.

Proof. Since $a^{-1}\mu_g(a)$ is a measure on $\mathbf{Z}_p^*$, there exists a power series f such that $f \in \mathfrak{o}[[X]]$,

$$a^{-1}\mu_g(a) = \mu_f(a) \quad \text{and} \quad \mathbf{U}f = f.$$

Then

$$\mu_g(x) = x\mu_f(x)$$

whence by **Meas 6,**

$$g = Df = D\mathbf{U}f.$$

We let $h = \mathbf{U}f$ to conclude the proof.

§3. The Mellin Transform and p-adic L-function

Let ω be the **Teichmuller character.** If p is odd, then

$$\omega\colon \mathbf{Z}_p^* \to \mu_{p-1}$$

is the character such that $\omega(a) \equiv a \bmod p$. If $p = 2$, then we define $\omega(a) = \pm 1$ such that

$$\omega(a) \equiv a \bmod 4.$$

Then we can write uniquely an element $a \in \mathbf{Z}_p^*$ as

$$a = \omega(a)\langle a\rangle,$$

where $\langle a\rangle \equiv 1 \bmod p$ if p is odd, and $\langle a\rangle \equiv 1 \bmod 4$ if $p = 2$.

Let μ be a measure. We define its **Gamma transform** as a function on $\mathbf{Z}_p$ by the integral

$$\Gamma_p\mu(s) = \int_{\mathbf{Z}_p^*} \langle a\rangle^s \, d\mu(a),$$

and we define its **Mellin transform**, also as function on $\mathbf{Z}_p$, by

$$\mathbf{M}_p\mu(s) = \int_{\mathbf{Z}_p^*} \langle a\rangle^s a^{-1} \, d\mu(a).$$

It is clear that $\Gamma_p\mu$ and $\mathbf{M}_p\mu$ are continuous in s. (For analyticity, see below.) Since the integral is taken on $\mathbf{Z}_p^*$, $\mathbf{M}_p\mu$ depends only on the restriction of μ to $\mathbf{Z}_p^*$, so if $\mu = \mu_f$, then

$$\mathbf{M}_p\mu_f = M_p\mu_{\mathbf{U}f}.$$

If $\mu = \mu_f$, we write sometimes $\mathbf{M}_p f$ instead of $\mathbf{M}_p\mu_f$, and similarly for the Gamma transform.

Note that $a^{-1}\,d\mu(a)$ for $a \in \mathbf{Z}_p^*$ is also the functional associated with a measure, so that the Mellin transform is actually a special case of the Gamma transform (of another measure).

Theorem 3.1. *Let $g \in \mathfrak{o}[[X]]$ be such that $\mathbf{U}g = g$, and let h be a power series such that $Dh = g$. Then $\mathbf{U}h \in \mathfrak{o}[[X]]$ and*

$$\Gamma_p \mathbf{U}h = \mathbf{M}_p\mu_g.$$

Proof. This is an immediate application of **Meas 7**, after integrating the function $\langle a\rangle^s$.

We now consider the analyticity properties.

Lemma. *Let μ be a measure on $\mathbf{Z}_p^*$. Then there exists a power series $h \in \mathbf{Z}_p[[s]]$,*

$$h(s) = \sum_{n=0}^{\infty} b_n s^n$$

such that $b_n \to 0$ as $n \to \infty$, with the property that for all $s \in \mathbf{Z}_p$,

$$h(s) = \int_{\mathbf{Z}_p^*} \langle a\rangle^s\,d\mu.$$

Proof. The integral can be written as a sum of integrals over cosets of $1 + p\mathbf{Z}_p$ (or $1 + 4\mathbf{Z}_2$ if $p = 2$). Changing the measure appropriately with respect to each coset, we are reduced to proving (say for odd p) that for any measure μ, the integral

$$\int_{1+p\mathbf{Z}_p} \langle a\rangle^s\,d\mu$$

has the desired analyticity property. We note that

$$\begin{aligned}\int_{1+p\mathbf{Z}_p} \langle a\rangle^s\,d\mu &= \int_{1+p\mathbf{Z}_p} \sum_{n=0}^{\infty} \binom{s}{n}(a-1)^n\,d\mu(a)\\ &= \int_{1+p\mathbf{Z}_p} \sum_{n=0}^{\infty} s(s-1)\cdots(s-n+1)\frac{(a-1)^n}{n!}\,d\mu(a).\end{aligned}$$

But $a - 1 \equiv 0 \bmod p$, and so $(a-1)^n/n!$ is p-integral for all n. Furthermore, $(a-1)^n/n!$ tends to 0 p-adically as $n \to \infty$. Hence we can interchange the sum and integral to yield

$$\int_{1+p\mathbf{Z}_p} \langle a \rangle^s \, d\mu = \sum_{n=0}^{\infty} P_n(s) c_n$$

where P_n is a polynomial of degree n with integral coefficients, and

$$c_n = \int_{1+p\mathbf{Z}_p} \frac{(a-1)^n}{n!} \, d\mu(a)$$

is p-integral, and $c_n \to 0$. It is then clear that $\sum P_n(s)c_n$ can be written as a power series $h(s)$ whose coefficients b_n tend to 0 as desired.

We had the measure $E_{1,c}$ in Chapter 2, with $c \in \mathbf{Z}_p^*$. Let s be a p-adic variable in $\mathbf{Z}_p$. For any c such that $\chi(c)\langle c \rangle^s$ is not identically 1 we define the **p-adic L-function L_p** by

$$\begin{aligned} L_p(1-s, \chi) &= \frac{-1}{1 - \chi(c)\langle c \rangle^s} \mathbf{M}_p(\chi E_{1,c})(s) \\ &= \frac{-1}{1 - \chi(c)\langle c \rangle^s} \int_{\mathbf{Z}_p^*} \langle a \rangle^s \chi(a) a^{-1} \, dE_{1,c}(a). \end{aligned}$$

By the lemma, the integral is analytic as a function of s. The factor in front is analytic except when

$$\chi(c)\langle c \rangle^s = 1.$$

If χ is non-trivial, we can select c such that $\chi(c) \neq 1$, and then the factor in front is also analytic at $s = 0$.

Theorem 3.2. *The value of $L_p(1-s, \chi)$ is independent of the choice of c, and for any positive integer k,*

$$L_p(1-k, \chi) = -\frac{1}{k} B_{k, \chi\omega^{-k}}.$$

In particular, if $k \equiv 0 \bmod p - 1$, and p is odd, then

$$L_p(1-k, \chi) = -\frac{1}{k} B_{k, \chi}.$$

Proof. Since the set of sufficiently large integers $k \equiv 0 \bmod p - 1$ is dense in $\mathbf{Z}_p$, we see that the first assertion follows from the explicit values given at integers of the form $1 - k$ as described. For these, we have:

$$\begin{aligned}
\mathbf{M}_p(\chi E_{1,c})(k) &= \int_{\mathbf{Z}_p^*} \langle a \rangle^{k-1} \chi(a)\omega(a)^{-1}\, dE_{1,c}(a) \\
&= \int_{\mathbf{Z}_p^*} a^{k-1}\omega(a)^{-k}\chi(a)\, dE_{1,c}(a) \\
&= (1 - \chi\omega^{-k}(c)c^k)\frac{1}{k} B_{k,\chi\omega^{-k}} \quad \text{by Theorem 2.4 of Chapter 2} \\
&= (1 - \chi(c)\langle c \rangle^k)\frac{1}{k} B_{k,\chi\omega^{-k}}.
\end{aligned}$$

This proves the theorem.

Theorem 3.3. *Let $g = \mathbf{U}g$ and let h be the power series such that*

$$Dh = g \quad \textit{and} \quad h(0) = 0.$$

Then

$$\mathbf{M}_p\mu_g(0) = -\frac{1}{p}\sum_{\xi^p = 1} h(\xi - 1).$$

Proof. By **Meas 7** we have

$$\mathbf{M}_p\mu_g(0) = \int a^{-1}\, d\mu_g(a) = \int d\mu_{\mathbf{U}h}(a) = \mathbf{U}h(0).$$

The formula is then clear from the definition of $\mathbf{U}$.

To compute $L_p(1, \chi)$ we have to work out the power series associated to $E_{1,c}$ and then apply the formalism of the preceding section systematically to get the answer, with $s = 0$ in $L_p(1 - s, \chi)$, using Theorem 3.3.

Proposition 3.4. *Let $c \in \mathbf{Z}_p^*$. The power series associated with the measure $E_{1,c}$ is*

$$f_{1,c} = \frac{1}{T-1} - \frac{c}{T^c - 1}, \qquad \textit{with } T = 1 + X.$$

Proof. It is immediate to verify that as power series in X the expression on the right-hand side is holomorphic at $X = 0$, and that its coefficients are p-integral because c is a p-unit. Let

$$f(T) = \frac{\log T}{T-1} - \frac{c \log T}{T^c - 1}.$$

Putting $T = e^Z$ we find

$$f(T) = f^*(Z) = \frac{Z}{e^Z - 1} - \frac{cZ}{e^{cZ} - 1}$$

$$= \sum (1 - c^k) B_k \frac{Z^k}{k!}.$$

On the other hand, let $f_{1,c}$ be the power series associated with $E_{1,c}$, and write

$$f_{1,c}(X) = f_{1,c}^*(Z) = \sum_{k=1}^{\infty} c_{k-1} \frac{Z^{k-1}}{(k-1)!}.$$

Since

$$\int_{\mathbf{Z}_p} x^{k-1}\, dE_{1,c} = \frac{1}{k} B_k(1 - c^k),$$

it follows from **Meas 6** that

$$c_{k-1} = D^{k-1} f_{1,c}^*(0) = \frac{1}{k} B_k(1 - c^k),$$

so

$$Z f_{1,c}^*(Z) = \sum (1 - c^k) B_k \frac{Z^k}{k!} = f^*(Z) = \frac{Z}{T - 1} - \frac{cZ}{T^c - 1}.$$

It follows that

$$f_{1,c} = \frac{1}{T - 1} - \frac{c}{T^c - 1}$$

as desired.

Proposition 3.5. *Let χ be a non-trivial character on $\mathbf{Z}_p^*$ with conductor N. The power series associated with $\chi E_{1,c}$ is*

$$g_{\chi,c} = G_\chi(T) - c\chi(c) G_\chi(T^c)$$

where

$$G_\chi(T) = \frac{S(\chi, \zeta)}{N} \sum_{a \in \mathbf{Z}(N)^*} \bar{\chi}(a) \frac{1}{\zeta^{-a} T - 1}.$$

Proof. Immediate from **Meas 5**.

We shall now assume that c is an integer > 1 prime to p.
Written in full, the power series for $g_{\chi,c}$ is

$$g_{\chi,c} = \frac{S(\chi, \zeta)}{N} \sum_a \bar{\chi}(a) \left[\frac{\zeta^a}{T - \zeta^a} - \frac{c\chi(c)\zeta^a}{T^c - \zeta^a} \right].$$

If we let

$$h_{\chi,c}(X) = \frac{-S(\chi,\zeta)}{N} \sum_{\lambda \neq 1} \sum_{a} \bar{\chi}(a) \log\left(1 + \frac{X}{1 - \lambda\zeta^a}\right)$$

where:

λ ranges over cth roots of unity $\neq 1$,
a ranges over $\mathbf{Z}(N)^*$,

then it is easy to see (and we carry out the computation below) that

$$Dh_{\chi,c} = g_{\chi,c} \quad \text{and} \quad h_{\chi,c}(0) = 0.$$

Furthermore,

$$-\mathbf{M}_p g_{\chi,c}(s) = (1 - \chi(c)\langle c\rangle^s)L_p(1 - s, \chi).$$

The situation is then set up to apply Theorem 3.3.

We now prove that $Dh_{\chi,c} = g_{\chi,c}$. Observe that since $\sum \bar{\chi}(a) = 0$, we have

$$G_\chi(T) = \frac{S(\chi,\zeta)}{N} \sum_a \bar{\chi}(a) \frac{\zeta^a}{T - \zeta^a} = \frac{S(\chi,\zeta)}{N} \sum_a \bar{\chi}(a) \frac{T}{T - \zeta^a}.$$

Lemma. $\displaystyle\frac{S(\chi,\zeta)}{N} \sum_{\lambda \neq 1} \sum_a \bar{\chi}(a) \frac{T}{T - \lambda\zeta^a} = c\chi(c)G_\chi(T^c) - G_\chi(T).$

Proof. Taking the logarithmic derivative of

$$T^c - \zeta^{ac} = \prod_\lambda (T - \lambda\zeta^a)$$

we obtain

$$\sum_{\lambda \neq 1} \sum_a \bar{\chi}(a) \frac{T}{T - \lambda\zeta^a} = \sum_a \frac{\bar{\chi}(a)cT^c}{T^c - \zeta^{ac}} - \sum_a \frac{\bar{\chi}(a)T}{T - \zeta^a}.$$

Multiplying by $S(\chi, \zeta)/N$ proves the lemma.

The assertion

$$Dh_{\chi,c} = g_{\chi,c}$$

follows by using

$$1 + \frac{X}{1 - \lambda\zeta^a} = \frac{T - \lambda\zeta^a}{1 - \lambda\zeta^a}$$

and differentiating naively using $D = TD_T$.

We shall recall below how it is possible to extend the definition of the p-adic logarithm uniquely to a continuous function on all of $\mathbf{C}_p^*$ such that $\log p = 0$. This is the log with which we deal in the next theorem, giving us Leopoldt's value of the L-function $L_p(s, \chi)$ at $s = 1$.

Theorem 3.6. *Let χ be a primitive Dirichlet character with conductor N equal to a power of p. Then*

$$L_p(1, \chi) = -\frac{S(\chi, \zeta)}{N} \sum_{a \in \mathbf{Z}(N)^*} \bar{\chi}(a) \log(1 - \zeta^a).$$

Proof. By Theorem 3.3, Proposition 3.5, and the definition of $L_p(1, \chi)$, we find:

$$\begin{aligned}(1 - \chi(c))L_p(1, \chi) &= \frac{1}{p} \frac{S(\chi, \zeta)}{N} \sum_{\xi} \sum_{a} \sum_{\lambda \neq 1} \bar{\chi}(a) \log\left(1 + \frac{\xi - 1}{1 - \lambda\zeta^a}\right) \\ &= \frac{1}{p} \frac{S(\chi, \zeta)}{N} \sum_{a} \bar{\chi}(a) \log \prod_{\lambda \neq 1} \prod_{\xi} \frac{\xi - \lambda\zeta^a}{1 - \lambda\zeta^a}.\end{aligned}$$

But

$$\prod_{\lambda \neq 1} \prod_{\xi} \frac{\xi - \lambda\zeta^a}{1 - \lambda\zeta^a} = \prod_{\lambda \neq 1} \frac{1 - \lambda^p\zeta^{ap}}{(1 - \lambda\zeta^a)^p} = \prod_{\lambda \neq 1} \frac{1 - \lambda\zeta^{ap}}{(1 - \lambda\zeta^a)^p}.$$

Using the fact that N is the conductor of χ and that the sum of a non-trivial character over a group is 0, we leave to the reader the verification that

$$\sum_{a} \bar{\chi}(a) \log(1 - \lambda\zeta^{ap}) = 0.$$

It follows that

$$\begin{aligned}(1 - \chi(c))L_p(1, \chi) &= \frac{S(\chi, \zeta)}{N} \sum_{a} \bar{\chi}(a) \log \prod_{\lambda \neq 1} (1 - \lambda\zeta^a) \\ &= \frac{S(\chi, \zeta)}{N} \sum_{a} \bar{\chi}(a) \log \frac{1 - \zeta^{ca}}{1 - \zeta^a} \\ &= \frac{S(\chi, \zeta)}{N} (\chi(c) - 1) \sum_{a} \bar{\chi}(a) \log(1 - \zeta^a),\end{aligned}$$

as was to be shown.

Appendix. The p-adic Logarithm

We recall briefly how to extend the p-adic log to the multiplicative group $\mathbf{C}_p^*$. The p-adic log is defined first by the usual series

$$\log_p(1 + x) = x - \frac{x^2}{2} + \cdots.$$

We shall omit the p as subscript. The series converges for $|x| < 1$ (the absolute value is that on $\mathbf{C}_p$, the completion of the algebraic closure of $\mathbf{Q}_p$). We extend the log to all units of $\mathbf{C}_p^*$ as follows. The units have a product decomposition

$$U = \mu_{[p]} \times U_1$$

where $\mu_{[p]}$ is the group of roots of unity in F of order prime to p, and U_1 is the group of units $\equiv 1 \pmod{\mathfrak{p}}$, and $\mathfrak{p}|p$ in $\mathbf{C}_p^*$. For each unit u we let $\langle u \rangle$ be its projection on U_1, and we define

$$\log u = \log \langle u \rangle.$$

Thus the log has been extended to all units, and it is clear that this extension is continuous, and is a homomorphism.

It is even possible to extend the log to the whole multiplicative group $\mathbf{C}_p^*$ (following Iwasawa). We let P be a subgroup of $\mathbf{C}_p^*$ containing the powers of p, and one tth root of p for each positive rational number t. Then the multiplicative group of $\mathbf{C}_p^*$ has the product decomposition

$$P \times \mu_{[p]} \times U_1.$$

Again we define the log of an element $\alpha \in \mathbf{C}_p^*$ to be the log of its projection on U_1.

We leave it as an exercise to the reader to verify that this extension is continuous. It is obviously a homomorphism. In particular, $\log p = 0$.

As for uniqueness, suppose the log has been extended to a continuous function on $\mathbf{C}_p^*$, which is a homomorphism into the additive group. Then the log has to vanish on all roots of unity. If $\log p = 0$ then $\log p^r = 0$ for all rational numbers r. Given $a \in \mathbf{C}_p^*$ there exists r such that ap^r is a unit. Hence the extension is determined by its values on units. Furthermore there exists a root of unity ζ such that $ap^r\zeta$ is $\equiv 1 \bmod \mathfrak{m}$, where $\mathfrak{m}$ is the maximal ideal of the integers of $\mathbf{C}_p$. Hence the log is determined by its values on elements $\equiv 1 \bmod \mathfrak{m}$, where it is defined by the usual power series. This proves uniqueness.

§4. The p-adic Regulator

Let K be a totally real number field, and let $E = E_K$ be the group of units of K. Let p be a prime number. Let $u_1, \ldots, u_r$ be a family of independent units in K, and let

$$\sigma_i : K \to \mathbf{C}_p, \quad i = 1, \ldots, r + 1$$

be the embeddings of K in the p-adic complex numbers (completion of the algebraic closure of $\mathbf{Q}_p$). We suppose $\sigma_{r+1} = id$. We define the **p-adic regulator** up to sign,

$$R_p(u_1, \ldots, u_r) = \pm \det \log \sigma_i u_j.$$

If $u_1, \ldots, u_r$ are a basis for the units (mod roots of unity), we simply call it the **p-adic regulator**, and write

$$R_p = R_{K,p} = R_p(E_K) = R_p(E).$$

If K is the real subfield of $\mathbf{Q}(\mu_m)$, and $\mathscr{E}$ is the group generated by the real cyclotomic units, then we let

$$R_p(\mathscr{E}) = R_p(u_1, \ldots, u_r)$$

where $u_1, \ldots, u_r$ generate these cyclotomic units, mod ± 1.

We leave it to the reader to verify:

Theorem 4.1. *Let K be the real subfield of $\mathbf{Q}(\mu_m)$. Then*

$$R_p(\mathscr{E}) = (E : \mathscr{E})R_p(E) = (E : \mathscr{E})R_p.$$

We know from Theorem 5.1 of the preceding chapter that

$$h^+ = (E : \mathscr{E}).$$

Let g_a (a prime to m) be the cyclotomic units, and g_a^+ the corresponding real cyclotomic units. From our definition of the p-adic log, we know that for any embedding $\sigma \colon \mathbf{Q}(\mu_m) \to \mathbf{C}_p$,

$$\log \sigma g_a = \log \sigma g_a^+.$$

Thus in writing down the regulator, we can use the usual form for the cyclotomic units, without bothering to write down the extra root of unity.

We may write the p-adic cyclotomic regulator by the Frobenius determinant formula,

$$R_p(\mathscr{E}) = \det_{a,b \neq 1} \log \sigma_a g_b = \prod_{\chi \neq 1} \sum_{a \in G} \bar{\chi}(a) \log g_a,$$

where $G = \mathbf{Z}(m)^*/\pm 1$. Since

$$g_a = \frac{\zeta^a - 1}{\zeta - 1},$$

and since $\sum_{a\in G} \bar{\chi}(a) = 0$, we may also write this formula in the form

$$\boxed{R_p(\mathscr{E}) = \prod_{\chi\neq 1} \sum_{a\in G} \bar{\chi}(a) \log(\zeta^a - 1).}$$

The product is taken over all non-trivial characters of $\mathbf{Z}(m)^*/\pm 1$.

Theorem 4.2 (Brumer). *We have $R_p \neq 0$ for the real cyclotomic field $\mathbf{Q}(\mu_m)^+$.*

Proof. The cyclotomic units are algebraic, and it is a known theorem from the theory of transcendental numbers that the logs (p-adic or otherwise) of multiplicatively independent algebraic numbers are linearly independent over the algebraic numbers. The proof is the p-adic analogue of Baker's proof for the corresponding result over the complex numbers, see Brumer [Br], or [L 4], before Chapter VIII, Introduction to the Baker method. The proof given there applies p-adically. The factorization of the regulator into a product of linear forms in logarithms then shows that the regulator is not 0.

Theorem 4.3 (Leopoldt p-adic Class Number-regulator Formula). *Let $m = p^n$ be a prime power, and $K^+ = \mathbf{Q}(\mu_m)^+$. Then*

$$\prod_{\substack{\chi\neq 1\\ \chi \text{ even}}} \frac{1}{2} L_p(1, \chi) = \frac{h^+}{\sqrt{d_{K^+}}} R_p.$$

Proof. From Theorem 4.1 and the *complexly derived* index

$$h^+ = (E : \mathscr{E})$$

of Theorem 5.1 in the preceding chapter, we find:

$$\begin{aligned}\pm h^+ R_p(E) = \pm R_p(\mathscr{E}) &= \prod_{\substack{\chi\neq 1\\ \chi \text{ even}}} \prod_{a\in G} \bar{\chi}(a) \log_p(\zeta^a - 1)\\ &= \prod_{\substack{\chi\neq 1\\ \chi \text{ even}}} - \frac{m(\chi)}{S(\chi, \lambda)} \frac{1}{2} L_p(1, \chi)\end{aligned}$$

by Theorem 3.6, the $\frac{1}{2}$ appearing because $G = \mathbf{Z}(m)^*/\pm 1$,

$$= \sqrt{d_{K^+}} \prod_{\substack{\lambda\neq 1\\ \lambda \text{ even}}} - \frac{1}{2} L_p(1, \chi)$$

by using the complex Theorem 3.1 of Chapter 3. Selecting the sign of the regulator R_p appropriately yields the desired formula.

Remark. The proof of the formula involves the *complex case*. Presumably there is a direct proof, which is valid for all totally real number fields K. Cf. the Appendix of Coates Durham lectures [Co 3], where such a proof is given for the characteristic polynomial of a certain Iwasawa module. The extent to which analogues of cyclotomic units will ultimately play a role in such proofs is not clear at present.

§5. The Formal Leopoldt Transform

Let K be a field of characteristic 0. Let T be the variable on the multiplicative group. We put:

$$T = e^Z, \qquad X = T - 1, \qquad T = X + 1.$$

Then Z is the corresponding variable on the additive group. Note that

$$X^n = (T - 1)^n = (e^Z - 1)^n.$$

Changing variables gives rise to the notation

$$f(X) = f(e^Z - 1) = f_{G_a}(Z) = f_{G_m}(T).$$

This last equality makes sense only when f is a rational function.

For any power series $f \in K[[X]]$ we define the **Leopoldt transform** Γf as a function on integers ≥ 0 by

$$f_{G_a}(Z) = \sum \Gamma f(k) \frac{Z^k}{k!}.$$

As before we let

$$D_X = d/dX, \qquad D_Z = d/dZ, \qquad D_T = d/dT.$$

Then

$$D_Z = (1 + X)D_X = TD_T,$$

and for any integer $k \geq 0$,

$$\Gamma f(k) = D_Z^k f_{G_a}|_{Z=0} = (TD_T)^k f_{G_m}|_{T=1}.$$

Define coefficients $\gamma_n(k)$ by

$$(e^Z - 1)^n = \sum_{k=1}^{\infty} \gamma_n(k) \frac{Z^k}{k!}.$$

Then

$$\gamma_n(k) = D_Z^k(e^Z - 1)^n|_{Z=0} = (TD_T)^k(T - 1)^n|_{T=1}$$
$$\gamma_n(k) = 0 \quad \text{if } k < n.$$

Lemma. (a) *We have*

$$\gamma_n(k) = \sum_{i=0}^{n} (-1)^{n-i}\binom{n}{i} i^k.$$

(b) *Each integer $\gamma_n(k)$ is divisible by $n!$.*

Proof. As to the first assertion, it is immediate by induction that

$$(TD_T)^k T^i = i^k T^i.$$

Hence

$$\gamma_n(k) = (TD_T)^k (T-1)^n|_{T=1} = (TD_T)^k \sum_{i=0}^{n} (-1)^{n-i}\binom{n}{i} T^i$$

has the expression as stated. On the other hand by induction it follows that given an integer n, for each integer k there exist integers $a_n, \ldots, a_{n-k}$ such that $a_i = 0$ if $i > 0$, and

$$(TD_T)^k (T-1)^n = a_n (T-1)^n + n a_{n-1}(T-1)^{n-1} + \cdots$$
$$+ n(n-1)\cdots(n-k+1)a_{n-k}(T-1)^{n-k}.$$

Putting $T = 1$ yields the second assertion.

In the light of the lemma, a power series $f(X)$ has the Z-expression

$$f(X) = \sum_{n=0}^{\infty}\sum_{k=0}^{\infty} a_n \frac{\gamma_n(k)}{n!}\frac{Z^k}{k!}.$$

Consequently,

$$ZD_Z f = \sum_{n=0}^{\infty}\sum_{k=0}^{\infty} a_n \frac{\gamma_n(k)}{n!}\frac{Z^k}{(k-1)!}$$
$$= \sum_{n=0}^{\infty}\sum_{k=0}^{\infty} a_n k \frac{\gamma_n(k)}{n!}\frac{Z^k}{k!}.$$

These formulas can be summarized in the following theorem. We let $C(\mathbf{Z}, K)$ denote the space of functions from $\mathbf{Z}$ into K.

Theorem 5.1. *There exists a unique linear map*

$$\Gamma: K[X] \to C(\mathbf{Z}, K)$$

satisfying any one of the following equivalent conditions:

Γ1. $\Gamma((1+X)^m)(k) = m^k$ *for all integers* $m \geq 0$.

Γ2. $$\Gamma\left(\frac{X^n}{n!}\right)(k) = \frac{\gamma_n(k)}{n!}.$$

Γ3. $$f_{G_a}(Z) = \sum \Gamma f(k) \frac{Z^k}{k!}.$$

This map also satisfies:

Γ4. $$\Gamma(ZD_Z f)(k) = k\Gamma f(k).$$

Observe that the Leopoldt transform is defined by **Γ3** for power series. The other two conditions **Γ1** and **Γ2** do not make sense for power series. However, in the next section, we shall work over a p-adic field K where these other conditions do make sense for a suitably restricted set of power series, with certain convergence conditions.

§6. The p-adic Leopoldt Transform

For simplicity, we suppose that p is an odd prime number. Let K be finite over $\mathbf{Q}_p$. Let $\mathbf{C}_p$ = completion of the algebraic closure of K. We denote the p-adic absolute value by $| \ | = | \ |_p$, normalized so that

$$|p| = 1/p.$$

We define the **Leopoldt space**:
$\mathscr{L} = \mathscr{L}_K$ = space of power series

$$f(X) = \sum a_n \frac{X^n}{n!}, \qquad a_n \in K,$$

such that

$$\lim_{n\to\infty} |a_n| = 0.$$

We define the **Leopoldt norm**

$$\|f\|_{\mathscr{L}} = \max_n |a_n|.$$

Then $\mathscr{L}$ is a Banach space, and a Banach algebra because $\|fg\|_{\mathscr{L}} \leq \|f\|_{\mathscr{L}}\|g\|_{\mathscr{L}}$.

Theorem 6.1. *If $f \in \mathscr{L}$ then f converges on the disc of elements*

$$x \in \mathbf{C}_p \quad \textit{and} \quad |x| \leq |p|^{1/(p-1)}.$$

For such x we have

$$|f(x)| \leq \|f\|_{\mathscr{L}}.$$

Proof. Obvious, because

$$|p|^{n/(p-1)} \leq |n!| \quad \text{and so} \quad \left| a_n \frac{x^n}{n!} \right| \leq |a_n|.$$

We let $C(\mathbf{Z}_p, K)$ = Banach space of continuous functions on $\mathbf{Z}_p$ with values in K, and the sup norm.

If $a \in \mathbf{Z}_p$ and $p|a$, we let $\langle a \rangle = 0$.

If $a \in \mathbf{Z}_p^*$ we write

$$a = \zeta\langle a \rangle = \omega(a)\langle a \rangle$$

where $\zeta \in \mu_{p-1}$ and $\langle a \rangle \equiv 1 \bmod p$. The **Teichmuller character** ω by definition is such that

$$\omega(a) = \zeta.$$

If $s \in \mathbf{Z}_p$, then

$$\langle a \rangle^s = \lim_{k \to s} \langle a \rangle^k$$

is defined in the usual way, where k ranges over positive integers approaching s p-adically. If a is not prime to p, then we let $\langle a \rangle^s = 0$ for all s.

If χ is a character on $\mathbf{Z}_p^*$, as usual we put $\chi(m) = 0$ if m is divisible by p, so

$$\chi(m)\langle m \rangle^s = 0 \quad \text{if } p|m.$$

If $\chi = \omega^\alpha$ where α is a residue class mod $p - 1$, and $k \equiv \alpha \bmod p - 1$, then for any positive integer i such that $p \nmid i$, we have

$$i^k = \omega^\alpha(i)\langle i \rangle^k.$$

Theorem 6.2. *Let α be a residue class* mod $p - 1$. *There exists a unique continuous linear map*

$$\Gamma_\alpha : \mathscr{L}_K \to C(\mathbf{Z}_p, K)$$

satisfying any one of the following three equivalent conditions:

Γα 1. $$\Gamma_\alpha((1 + X)^m)(s) = \omega^\alpha(m)\langle m \rangle^s,$$

for any integer $m \geq 0$.

Γα 2. $$\Gamma_\alpha\left(\frac{X^n}{n!}\right)(s) = \frac{1}{n!} \sum_{i=0}^{\infty} (-1)^{n-i} \binom{n}{i} \omega^\alpha(i)\langle i \rangle^s$$

Γα 3. $$\Gamma_\alpha f(s) = \lim_m \Gamma f(m) = \lim_m \sum_n a_n \Gamma\left(\frac{X^n}{n!}\right)(m)$$

where the limit is taken over positive integers m satisfying:

$$(*) \qquad m \to \infty, \qquad m \to s \text{ } p\text{-adically}, \qquad m \equiv \alpha \bmod p - 1.$$

This map also satisfies

$$\|\Gamma_\alpha f\| \leq \|f\|_{\mathscr{L}},$$

and for $s \in \mathbf{Z}_p$,

$$\mathbf{\Gamma_\alpha\ 4.} \qquad \Gamma_\alpha((ZD_Z)f)(s) = s\Gamma_\alpha f(s).$$

Proof. Any continuous linear map on the space of polynomials (with Leopoldt norm) extends uniquely by continuity to the Leopoldt Banach algebra. We shall prove that the linear map

$$\Gamma_\alpha\colon K[X] \to C(\mathbf{Z}_p, K)$$

with values

$$\Gamma_\alpha\left(\frac{X^n}{n!}\right)(s) = \frac{1}{n!}\sum_{i=0}^{n}(-1)^{n-i}\binom{n}{i}\omega^\alpha(i)\langle i\rangle^s$$

is continuous, and has the other properties. Uniqueness is obvious.

If $f(X) = \sum a_n(X^n/n!)$ lies in $\mathscr{L}$, then $|a_n| \to 0$ as $n \to \infty$. To prove that Γ_α is continuous, it will therefore suffice to prove that the values

$$\Gamma_\alpha\left(\frac{X^n}{n!}\right)(s)$$

are bounded. In fact, we shall see that $\Gamma_\alpha(X^n/n!)(s)$ is p-integral. This will also prove that

$$\|\Gamma_\alpha f\| \leq \|f\|_{\mathscr{L}}.$$

Fix the integer n. Let m range over integers as in $\mathbf{\Gamma_\alpha\ 3}$. Such integers are dense in $\mathbf{Z}_p$, so it suffices to prove that

$$\Gamma_\alpha\left(\frac{X^n}{n!}\right)(m) \text{ is } p\text{-integral for such } m.$$

If $i \equiv 0 \bmod p$, then $i^m/n!$ is p-integral for large m. If $i \not\equiv 0 \bmod p$, and

$$i = \omega(i)\langle i\rangle,$$

then for m close to s p-adically,

$$i^m = \omega^\alpha(i)\langle i\rangle^m \equiv \omega^\alpha(i)\langle i\rangle^s \quad \text{mod high power of } p.$$

The Lemma (b) of §2 then concludes the proof that $\|\Gamma_\alpha f\| \le \|f\|_{\mathscr{L}}$, and the arguments also show that $\boldsymbol{\Gamma}_\alpha$ **1** and $\boldsymbol{\Gamma}_\alpha$ **3** are satisfied. It is clear that Γ_α also satisfies $\boldsymbol{\Gamma}_\alpha$ **4**, thereby concluding the proof of the theorem.

We call Γ_0 the **p-adic Leopoldt transform.**

The Leopoldt transform p-adically with characters ω^α was first used by Lichtenbaum [Li 3] to deal with elliptic curves.

In Theorem 6.5 below we shall prove that Γ_0 is the p-adic Gamma transform already mentioned in §3 when applied to a power series with coefficients in $\mathfrak{o}$. Hence we may then write

$$\Gamma_0 f = \Gamma_p f.$$

We recall the operator

$$\mathbf{U}f(X) = f(X) - \frac{1}{p} \sum_{\zeta^p = 1} f(\zeta(X+1) - 1),$$

where the sum is taken over all pth roots of unity ζ. Then for the special polynomial $(1 + X)^n$ we have

$$\begin{aligned} \mathbf{U}((1+X)^n) &= (1+X)^n - \frac{1}{p}\sum_\zeta \zeta^n (1+X)^n \\ &= (1+X)^n\left(1 - \frac{1}{p}\sum_\zeta \zeta^n\right), \end{aligned}$$

and

$$\sum \zeta^n = \begin{cases} p & \text{if } p \mid n \\ 0 & \text{if } p \nmid n. \end{cases}$$

Hence

$$\mathbf{U}((1+X)^n) = \begin{cases} 0 & \text{if } p \mid n \\ (1+X)^n & \text{if } p \nmid n, \end{cases}$$

and in particular, $\mathbf{U}$ is a projection operator, i.e.,

$$\mathbf{U}^2 = \mathbf{U}.$$

The next lemma describes the continuity property of the operator $\mathbf{U}$ for the Leopoldt norm.

Lemma. $$\|\mathbf{U}f\|_{\mathscr{L}} \le \|f\|_{\mathscr{L}}.$$

Proof.

$$\begin{aligned}\sum_{\zeta\neq 1} f(\zeta X + \zeta - 1) &= \sum_{\zeta\neq 1} a^n \frac{(\zeta X + \zeta - 1)^n}{n!}\\ &= \sum_{\zeta\neq 1} \frac{a_n}{n!} \sum_{k=0}^{n} \binom{n}{k} (\zeta X)^k (\zeta - 1)^{n-k}\\ &= \sum_{\zeta\neq 1} \left(\sum_n \frac{a_n}{n!} \binom{n}{k} (\zeta - 1)^{n-k} \zeta^k\right) X^k.\end{aligned}$$

The coefficient of $X^k/k!$ in the above sum is either 0 or

$$\sum_n \sum_{\zeta\neq 1} \frac{a^n}{n!} \frac{n!}{(n-k)!} (\zeta - 1)^{n-k} \zeta^k,$$

and $|\zeta - 1|^{n-k} < |(n-k)!|$, so the coefficient of a_n is not a unit at p. But

$$\sum_{\zeta\neq 1} (\zeta - 1)^{n-k} \zeta^k$$

is a rational integer, and is therefore $\equiv 0 \bmod p$. Hence

$$\frac{1}{p} \sum_{\zeta\neq 1} f(\zeta X + \zeta - 1) = \frac{1}{p} \sum_k \left(\sum_{n=0}^{\infty} a_n b_{n,k}\right) \frac{X^k}{k!}$$

where $b_{n,k} \in \mathbf{Z}$ and $b_{n,k} \equiv 0 \bmod p$. It is then immediate that

$$\|\mathbf{U}f\|_{\mathscr{L}} \le \|f\|_{\mathscr{L}},$$

as desired.

The next theorems prove for the Leopoldt transform on the Leopoldt space results which have already been proved for measures.

Theorem 6.3. *Let m be an integer ≥ 0, and $m \equiv \alpha \bmod p - 1$. Then*

$$\Gamma_\alpha f(m) = \Gamma \mathbf{U} f(m).$$

Proof. The two maps

$$f \mapsto \Gamma_\alpha f \quad \text{and} \quad f \mapsto \Gamma \mathbf{U} f$$

of $K[X] \to C(\mathbf{Z}_p, K)$ are equal on the polynomials $(1 + X)^\nu$. For a fixed m the maps

$$f \mapsto \Gamma_\alpha f(m) \quad \text{and} \quad f \mapsto \Gamma \mathbf{U} f(m)$$

are continuous, so the theorem follows by continuity.

Theorem 6.4. *For $f \in \mathscr{L}_K$ we have*

$$\Gamma_0 f(0) = \mathbf{U}f(0) = f(0) - \frac{1}{p} \sum_{\zeta^p = 1} f(\zeta - 1).$$

Proof. The power series for $\mathbf{U}f$ in terms of X or Z have the same constant term. Hence

$$\Gamma \mathbf{U}f(0) = \mathbf{U}f(0).$$

Taking $\alpha = 0$, the theorem is obvious from Theorem 6.3, and the fact that $\zeta - 1$ lies in the domain of convergence of f by Theorem 6.1.

The next theorem resulted from a conversation with Ribet.

Theorem 6.5. *For $s \in \mathbf{Z}_p$ and $f \in \mathfrak{o}[[X]]$ we have*

$$\int_{\mathbf{Z}_p^*} \langle a \rangle^s \, d\mu_f(a) = \Gamma_0 f(s).$$

Proof. By continuity in s, it suffices to prove the theorem when $s = k$ is an integer ≥ 1 and $k \equiv 0 \bmod p - 1$. Let φ be the characteristic function of $\mathbf{Z}_p^*$. Then

$$\begin{aligned}
\int_{\mathbf{Z}_p^*} \langle a \rangle^k \, d\mu_f(x) &= \int_{\mathbf{Z}_p} x^k \varphi(x) \, d\mu_f(x) \\
&= \int_{\mathbf{Z}_p} x^k \, d\mu_{\mathbf{U}f}(x) \\
&= D^k \mathbf{U}f(0) \\
&= \Gamma \mathbf{U}f(k) \\
&= \Gamma_0 f(k).
\end{aligned}$$

This proves the theorem.

We now see that the Leopoldt transform is an extension of the Gamma transform to the Leopoldt space.

Iwasawa Theory and Ideal Class Groups 5

We shall now study Iwasawa's theory concerning projective limits in $\mathbf{Z}_p$-extensions.

The first three sections establish purely algebraic facts about projective limits, and finitely generated modules over the power series ring $\mathbf{Z}_p[[X]]$ which appears as the limit of p-adic group rings of cyclic groups. The situation is quite similar to modules over principal rings when considering finitely generated modules over integrally closed Noetherian domains. Cf. Bourbaki, *Commutative Algebra*, Chapter VII, §4, where a general structure theorem is given. For 2-dimensional local rings, this was complemented by Serre [Se 1] who showed that reflexive modules in that case are free, thus getting a complete result for $\mathbf{Z}_p[[X]]$. Here we shall follow Paul Cohen's proof analogous to finding elementary divisors by row and column operations.

We shall also follow Serre's exposition [Se 1], giving the asymptotic estimate for the orders of the factor modules. This is applied afterwards to the orders of ideal class groups. Iwasawa's original proofs were rather complicated, and his point of view was that of projective limits of finite abelian p-groups on which $\Gamma \approx \mathbf{Z}_p$ operates continuously, and which are of topologically finite type for this action. See [Iw 1], [Iw 6]. Serre [Se 1] saw that there was an isomorphism of categories between these objects and $\mathbf{Z}_p[[X]]$-modules of finite type. He introduced this point of view which simplified the proofs and also proved successful in subsequent applications.

The next three sections deal with arithmetic situations arising as special cases (but which historically motivated the general results). We consider several modules over the Iwasawa algebra. First, we deal with the projective limit of ideal class groups. Class field theory identifies this projective limit with a Galois group. The reader unacquainted with class field theory can simply take for granted the isomorphism, which is described as we need it.

We follow mostly Serre's exposition [Se 1]. The results are valid for arbitrary $\mathbf{Z}_p$-extensions, not necessarily cyclotomic ones.

The final two sections go further into certain Galois groups as modules over the Iwasawa algebra, and also describe all possible $\mathbf{Z}_p$-extensions of a given number field in class field theoretic terms. The Leopoldt conjecture would imply that there are precisely $r_2 + 1$ independent ones. This depends on the $\mathbf{Z}_p$-rank of the closure of the global units in the local units. See §5, Theorem 5.2.

§1. The Iwasawa Algebra

Let Γ be a topological group isomorphic to $\mathbf{Z}_p$. We write Γ multiplicatively, and let γ be a fixed generator, so that the isomorphism may be written

$$x \mapsto \gamma^x \text{ for } x \in \mathbf{Z}_p.$$

Let

$$\Gamma_n = \Gamma/\Gamma^{p^n} \approx \mathbf{Z}(p^n).$$

Then Γ_n is cyclic of order p^n, generated by the image of γ. Conversely, a compatible system $\{\gamma_n\}$ of generators in a projective system $\{\Gamma_n\}$ of cyclic groups of order p^n would give rise to a generator γ in their projective limit.

We have a commutative diagram

$$\begin{array}{ccc} \mathbf{Z}_p[\Gamma_{n+1}] & \to & \mathbf{Z}_p[T]/(T^{p^{n+1}} - 1) \\ \downarrow & & \downarrow \\ \mathbf{Z}_p[\Gamma_n] & \to & \mathbf{Z}_p[T]/(T^{p^n} - 1) \end{array}$$

where T is a variable. Let $X = T - 1$, $T = X + 1$. Then $\mathbf{Z}_p[T] = \mathbf{Z}_p[X]$, and

$$\mathbf{Z}_p[T]/(T^{p^n} - 1) \approx \mathbf{Z}_p[X]/((X + 1)^{p^n} - 1).$$

Let

$$h_n = h_n(X) = (1 + X)^{p^n} - 1.$$

Then

$$h_n = X^{p^n} + \cdots,$$

and all coefficients other than the leading coefficient are divisible by p. Such a polynomial is called **distinguished**.

We wish to establish an isomorphism

$$\mathbf{Z}_p[[X]] \xrightarrow{\approx} \lim \mathbf{Z}_p[\Gamma_n] = \lim \mathbf{Z}_p[X]/(h_n).$$

Let

$$\Lambda = \mathbf{Z}_p[[X]].$$

We first note that if h is any distinguished polynomial, then

$$\mathbf{Z}_p[X]/(h) \approx \Lambda/h\Lambda.$$

This is immediate from the Euclidean algorithm (see Theorem 3.1), which shows that $\Lambda/h\Lambda$ is free of rank deg h over $\mathbf{Z}_p$, and similarly $\mathbf{Z}_p[X]/(h)$ is free of the same rank over $\mathbf{Z}_p$. Furthermore this same algorithm shows that the natural map

$$\mathbf{Z}_p[X]/(h) \to \Lambda/h\Lambda$$

is surjective, so is an isomorphism.

We thus obtain a natural map for each n,

$$\mathbf{Z}_p[[X]] \to \mathbf{Z}_p[\Gamma_n] = \mathbf{Z}_p[X]/(h_n),$$

whence a homomorphism

$$\varepsilon: \Lambda = \mathbf{Z}_p[[X]] \to \lim \mathbf{Z}_p[X]/(h_n).$$

Theorem 1.1. *The homomorphism* ε *is an isomorphism.*

Proof. A trivial induction shows that

$$h_n = (1 + X)^{p^n} - 1 \in (p, X)^{n+1}$$

where (p, X) denotes the maximal ideal of $\mathbf{Z}_p[X]$, generated by p and X. It follows that the intersection of the ideals $h_n\Lambda$ must be 0, whence the kernel of ε is 0. Since ε is clearly surjective, this proves the theorem.

Note that the isomorphism ε depends on the original choice of generator γ.

The projective limit

$$\varprojlim \mathbf{Z}_p[\Gamma_n]$$

is called the **Iwasawa algebra**. Given a choice of generator γ, it is identified with $\mathbf{Z}_p[[X]]$ by Theorem 1.1, and then we also call $\mathbf{Z}_p[[X]]$ the **Iwasawa algebra**.

We now consider modules over the Iwasawa algebra.

For each n let V_n be a module over $\mathbf{Z}_p[\Gamma_n]$, and suppose we have homomorphisms

$$V_{n+1} \to V_n$$

compatible with the action of the group rings $\mathbf{Z}_p[\Gamma_{n+1}]$ and $\mathbf{Z}_p[\Gamma_n]$ respectively. We may form the projective limit

$$V = \lim V_n,$$

which is then a Λ-module.

Examples. In §4 of this chapter, $V_n = C_n$ is the p-primary part of the ideal class group, and so the projective limit C is a module over $\Lambda = \mathbf{Z}_p[[X]]$. In Chapter 7, we shall consider projective systems of local units as modules over the Iwasawa algebra.

If each V_n is a finite abelian group, or is compact, then the projective limit V is compact, and $\mathbf{Z}_p[[X]]$ operates continuously on V, which is then what we call a topological module over $\mathbf{Z}_p[[X]]$. (Here and in the sequel, compact means compact Hausdorff.) Note that $\mathbf{Z}_p[[X]]$ itself is compact.

Nakayama's lemma. *Let $\mathfrak{o}$ be a local ring with maximal ideal $\mathfrak{m}$, and $\mathfrak{m}$-adic topology. Let V be a compact topological $\mathfrak{o}$-module.*

(i) *If $\mathfrak{m}V = V$ then $V = 0$.*
(ii) *If $\mathfrak{o}$ is compact, and $V/\mathfrak{m}V$ is finitely generated, then V is finitely generated by any set of representatives of $V/\mathfrak{m}V$.*

Proof. Let U be a neighborhood of 0 in V. Since V is a topological $\mathfrak{o}$-module, for each $x \in V$ there exists an open neighborhood U_x of x and a positive integer $n(x)$ such that

$$\mathfrak{m}^{n(x)}U_x \subset U.$$

A finite number of neighborhoods U_x cover V. Hence there exists an integer n such that $\mathfrak{m}^nV \subset U$. But $\mathfrak{m}V = V$ implies $\mathfrak{m}^nV = V$, and hence $V \subset U$ for all U. Since V is Hausdorff, it follows that $V = 0$, which proves (i).

For (ii), let $x_1, \ldots, x_s$ be representatives of $V/\mathfrak{m}V$, and let W be the $\mathfrak{o}$-submodule generated by them. Then W is a continuous image of $\mathfrak{o}^{(s)}$, and is therefore compact, and closed. Then V/W is compact, and we have

$$\mathfrak{m}(V/W) = V/W.$$

Hence $V/W = 0$, and $V = W$, thereby proving (ii).

Next we pass to certain results concerning finitely generated modules over the Iwasawa algebra. These will be applied to computing orders of certain factor groups (which in §4 will be ideal class groups). The reader may omit the rest of this section if he wishes to disregard such computations for the moment and merely wishes to concentrate on general structural results.

Two modules V, V' are said to be **quasi-isomorphic** if there is a homomorphism

$$V \to V'$$

with finite kernel and cokernel. It will be shown in §3 that any finitely generated V has a quasi-isomorphism with a finite product

$$(*) \qquad V \to \Lambda^{(r)} \oplus \prod \Lambda/(p^{m_i}) \oplus \prod \Lambda/(f_j),$$

where the f_j are distinguished. The first factor $\Lambda^{(r)}$ is the free part. The other factors are Λ-torsion modules.

Suppose now that V is a torsion module such that $V/h_n V$ is finite for all n. We wish to get an asymptotic formula for the order of $V/h_n V$. Such a formula does not change under a quasi-isomorphism, so we are reduced to consider the two cases when

$$V = \Lambda/p^m \quad \text{and} \quad V = \Lambda/f$$

for some positive integer m, and f is distinguished.

In the first case we have

$$\Lambda/p^m = \mathbf{Z}(p^m)[[X]],$$

the power series ring over $\mathbf{Z}(p^m)$. In the second case, Λ/f is a free module over $\mathbf{Z}_p$, whose rank is $\deg f$. It may happen in this second case that

$$V_n = V/(\gamma^{p^n} - 1)V$$

is not finite. We shall first make the assumption of finiteness to get the formula for the order, which is a power of p, so we put

$$\text{card } V_n = p^{e_n} \quad \text{where} \quad e_n = e_n(V).$$

Theorem 1.2. (i) *If $V = \Lambda/p^m$ then $e_n = mp^n$.*

(ii) *Let $V = \Lambda/f$ where f is distinguished of degree d, and assume V_n finite for all n. Then there exists a constant c_0 such that for all n sufficiently large,*

$$e_n = dn + c_0.$$

(iii) *If V is finitely generated over Λ such that V_n is finite for all n, then there exists a constant c such that*

$$e_n(V) = mp^n + dn + c$$

for all n sufficiently large. In the representation of V as in (*) with $r = 0$, we have

$$m = \sum m_i \quad \text{and} \quad d = \sum \deg f_j.$$

Proof. In case (i) we have

$$\mathbf{Z}(p^m)[[X]]/((X+1)^{p^n}-1) \approx \mathbf{Z}(p^m)[X]/((X+1)^{p^n}-1),$$

and this is just $\mathbf{Z}(p^m)[T]/(T^{p^n}-1)$, which is a free module of rank p^n over $\mathbf{Z}(p^m)$. Thus the computation of the order is obvious.

Let us now look at case (ii). For any $h \in \Lambda$ we let h_V be the endomorphism of V induced by h. We let

$$\gamma_n = \gamma_V^{p^n}, \qquad \gamma_V = X_V + 1.$$

We have

$$\gamma^{p^n} - 1 = (X+1)^{p^n} - 1 \equiv X^{p^n} \pmod p,$$
$$f \equiv X^d \pmod p.$$

Hence there exists n_0 such that for $n > n_0$ we have

$$X^{p^{n-1}} \equiv 0 \bmod (f, p) \quad \text{and} \quad X_V^{p^{n-1}} \equiv 0 \bmod p,$$

and therefore

$$\gamma_{n-1} = \gamma_V^{p^{n-1}} \equiv 1 \pmod p.$$

It follows that

$$\gamma_n = \gamma_{n-1}^p \equiv 1 \pmod{p^2}.$$

Now

$$\begin{aligned}\gamma_{n+1} - 1 &= (1 + \gamma_n + \cdots + \gamma_n^{p-1})(\gamma_n - 1)\\ &= (1 + 1 + \cdots + 1 + O(p^2))(\gamma_n - 1)\\ &= pu(\gamma_n - 1)\end{aligned}$$

where u is invertible. We have therefore shown that

$$(\gamma_n - 1)V = p^{n-n_0}(\gamma_{n_0} - 1)V.$$

Furthermore, $(\gamma_{n_0} - 1)V$ is of finite index in V, and is therefore a free module over $\mathbf{Z}_p$ of the same rank d as V. This proves (ii). Case (iii) is then obvious, thus proving the theorem.

Next we consider the case when $V/h_n V$ is not necessarily finite, but make additional hypotheses which still allow us to compute the orders of certain factor groups asymptotically, and which are satisfied in the application to ideal class groups.

We let

$$g_n = 1 + \gamma + \cdots + \gamma^{p^n - 1}.$$

We say that V is of **Iwasawa type** if there exist elements $v_1, \ldots, v_s \in V$ such that, if we put

$$U_0 = \mathbf{Z}_p\text{-submodule of } V \text{ generated by } (\gamma - 1)V \text{ and } v_1, \ldots, v_s,$$
$$U_n = g_n U_0,$$

then V/U_n is finite for all n. In particular, V/U_0 is finite. For a module of Iwasawa type, we let

$$V_n = V/U_n.$$

Theorem 1.3. *Assume that V is of Iwasawa type. Then the conclusions of Theorem* 1.2(i), (ii), (iii) *remain valid.*

Proof. Note that Case (i) is unchanged, only Case (ii) is now slightly different, but the proof runs along entirely similar lines as follows. In this case, V is $\mathbf{Z}_p$-free of rank d. An argument similar to that of Theorem 1.2(ii) shows that

$$g_n V = p^{n-n_0} g_{n_0} V$$

for all $n \geq n_0$. Let $W = g_{n_0} V$. Then

$$e(V/g_n V) = e(V/W) + e(W/p^{n-n_0} W) = c_1 + d(n - n_0)$$

for some constant c_1, since W has the same $\mathbf{Z}_p$-rank as V. This proves the theorem since $e(g_n V/g_n U_0) \leq e(V/U_0)$ and stabilizes for large n.

§2. Weierstrass Preparation Theorem

The proof of the Weierstrass theorem in this section is due to Manin [Man 1]. We start with the **Euclidean algorithm.**

Theorem 2.1. *Let $\mathfrak{o}$ be a complete local ring with maximal ideal $\mathfrak{m}$. Let*

$$f(X) = \sum_{i=0}^{\infty} a_i X^i$$

be a power series in $\mathfrak{o}[[X]]$, such that not all a_i lie in $\mathfrak{m}$. Say $a_0, \ldots, a_{n-1} \in \mathfrak{m}$, and $a_n \in \mathfrak{o}^*$ is a unit. Given $g \in \mathfrak{o}[[X]]$ we can solve the equation uniquely

$$g = qf + r,$$

with $q \in \mathfrak{o}[[X]]$, $r \in \mathfrak{o}[X]$, and $\deg r \leq n - 1$.

Proof. Let α and τ be the projections on the beginning and tail end of the series, given by

$$\alpha\colon \sum b_i X^i \mapsto \sum_{i=0}^{n-1} b_i X^i = b_0 + b_1 + \cdots + b_{n-1}X^{n-1}$$

$$\tau\colon \sum b_i X^i \mapsto \sum_{i=n}^{\infty} b_i X^{i-n} = b_n + b_{n+1}X + b_{n+2}X^2 + \cdots.$$

Note that $\tau(X^n h) = h$ for any $h \in \mathfrak{o}[[X]]$, and h is a polynomial of degree $<n$ if and only if $\tau(h) = 0$.

The existence of q, r is equivalent with the condition that there exists q such that

$$\tau(g) = \tau(qf).$$

But

$$f = \alpha f + X^n \tau(f).$$

Hence our problem is equivalent with solving

$$\tau(g) = \tau(q\alpha(f)) + \tau(qX^n\tau(f)) = \tau(q\alpha(f)) + q\tau(f).$$

Note that $\tau(f)$ is invertible. Put $Z = q\tau(f)$. Then the above equation is equivalent with

$$\tau(g) = \tau\left(Z\frac{\alpha(f)}{\tau(f)}\right) + Z = \left(I + \tau \circ \frac{\alpha(f)}{\tau(f)}\right)Z.$$

Note that

$$\tau \circ \frac{\alpha(f)}{\tau(f)}\colon \mathfrak{o}[[X]] \to \mathfrak{m}\mathfrak{o}[[X]],$$

because $\alpha(f)/\tau(f) \in \mathfrak{m}\mathfrak{o}[[X]]$. We can therefore invert to find Z, namely

$$Z = \left(I + \tau \circ \frac{\alpha(f)}{\tau(f)}\right)^{-1} \tau(g),$$

which proves both existence and uniqueness and concludes the proof.

Theorem 2.2 (Weierstrass Preparation). *The power series f in the previous theorem can be written in the form*

$$f(X) = (X^n + b_{n-1}X^{n-1} + \cdots + b_0)u,$$

where $b_i \in \mathfrak{m}$, and u is a unit in $\mathfrak{o}[[X]]$.

Proof. Write

$$X^n = qf + r,$$

by the Euclidean algorithm. Then q is invertible because

$$q = c_0 + c_1 X + \cdots$$
$$f = \cdots + a_n X^n + \cdots$$

so that

$$1 \equiv c_0 a_n \pmod{\mathfrak{m}},$$

and c_0 is a unit in $\mathfrak{o}$. We obtain $qf = X^n - r$, and

$$f = q^{-1}(X^n - r),$$

with $r \equiv 0 \pmod{\mathfrak{m}}$. This proves the theorem.

The integer n in Theorems 2.1 and 2.2 is called the **Weierstrass degree** of f, and is denoted by

$$\deg_W f.$$

We see that a power series not all of whose coefficients lie in $\mathfrak{m}$ can be expressed as a product of a polynomial having the given Weierstrass degree, times a unit in the power series ring. Furthermore, all the coefficients of the polynomial except the leading one lie in the maximal ideal. Such a polynomial is called **distinguished**.

§3. Modules over $\mathbf{Z}_p[[X]]$

The structure of finitely generated modules over $\mathbf{Z}_p[[X]]$ was first determined by Serre [Se 1] who introduced this point of view in Iwasawa theory. As already mentioned, cf. Bourbaki for general structure theorems over integrally closed Noetherian domains. Paul Cohen showed how one could give a proof along the standard lines of row and column operations, cf. [L 3]. Robert Coleman pointed out to me that the inductive step as given in [L 3] had to be modified, and I am indebted to him for the exposition given in the lemma and Theorem 3.2 below.

We let $\Lambda = \mathfrak{o}[[X]]$, where $\mathfrak{o}$ is a complete discrete valuation ring. We denote by p a prime element of $\mathfrak{o}$. By a finite module over $\mathfrak{o}$ we mean a finitely generated module annihilated by some power p^k and some distinguished element λ. If $\mathfrak{o} = \mathbf{Z}_p$, then "finite" has the usual meaning.

By a **quasi-isomorphism** we mean a homomorphism with finite kernel and cokernel. We denote a quasi-isomorphism by the sign

$$M \sim M'.$$

Theorem 3.1. *Let M be a finitely generated Λ-module. There exists a quasi-isomorphism*

$$M \sim \Lambda^{(r)} \oplus \prod \Lambda/p^{n_i} \oplus \prod \Lambda/(f_j^{m_j})$$

where each f_j is a distinguished polynomial, irreducible in $\mathfrak{o}[X]$, i, j range over finite sets of indices, and $\Lambda^{(r)}$ is the product of Λ taken r times, for some integer r.

The rest of this section is devoted to the proof.

Suppose that M has generators $u_1, \ldots, u_n$. Relative to such generators we can form the matrix of relations, whose rows are vectors

$$(\lambda_1, \ldots, \lambda_n)$$

such that

$$\lambda_1 u_1 + \cdots + \lambda_n u_n = 0.$$

Since Λ is Noetherian, a finite number of the rows generate all of them.

Performing the usual row and column operations on the matrix amounts to changing the generators of the module. We shall describe other operations, corresponding to embedding the module in a bigger one with finite cokernel.

An element $\lambda \in \Lambda$ is called **p-free** if λ does not lie in $p\Lambda$, in other words, if we can apply the Weierstrass preparation theorem to it.

Suppose that there is a relation of the form

$$\lambda_1 u_1 + p(\lambda_2 u_2 + \cdots + \lambda_n u_n) = 0,$$

where λ_1 is p-free. We can form the new module M' obtained by adjoining a new generator v with the relations

$$pv = u_1, \qquad \lambda_1 v = -(\lambda_2 u_2 + \cdots + \lambda_n u_n).$$

This can be formalized by considering a direct sum

$$M \oplus (v)$$

modulo the desired relations, i.e., modulo the submodule generated by the elements

$$(0, pv) - (u_1, 0) \quad \text{and} \quad (0, \lambda_1 v) - (\lambda_2 u_2 + \cdots + \lambda_n u_n, 0).$$

It is then immediately verified that the canonical map of M into the factor module is injective. The factor module M'/M is annihilated by p and λ_1, whence is finite. Furthermore, the elements $v, u_2, \ldots, u_n$ generate M', and have the relation

$$\lambda_1 v + \lambda_2 u_2 + \cdots + \lambda_n u_n = 0.$$

In terms of the relation matrix, this means that we shall allow the following operations, replacing the matrix R by a matrix R'.

O 1. *If R contains a row $(\lambda_1, p\lambda_2, \ldots, p\lambda_n)$ with λ_1 not divisible by p, then we let R' be the matrix whose rows consist of*

$$(\lambda_1, \ldots, \lambda_n)$$

and the rows of R with first element multiplied by p.

Observe that in this first operation, we may have $\lambda_2 = \cdots = \lambda_n = 0$.

Next suppose that some power p^k $(k \geq 1)$ divides all elements of R, but that there exists one relation

$$p^k(\lambda_1, \ldots, \lambda_n)$$

such that λ_1 is distinguished (or equivalently, λ_1 is not divisible by p). We may then form the module M' obtained by adjoining a new element v with the relations

$$p^k v = p^k u_1 \quad \text{and} \quad \lambda_1 v = -(\lambda_2 u_2 + \cdots + \lambda_n u_n).$$

Again, it is easily verified that M is embedded in M' and that M'/M is finite. Note that $p^k(v - u_1) = 0$. The relations of the submodule

$$(v, u_2, \ldots, u_n)$$

are generated by R and the additional relation

$$(\lambda_1, \ldots, \lambda_n).$$

We have a direct sum decomposition

$$M' = (v, u_2, \ldots, u_n) \oplus (v - u_1),$$

and the relations of $v - u_1$ are generated by p^k. To prove the theorem, it suffices to consider the first component of M'. Thus our second operation is described as follows.

O 2. *If all elements of the first column in R are divisible by p^k, and if there exists one relation $(p^k\lambda_1, \ldots, p^k\lambda_n)$ such that λ_1 is not divisible by p, then we let R' consist of R and the new row*

$$(\lambda_1, \ldots, \lambda_n).$$

Finally we allow one more operation:

O 3. *If R has a relation of the form*

$$p^k(\lambda_1, \ldots, \lambda_n), \quad k \geq 0,$$

and there exists an element λ not divisible by p such that

$$(\lambda\lambda_1, \ldots, \lambda\lambda_n)$$

is also a relation, then we may replace R by the matrix R′ having the same rows as R, except that the row $p^k(\lambda_1, \ldots, \lambda_n)$ is replaced by

$$(\lambda_1, \ldots, \lambda_n).$$

This operation corresponds to the surjection with finite kernel

$$M \to M/(\lambda_1 u_1 + \cdots + \lambda_n u_n).$$

Row or column operations, or **O 1, O 2, O 3** will be called **admissible operations.**

Given a matrix R over Λ, we define

$$\deg^{(k)}(R) = \min \deg_W(a'_{ij}) \quad \text{for } i, j \geq k,$$

where (a'_{ij}) ranges over all admissible transformations of R which leave unaltered the components of the first $k - 1$ rows.

Remark. If R' is obtained from R by admissible operations leaving the values in the first $k - 1$ rows unaltered, then

$$\deg^{(k)}(R) \leq \deg^{(k)}(R').$$

Let $r \geq 1$ be an integer. Suppose that R has the form

$$\begin{pmatrix} \lambda_{11} & 0 & \cdots & 0 & 0 & \cdots & 0 \\ \vdots & & & \vdots & \vdots & & \vdots \\ 0 & & \cdots & \lambda_{r-1,r-1} & 0 & \cdots & 0 \\ * & & \cdots & * & * & \cdots & * \end{pmatrix}$$

and

$$\begin{pmatrix} \lambda_{rr} & \cdots & \lambda_{rn} \\ * & \cdots & * \end{pmatrix} \neq 0.$$

Assume also that λ_{ii} for $i = 1, \ldots, r - 1$ is a distinguished polynomial with the property that

$$\deg^{(k)}(R) = \deg \lambda_{kk} \quad \text{for } k = 1, \ldots, r - 1.$$

Then we shall say that R is in $(r - 1)$-**normal form.** If $r = 1$ then this condition is vacuously satisfied, and is the starting point for the induction of the following lemma.

Lemma. *Suppose that R is in $(r - 1)$-normal form, with its first $r - 1$ diagonal elements $\lambda_{11}, \ldots, \lambda_{r-1,r-1}$. Then by admissible transformations, we can transform R into a matrix which is in r-normal form, and has the same first $r - 1$ diagonal elements.*

Proof. Using **O 1** with respect to each of the first $r - 1$ rows, we may assume without loss of generality that any given power p^k $(k \geq 0)$ divides all components λ_{ij} with $i \geq r$ and $j = 1, \ldots, r - 1$, that is all components lying below the portion of the matrix which has already been diagonalized. Using **O 2**, we may then arrange that p does not divide some λ_{ij} with $i \geq r$, and $j \geq r$. After a succession of admissible transformations on the lower right matrix

$$\begin{pmatrix} \lambda_{rr} & \cdots & \lambda_{rn} \\ & {*}{*}{*}{*}{*}{*}{*}{*}{*}{*} & \end{pmatrix}$$

induced by admissible transformations of R which leave the first $r - 1$ rows fixed elementwise, we may then find some element λ_{ij} with $i \geq r$ and $j \geq r$ such that

$$\deg_W \lambda_{ij} = \deg^{(r)}(R).$$

The Weierstrass preparation theorem allows us to assume that this element λ_{ij} is a distinguished polynomial, and

$$\deg \lambda_{ij} = \deg^{(r)}(R).$$

Finally, row and column interchanges which do not involve the elements λ_{ii} $(i = 1, \ldots, r - 1)$ allow us to assume that $\lambda_{ij} = \lambda_{rr}$.

There remains to show that we can make all other elements on the rth row equal to 0 after appropriate transformations. By the Euclidean Algorithm, we may assume that

$$\deg \lambda_{rj} < \deg \lambda_{rr} \text{ for } r \neq j$$

$$\deg \lambda_{rj} < \deg \lambda_{jj} \text{ for } j < r.$$

We first deal with the elements to the right of λ_{rr} on the rth row. We may assume that λ_{rj} with $j > r$ is divisible by p, otherwise we contradict the minimality of the degree of λ_{rr}. Using **O 1** repeatedly as before with respect to the first $r - 1$ rows, we may then assume that all elements λ_{rj} with $j < r$ are divisible by a high power p^k. We then use **O 1** with respect to the rth row, to divide all elements λ_{rj} $(j \neq r)$ by successive powers of p, thus leading to some element $\lambda_{rj'}$ with $j' > r$ not divisible by p, a contradiction of $\deg \lambda_{rr} = \deg^{(r)}(R)$. Thus $\lambda_{rj} = 0$ for $j > r$.

For elements λ_{rj} to the left of λ_{rr}, that is with $j < r$, if some such element is not 0, then we may use **O 1** with respect to the rth row to divide by p, until we are in the situation where there exists $j < r$ such that λ_{rj} is not divisible by p, contradicting the facts that

$$\deg \lambda_{rj} < \deg \lambda_{jj} \quad \text{and} \quad \deg \lambda_{jj} = \deg^{(j)}(R).$$

Thus we have put the matrix in r-normal form, and proved the lemma.

Theorem 3.2. *If R is a matrix of relations, we can transform R with a finite number of admissible operations into a matrix R' of the form*

$$\begin{pmatrix} \lambda_{11} & 0 & \cdots & 0 & 0 & \cdots & 0 \\ \vdots & \vdots & & \vdots & \vdots & & \vdots \\ 0 & 0 & \cdots & \lambda_{rr} & 0 & \cdots & 0 \\ 0 & 0 & \cdots & 0 & 0 & \cdots & 0 \end{pmatrix}$$

where λ_{ii} are distinguished polynomials.

Proof. By the lemma, we can replace R by a matrix R' of the form

$$\begin{pmatrix} \lambda_{11} & 0 & \cdots & 0 & 0 & \cdots & 0 \\ \vdots & \vdots & & \vdots & \vdots & & \vdots \\ 0 & 0 & \cdots & \lambda_{rr} & 0 & \cdots & 0 \\ * & * & \cdots & * & 0 & \cdots & 0 \end{pmatrix}$$

where λ_{ii} are distinguished polynomials, and

$$\deg \lambda_{ii} = \deg^{(i)}(R) \text{ for } i = 1, \ldots, r.$$

By the Euclidean Algorithm, we may assume that $\lambda_{ji} = 0$ or

$$\deg \lambda_{ji} < \deg \lambda_{ii} \text{ for } j \neq i.$$

In fact, we contend that $\lambda_{ji} = 0$ for all $j \neq i$. Suppose otherwise, so that $\lambda_{ji} \neq 0$ for some $j > r > i$, so we have a relation

$$(\lambda_{j1}, \ldots, \lambda_{jr}, 0, \ldots, 0)$$

not identically 0. Let

$$\lambda = \lambda_{11} \cdots \lambda_{rr}.$$

Then λ is not divisible by p, and $\lambda u_i = 0$ for $i = 1, \ldots, r$, so

$$(\lambda\lambda_{j1}, \ldots, \lambda\lambda_{jr}, 0, \ldots, 0)$$

is also a relation. By **O 3** we may assume without loss of generality that some $\lambda_{j1}, \ldots, \lambda_{jr}$ is not divisible by p, and then contradict the minimality condition on the λ_{ii}. This proves the theorem.

We return to the module interpretation, to see that Theorem 3.2 implies the theorem. Indeed, any module with matrix of relations R' as in Theorem 3.2 is isomorphic to

$$\Lambda^{n-r} \oplus \bigoplus_{i=1}^{r} \Lambda/(\lambda_{ii}).$$

Finally, if f, g are distinguished and relatively prime, the map

$$\Lambda(fg) \to \Lambda/f \oplus \Lambda/g$$

is an embedding with finite cokernel. This allows us to decompose the factors Λ/λ_{ii} into a direct sum of factors

$$\Lambda/(f_j^{m_j})$$

where f_j is distinguished and irreducible, thereby concluding the proof of Theorem 3.1.

§4. $\mathbf{Z}_p$-extensions and Ideal Class Groups

Let K_0 be a number field. An extension K_∞ of K_0 is called a **$\mathbf{Z}_p$-extension** if it is abelian, and its Galois group is isomorphic to $\mathbf{Z}_p$. To give such an extension is the same as to give a tower of fields

$$K_\infty = \bigcup_{n=0}^{\infty} K_n \supset \cdots \supset K_n \supset \cdots \supset K_0$$

such that K_n is cyclic over K_0 of degree p^n.

Examples. Let p be a prime number. Let

$$K_n = \mathbf{Q}(\mu_{p^{n+1}}) \text{ if } p \text{ is odd}$$
$$K_n = \mathbf{Q}(\mu_{p^{n+2}}) \text{ if } p \text{ is even.}$$

This gives the cyclotomic $\mathbf{Z}_p$-extension over the field K_0.

More generally, let K be any number field, let

$$K^{(p)} = K(\mu^{(p)})$$

be the extension obtained by adjoining all p-power roots of unity. Then $K^{(p)}$ is abelian over K, and it is easy to see that the fixed field of the torsion subgroup of $\mathrm{Gal}(K^{(p)}/K)$ is a $\mathbf{Z}_p$-extension of $K_0 = K$, called the **cyclotomic** **$\mathbf{Z}_p$-extension**. We study it later in the book. Note that a non-totally real field K always has non-cyclotomic $\mathbf{Z}_p$-extensions, cf. §5. Natural examples can be constructed with elliptic curves having complex multiplication, cf. [C–W].

We say that a prime ideal $\mathfrak{p}_0$ of K_0 is **almost totally ramified** in a Galois extension K' if the inertia group of a prime $\mathfrak{p}$ in K' over $\mathfrak{p}_0$ is of finite index in $\mathrm{Gal}(K'/K_0)$. We say $\mathfrak{p}_0$ is **almost unramified** if its inertia group is finite.

We consider the following condition of Iwasawa.

IW. *K_∞ is totally ramified over K_0 over a finite number of prime ideals $\mathfrak{p}_1, \ldots, \mathfrak{p}_s$ lying above p, and is unramified over all other prime ideals.*

Lemma. *Let K_∞/K_0 be a $\mathbf{Z}_p$-extension. Then:*

(i) *Only a finite number of prime ideals of K_0 ramify in K_∞, they lie above p, and they are almost totally ramified.*
(ii) *For some positive integer d, the extension K_∞/K_d is a $\mathbf{Z}_p$-extension satisfying* **IW**.

Proof. Some prime ideal $\mathfrak{p}$ of K_0 must ramify in K_∞ because class field theory says the maximal unramified abelian extension of K_0 is finite. Let I be the inertia group. It is a closed subgroup of Γ, and $\neq 0$, hence equal to $p^m\mathbf{Z}_p$ for some m, so that $\mathfrak{p}$ is almost totally ramified. Over the completion $K_{0,\mathfrak{p}}$, the maximal tamely ramified abelian extension is finite. Hence the wild ramification group is of finite index in Γ, thus showing that $\mathfrak{p}$ lies above p. This proves (i). If we let $\mathfrak{p}_1, \ldots, \mathfrak{p}_s$ be the finite number of primes which are almost totally ramified; and let

$$I_j \cong p^{d_j}\mathbf{Z}_p$$

be the inertia groups, and $d = \max d_j$, then K_∞/K_d satisfies condition **IW** as desired.

Assume that condition **IW** is satisfied.

The same lemma as in Chapter 3, §4 shows that the norm map between any two successive steps in the tower is surjective on the ideal class groups. We let $C_n = C_n^{(p)}$ be the **p-primary part of the ideal class group in K_n**. Then we have a surjective sequence

$$C_0 \leftarrow C_1 \leftarrow C_2 \leftarrow \cdots$$

and we let

$$\boxed{C = \operatorname{lim\,proj} C_n}$$

be the projective limit. We may view C as consisting of all sequences

$$(c_0, c_1, c_2, \ldots)$$

with $c_n \in C_n$ and c_{n+1} mapping on c_n under the norm map.

Let M_n be the maximal p-primary abelian unramified extension of K_n, in other words the p-primary part of the Hilbert class field of K_n. There is an isomorphism given by class field theory

$$C_n \approx \mathrm{Gal}(M_n/K_n)$$

such that the following diagram is commutative.

$$\begin{array}{ccc} C_{n+1} & \rightarrow & \mathrm{Gal}(M_{n+1}/K_{n+1}) \\ {\scriptstyle \text{Norm}}\downarrow & & \downarrow{\scriptstyle \text{Restriction}} \\ C_n & \longrightarrow & \mathrm{Gal}(M_n/K_n) \end{array}$$

Since K_∞ is totally ramified over K_n, it follows that M_n is linearly disjoint from K_∞ over K_n. The lattice of fields looks as follows. We let $M_\infty = \bigcup M_n$.

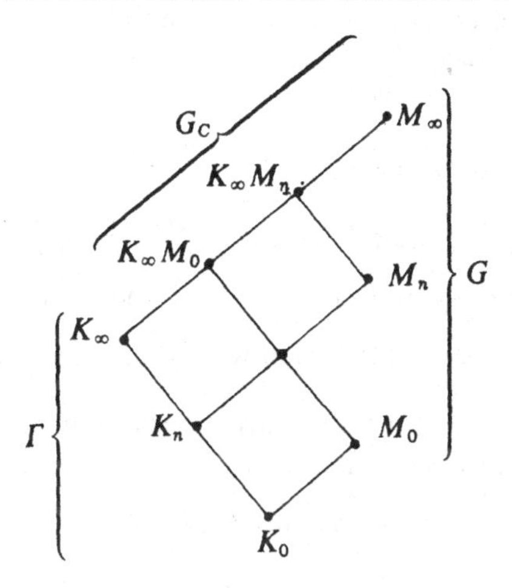

We let

$$G = \mathrm{Gal}(M_\infty/K_0) \quad \text{and} \quad G_C = \mathrm{Gal}(M_\infty/K_\infty) \approx C.$$

Remark. If we replace K_0 by K_1 then K_∞ over K_1 satisfies the same condition **IW**, so a number of results proved for K_∞ over K_0 apply *a fortiori* to K_∞ over K_1. Observe that if γ is a topological generator for Γ, then

$$\mathrm{Gal}(K_\infty/K_n) = \Gamma^{p^n} = \{\gamma^{p^n}\} \approx p^n\mathbf{Z}_p.$$

Theorem 4.1. *Assume first that* **IW** *is satisfied with* $s = 1$. *Let* I *be the inertia group of any prime above* $\mathfrak{p}$ *in* G. *Then:*

(i) $G = IG_C$ *is a semidirect product, and the restriction of* I *to* K_∞ *gives an isomorphism of* I *and* Γ.
(ii) *The commutator group* $G' = G_C^{\gamma-1}$.
(iii) *We have isomorphisms*

$$C/C^{\gamma-1} \approx C_0 \approx \mathrm{Gal}(M_0/K_0) \approx \mathrm{Gal}(K_\infty M_0/K_\infty) \approx G_C/G_C^{\gamma-1}.$$

Proof. We have an exact sequence

$$1 \to G_C \to G \to \Gamma \to 1.$$

The image of I in Γ by restriction to K_∞ is surjective because K_∞ is totally ramified over K_0. It is injective because M_∞ is unramified over K_∞ and so

$$I \cap G_C = \{1\}.$$

This proves (i). If $\sigma \in G_C$ then $\sigma^{\gamma - 1}$ is a commutator because Γ operates on G_C by conjugation. Hence $G_C^{\gamma-1} \subset G'$. On the other hand, $G/G_C^{\gamma-1}$ is abelian, so the reverse inclusion also holds and (ii) is proved. Finally, M_0 is the maximal p-primary abelian unramified extension of K_0 and so $\mathrm{Gal}(M_\infty/M_0)$ is the smallest subgroup of G containing G' and the inertia group I. Since G is the semidirect product of I and G_C, we see that (iii) follows from (ii), and conclude the proof of the theorem.

Corollary. *We have an isomorphism*

$$C_n \approx \mathrm{Gal}(M_n/K_n) \approx C/C^{\gamma^{p^n}-1} \approx G_C/G_C^{\gamma^{p^n}-1}.$$

Proof. Apply the theorem to the situation where K_0 is replaced by K_n.

Next consider the general situation with a finite number of primes.

Theorem 4.2. *Assume that* **IW** *is satisfied, with primes $\mathfrak{p}_1, \ldots, \mathfrak{p}_s$. Let I_j be the inertia group of $\mathfrak{p}_j$ in G. Then:*

(i) *There is a semidirect product decomposition*

$$G = I_1 G_C,$$

and $G' = G_C^{\gamma-1}$.

(ii) *Let σ_j be a generator for I_j, and write*

$$\sigma_j = \tau_j \sigma_1 \text{ with } \tau_j \in G_C.$$

Then

$$C_0 \approx G_C/(\tau_1, \ldots, \tau_s, G_C^{\gamma-1}).$$

Proof. Identical with that of Theorem 4.1, except that in the present more general situation, we have to look at the smallest subgroup of G containing the commutator group G' and all the inertia groups I_j instead of a single inertia group I.

Corollary. *Let U_0 be the $\mathbf{Z}_p$-submodule of G_C generated by the elements $\tau_1, \ldots, \tau_s$ and $G_C^{\gamma-1}$. Let*

$$U_n = U_0^{g_n} \quad \text{where } g_n = 1 + \gamma + \gamma^2 + \cdots + \gamma^{p^n-1}.$$

Then

$$C_n \approx G_C/U_n.$$

Proof. We apply the theorem to K_∞ as $\mathbf{Z}_p$-extension of K_n. This has the effect of replacing γ by γ^{p^n} and σ_i by $\sigma_i{}^{p^n}$. Then τ_i is replaced by $(\tau_i)^{g_n}$, because for every positive integer k, we have

$$\sigma_i^k = (\tau_i\sigma_1)^k = \tau_i\sigma_1\tau_i\sigma_1^{-1}\cdot\sigma_1^2\tau_i\sigma_1^{-2}\cdots\sigma_1^{k-1}\tau_i\sigma_1^{-k+1}\cdot\sigma_1^k,$$

whence for $k = p^n$ we obtain

$$\sigma_i^{p^n} = (\tau_i)^{g_n}\cdot\sigma_1^{p^n}.$$

Then

$$U_n = G_C^{\gamma^{p^n}-1}(\tau_1, \ldots, \tau_s)^{g_n}$$

where $(\tau_1, \ldots, \tau_s)$ is the group generated over $\mathbf{Z}_p$ by $\tau_1, \ldots, \tau_s$. Since

$$g_n(\gamma - 1) = \gamma^{p^n} - 1,$$

we find $U_n = U_0^{g_n}$, which proves the corollary.

It will be easily shown below in Theorem 4.4 that C is finitely generated over the Iwasawa algebra. Then Theorem 4.2 and its corollary show that C *is of Iwasawa type* as defined in §1, so that one can apply the counting procedure given there, to get an asymptotic formula for the orders of the groups C_n.

Iwasawa has conjectured that $m = 0$ in the case of the cyclotomic tower $\mathbf{Q}(\mu_{p^n})$, so that in this case, the order of the ideal class group (p-primary part) would have the form

$$\text{Card } C_n = p^{dn+c}$$

for n sufficiently large, in analogy with the orders of points of p-power order on abelian varieties. This conjecture has recently been proved by Ferrero and Washington. On the other hand, he has given examples of non-cyclotomic $\mathbf{Z}_p$-extensions K_∞ over K_0 for which $m > 0$.

Theorem 4.3. *Assume that* **IW** *is satisfied with one prime. If* $C_0 = \{1\}$, *then* $C_n = \{1\}$ *for all* n.

Proof. If $C_0 = 1$, then Theorem 4.1 shows that $C = C^{\gamma-1}$. Viewing C as module over $\mathbf{Z}_p[[X]]$, this means that $C = XC$. But X is contained in the maximal ideal of $\mathbf{Z}_p[[X]]$. By Nakayama's lemma, it follows that $C = \{1\}$, as desired.

Remark. We could let $K_n = \mathbf{Q}(\mu_{p^{n+1}})^+$ be the real subfield of the cyclotomic field. It is a conjecture of Kummer–Vandiver in that case that C_0 is trivial. [Remember: By definition, $C_0 = C_0^{(p)}$ in this chapter.] Thus the **Kummer–Vandiver conjecture** may also be formulated by saying that

$$h_0^+ \text{ is prime to } p,$$

and Theorem 4.3 shows that if this is the case, then h_n^+ is also prime to p for all n.

For historical comments on the Kummer–Vandiver conjecture, see the Introduction. Kummer usually denoted h^+ by D/Δ, where D and Δ denote the regulators of the group of units and cyclotomic units respectively, so their quotient is the index equal to h^+ by Theorem 5.1 of Chapter 3. Cf. also the discussion in [L 5].

Theorem 4.4. *For any $\mathbf{Z}_p$-extension the module C over $\mathbf{Z}_p[[X]]$ is a finitely generated torsion module.*

Proof. Suppose first for simplicity that condition **IW** is satisfied with only one prime. Then $C/\mathfrak{m}C$ is a factor group of $C/C^{\gamma-1}$, which is none other than C_0 by Theorem 4.1, and is therefore finite. That C is finitely generated is a special case of Nakayama's lemma.

In general, when **IW** is satisfied but with several primes, then we have to use another argument. By Theorem 4.2 we know that

$$G_C/G_C^{\gamma-1} \approx C/C^{\gamma-1}$$

is finitely generated over $\mathbf{Z}_p$ of rank uniformly bounded by s. Nakayama's lemma again shows that C is finitely generated over the Iwasawa algebra. Furthermore, applying Theorem 4.2 to K_∞/K_n, that is replacing γ by γ^{p^n}, shows that

$$C/(\gamma^{p^n} - 1)C$$

is also finitely generated over $\mathbf{Z}_p$ with a similar bound s for the rank. By the structure theorem of §3, if V is finitely generated over Λ, then there is a quasi-isomorphism

$$(*) \qquad V \to \Lambda^{(r)} \oplus \prod \Lambda/(p^{m_i}) \oplus \prod \Lambda/(f_j),$$

where f_j are distinguished. We use this with $V = C$, writing V additively. The uniform bound on the rank immediately shows that there cannot be any free part, i.e., $r = 0$. This proves Theorem 4.4.

§5. The Maximal p-abelian p-ramified Extension

The next two sections describe in class-field theoretic terms some properties of the Galois group of the maximal p-abelian p-ramified extension of a number field, and describe its $\mathbf{Z}_p$-extensions.

Let K be a number field. We let:

$M_p(K)$ = the maximal p-abelian p-ramified extension of K.

$M_p^{\mathrm{nr}}(K)$ = the maximal p-abelian unramified extension of K.

We fix the prime number p and the field K, so we sometimes omit reference to them in the notation.

$J = J_K$ = ideles of K, and U is the group of unit ideles,

$$U = \prod_{\mathfrak{p}} U_{\mathfrak{p}} \quad \text{and} \quad J^{\infty} = \prod_{v \in S_{\infty}} K_v^*.$$

In the first product, $\mathfrak{p}$ ranges over the prime ideals of K. We write

$$U_p = \prod_{\mathfrak{p}|p} U_{\mathfrak{p}}, \quad \text{and} \quad U_{[p]} = \prod_{l \neq p} U_l.$$

$E = E_K$ = units in K. We have an embedding on the diagonal:

$$\sigma_p : E \to U_p.$$

$G_p^{\mathrm{ab}}(K) = \mathrm{Gal}(M_p(K)/K)$.

By class field theory, an abelian extension of K is unramified at primes dividing l if and only if its associated group in the ideles contains U_l. Consequently we have an isomorphism

$$\boxed{G_p^{\mathrm{ab}}(K) \approx p\text{-part of } J_K/\overline{U_{[p]}J^{\infty}K^*}}$$

where the bar denotes closure in the idele topology. We have the inclusions

$$J \supset U_p U_{[p]} J^{\infty} K^* \supset \overline{U_{[p]} J^{\infty} K^*}.$$

The first factor group

$$J/U_p U_{[p]} J^{\infty} K^* = J/UK^*$$

is isomorphic to the Galois group of the Hilbert class field, and is finite. The second factor group is equal to

$$U_p U_{[p]} J^{\infty} K^* / \overline{U_{[p]} J^{\infty} K^*} \approx U_p / U_p \cap \overline{U_{[p]} J^{\infty} K^*}.$$

Lemma. $$U_p \cap \overline{U_{[p]}J^\infty K^*} = \overline{\sigma_p E}.$$

Proof. Let $U_p^{(n)}$ be the group of units in U_p which are $\equiv 1 \bmod p^n$. Then the groups

$$U_p^{(n)} U_{[p]} J^\infty K^*$$

form a fundamental system of neighborhoods for $U_{[p]}J^\infty K^*$, and their intersection is this closure. Intersecting with U_p (whose elements have component 1 at all primes not dividing p) shows that

$$U_p^{(n)} U_{[p]} J^\infty K^* \cap U_p = EU_p^{(n)}.$$

Taking the intersection for all n proves the lemma.

Theorem 5.1. *Let H be the p-Hilbert class field of K. Then we have an isomorphism*

$$\begin{aligned}\mathrm{Gal}(M_p(K)/H) &\approx p\text{-part of } U_p/\overline{\sigma_p E}\\ &= U_p^{(1)}/(U_p^{(1)} \cap \overline{\sigma_p E}).\end{aligned}$$

Again, as p is fixed, we write simply $U_p/\bar{E}$. By a **quasi-isomorphism**, we shall mean a homomorphism with finite kernel and cokernel. We denote a quasi-isomorphism by a single $\sim$. The theorem yields a quasi-isomorphism

$$G_p^{\mathrm{ab}}(K) \sim U_p/\bar{E}.$$

Furthermore, since U_p contains an open subgroup of finite index isomorphic to $\mathbf{Z}_p^{[K:\mathbf{Q}]}$, by means of the exponential map, say, we have a quasi-isomorphism

$$G_p^{\mathrm{ab}}(K) \sim \mathbf{Z}_p^{[K:\mathbf{Q}]-r_p}, \quad \text{where } r_p = \mathrm{rank}_{\mathbf{Z}_p}\bar{E} = r_p(E).$$

The **Leopoldt conjecture** states that $r_p = r = r_1 + r_2 - 1$.

Let $Z_p(K)$ = composite of all $\mathbf{Z}_p$-extensions of K. From the quasi-isomorphism we find:

$$[M_p(K) : Z_p(K)] < \infty.$$

Theorem 5.2. *Assume the Leopoldt conjecture for K. Then we have a quasi-isomorphism*

$$G_p^{\mathrm{ab}}(K) \sim \mathbf{Z}_p^{r_2+1} \approx \mathrm{Gal}(Z_p(K)/K).$$

Proof. The first statement comes from the definitions and

$$[K : \mathbf{Q}] = r_1 + 2r_2.$$

For the second statement, we note that the composite of all $\mathbf{Z}_p$-extensions of

K has a Galois group embedded in the product of $\mathbf{Z}_p$ with itself, and as such is a torsion free finitely generated module over $\mathbf{Z}_p$, whose rank is exactly $r_2 + 1$ by the first statement.

Example. Let $K_\infty = \mathbf{Q}(\mu^{(p)})$. Let U_n be the local units in the completion of K_n, congruent to 1 mod $\mathfrak{p}_n$. Let U_n' be the subgroup of units whose norm to $\mathbf{Q}_p$ is 1. Assume the Kummer–Vandiver conjecture. Let the notation be as in §4 of the next chapter. Then we obtain an isomorphism

$$\mathrm{Gal}(\Omega/\Omega^{\mathrm{nr}}) \approx \varprojlim U_n'/\bar{E}_n.$$

Without assuming the Kummer–Vandiver conjecture, we shall study the projective limit of the local groups in Chapter 7.

§6. The Galois Group as Module over the Iwasawa Algebra

Let K_0 be a number field, K_∞/K_0 any $\mathbf{Z}_p$-extension, with Galois group

$$\Gamma = \{\gamma\},$$

with topological generator γ. Let Ω be a p-abelian extension of K_∞ which is also Galois over K_0. For each n we let Ω_n be the maximal subfield of Ω which is abelian over K_n.

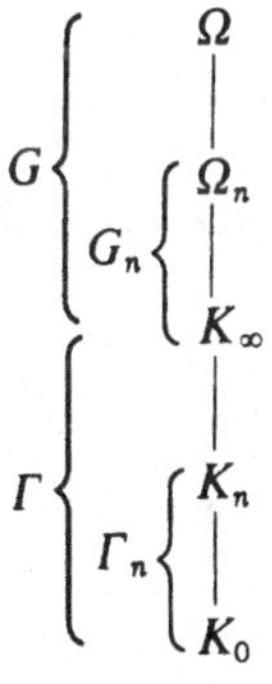

$$\begin{array}{ll} & \Omega \\ G & \;| \\ & \Omega_n \\ G_n & \;| \\ & K_\infty \\ & \;| \\ \Gamma & K_n \\ \Gamma_n & \;| \\ & K_0 \end{array}$$

The Galois groups are denoted by the letters shown on the diagram. Since Ω is assumed Galois over K_0 and is abelian over K_∞, it follows that the commutator subgroup is

$$\mathrm{Gal}(\Omega/K_0)^c = G^{\gamma - 1},$$

in other words, it consists of all elements

$$\sigma^{\gamma-1} = \sigma\gamma\sigma^{-1}\gamma^{-1} \text{ with } \sigma \in G.$$

It is frequently useful to view G as an additive module over the Iwasawa algebra. Indeed, Γ_n operates by conjugation on $\mathrm{Gal}(\Omega/K_n)$, and hence on the commutator group

$$\mathrm{Gal}(\Omega/K_n)^{\gamma^{p^n}-1}, \text{ also written } (\gamma^{p^n} - 1)\,\mathrm{Gal}(\Omega/K_n).$$

Hence

$$\lim G_n = G \text{ is a compact module over } \Lambda = \mathbf{Z}_p[[X]] = \lim \mathbf{Z}_p[\Gamma_n].$$

Taking K_n as ground field instead of K_0, we obtain *mutatis mutandis*

$$\mathrm{Gal}(\Omega/\Omega_n) = (\gamma^{p^n} - 1)G = ((1 + X)^{p^n} - 1)G.$$

Thus in terms of the Iwasawa algebra, we find

$$G_n = G/(\gamma^{p^n} - 1)G.$$

We denote by the sign $\sim$ a quasi-isomorphism of Λ-modules.

Theorem 6.1. *Let Ω be the maximal p-abelian p-ramified extension of K_∞. Then:*

(i) $G = \mathrm{Gal}(\Omega/K_\infty)$ *is finitely generated over the Iwasawa algebra, and in fact*

$$G/G^{\gamma-1} \sim \mathbf{Z}_p^\rho \qquad \text{where } \rho = [K_0 : \mathbf{Q}] - r_p - 1.$$

(ii) *If K_0 satisfies the Leopoldt conjecture, then $\rho = r_2$, and*

$$G/XG \sim \mathbf{Z}_p^{r_2}.$$

Proof. By definition,

$$\Omega_0 = M_p(K_0),$$

and the rank over $\mathbf{Z}_p$ of a subgroup of finite index in its Galois group was determined to be $[K_0 : \mathbf{Q}] - r_p$ in Theorem 5.2. Taking into account Γ itself shows that $G/XG \sim \mathbf{Z}_p^\rho$ where ρ is as stated. Nakayama's lemma then proves the first assertion, and (i). Part (ii) is then a matter of definitions.

Theorem 6.2. *Assume that K_0 is totally imaginary, and that each K_n satisfies the Leopoldt conjecture (namely*

$$r_p(K_n) = r_2(K_n).$$

Then there is a quasi-isomorphism

$$G \sim \Lambda^{r_2} \times G_{\mathrm{tor}},$$

where G_{tor} is the Λ-torsion submodule of G.

Proof. From the structure theorem, we know that

$$G \sim \Lambda^t \times G_{\text{tor}}.$$

On the other hand,

$$r_2(K_n) = r_2 p^n.$$

By Theorem 5.2 we know that

$$\text{Gal}(\Omega_n/K_n) \sim \mathbf{Z}_p^{r_2 p^n + 1}.$$

From the structure theorem, one sees easily that this is possible only if $t = r_2$, as desired.

The above theorems give a sample of Iwasawa's results [Iw 12]. It is possible to vary some of the hypotheses to obtain variants. For instance, one need not assume the full Leopoldt conjecture in Theorem 6.2, merely assume that the defect in that conjecture is bounded as function of n. For the cyclotomic $\mathbf{Z}_p$-extension, this can be proved easily, see for instance Greenberg [Gr 4].

6 Kummer Theory over Cyclotomic $\mathbf{Z}_p$-extensions

In the last chapter we studied the ideal class groups in a $\mathbf{Z}_p$-extension of a number field. Here we shall consider especially the cyclotomic $\mathbf{Z}_p$-extension, and then Kummer extensions above it, as in Iwasawa [Iw 12], obtained by adjoining p^nth roots of units, p-units, and ideal classes of p-power order.

We give Leopoldt–Iwasawa's theorem that the Kummer–Vandiver conjecture implies that C^- is cyclic, in a precise version for the cyclotomic extension of $\mathbf{Q}$, following Kubert–Lang [KL 9]. We prove that the Galois group of the Kummer extension obtained by adjoining p-power roots of p-units is 1-dimensional free over the Iwasawa algebra. As a consequence, we see that C^- is a quotient of this free module. See Leopoldt [Le 5] and the last Satz in [Le 10], as well as Iwasawa [Iw 7], Theorem 2. In the limit, there is an analogous (but less precise) statement of Greenberg [Gr 4], see also Coates [Co 3], Theorem 5.7.

For a discussion of the case of totally real number fields, cf. Coates [Co 3], [Co 4]. In this connection it is likely that the units conjectured by Stark [St] (see also Lichtenbaum's conjectures [Li 2]) will play a significant role similar to the one played by the cyclotomic units, to clarify the situation.

§1. The Cyclotomic $\mathbf{Z}_p$-extension

Let $\mu^{(p)}$ be the group of p-power roots of unity. Then $\mathbf{Q}(\mu^{(p)})$ is the composite of an extension of degree $p - 1$ if p is odd, $\mathbf{Q}(i)$ if $p = 2$, and a $\mathbf{Z}_p$-extension which is uniquely determined as the fixed field of the (finite) torsion group of the Galois group, and will be called the **cyclotomic $\mathbf{Z}_p$-extension**. We denote it by $Z_p(\mathbf{Q})$. **It is real**. If K is a number field, we let

$$\mathrm{Cyc}_p(K) = KZ_p(\mathbf{Q})$$

be the composite of K and the cyclotomic $\mathbf{Z}_p$-extension. Then $\text{Cyc}_p(K)$ is a $\mathbf{Z}_p$-extension of K. If K is totally real, then this cyclotomic $\mathbf{Z}_p$-extension is also totally real.

Suppose on the other hand that K contains the pth roots of unity if p is odd, and contains i if $p = 2$. Let q_0 be the power of p such that the q_0th roots of unity lie in K. Let

$$q_n = q_0 p^n \quad \text{and} \quad K_n = K(\mu_{q_n}).$$

Then $[K_{n+1} : K_n] = p$, and

$$K_\infty = \bigcup K_n$$

is $\mathbf{Z}_p$-extension of K. Let $\Gamma = \text{Gal}(K_\infty/K_0)$ and let

$$\varkappa : \Gamma \to 1 + q_0 \mathbf{Z}_p$$

be the canonical representation such that for any p^nth root of unity ζ we have

$$\zeta^\gamma = \zeta^{\varkappa(\gamma)}.$$

A Galois extension is called p-**abelian** if its Galois group is a projective limit of finite p-abelian groups. We now discuss properties of such extensions of K_∞ which are Galois over K_0.

For the rest of this section, we assume that K_0 contains the pth roots of unity if p is odd and i if $p = 2$. Let A_n be a subgroup of K_n^*, and let

$$\Gamma_n = \text{Gal}(K_n/K_0), \qquad \Lambda_n = \mathbf{Z}(p^n)[\Gamma_n].$$

We assume that A_n is stable under Γ_n.

Ordinary Kummer theory gives a pairing

$$\text{Gal}(K_n(A_n^{1/p^n})/K_n) \times A_n^{1/p^n}/(A_n^{1/p^n} \cap K_p^*) \to \mu_{p^n}$$

expressed by the symbol

$$(\sigma, \alpha) \mapsto \langle \sigma, \alpha \rangle_n = \sigma\alpha/\alpha$$

for σ in the Galois group and $\alpha \in A_n^{1/p^n}$. If $\gamma \in \Gamma_n$ then

$$\boxed{\langle \sigma^\gamma, \alpha^\gamma \rangle_n = \langle \sigma, \alpha \rangle_n^\gamma = \langle \sigma, \alpha \rangle_n^{\varkappa(\gamma)},}$$

where $\sigma^\gamma = \tilde{\gamma}\sigma\tilde{\gamma}^{-1}$, and $\tilde{\gamma}$ is any extension of γ to $K_n(A_n^{1/p^n})$. Indeed,

$$\tilde{\gamma}\sigma\tilde{\gamma}^{-1}(\tilde{\gamma}\alpha)/\tilde{\gamma}\alpha = \tilde{\gamma}\left(\frac{\sigma\alpha}{\alpha}\right) = \gamma\left(\frac{\sigma\alpha}{\alpha}\right).$$

We may also write

$$\langle \sigma, \alpha \rangle_n^{\varkappa(\gamma)} = \langle \sigma, \alpha^{\varkappa(\gamma)} \rangle_n.$$

The group A_n^{1/p^n} mod $A_n^{1/p^n} \cap K_n^*$ has exponent p^n, so exponentiating with a p-adic integer is well defined. In particular, we may rewrite the functorial formula in the form

$$\langle \sigma^\gamma, \alpha \rangle_n = \langle \sigma, \alpha^{\gamma^*} \rangle_n, \text{ where } \gamma^* = \gamma^{-1}\varkappa(\gamma).$$

We wish to pass to the limit. We could have taken the Kummer pairing on

$$\mathrm{Gal}(K_\infty(A_n^{1/p^n})/K_\infty) \times A_n^{1/p^n}/(A_n^{1/p^n} \cap K_\infty{}^*) \to \mu_{p^n},$$

writing the symbol $\langle \sigma, \alpha \rangle$ without an index n, defined by the same formula. The Galois group on the left can be identified with a subgroup of $\mathrm{Gal}(K_n(A_n^{1/p^n})/K_n)$, arising from the change of base of the Kummer extension from K_n to K_∞. Let G_n be the Galois group on the left, so

$$G_n = \mathrm{Gal}(K_\infty(A_n^{1/p^n})/K_\infty).$$

The field diagram is as follows.

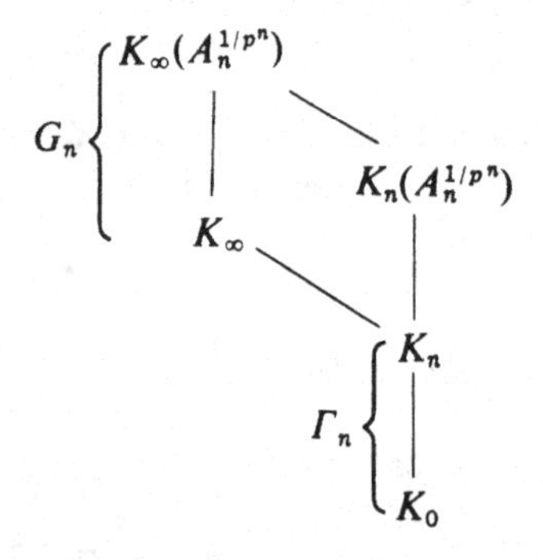

The group G_n is a Γ_n-module, hence a $\mathbf{Z}(p^n)[\Gamma_n] = \Lambda_n$-module. Hence via the natural homomorphism, it is a Λ-module, where

$$\Lambda = \lim \Lambda_n$$

is the Iwasawa algebra, isomorphic to $\mathbf{Z}_p[[X]]$, and $X = \gamma_0 - 1$, where γ_0 is a fixed generator of Γ.

Let

$$\lambda = \sum m_i X^i,\ m_i \in \mathbf{Z}_p$$

be an element of Λ. We define the **Iwasawa involution**

$$\lambda^* = \sum m_i X^{*i}, \quad \text{where } X^* = \varkappa(\gamma_0)(1 + X)^{-1} - 1.$$

Then X^* is also in the maximal ideal (p, X) of Λ, and

$$\lambda \mapsto \lambda^*$$

is an automorphism of Λ. The functorial formula for the action of $\gamma \in \Gamma$ on the Kummer symbol can then be expressed in terms of the involution by

$$\boxed{\langle \sigma^\lambda, \alpha \rangle = \langle \sigma, \alpha^{\lambda^*} \rangle,}$$

for $\sigma \in G_n$ and $\alpha \in A_n^{1/p^n}/(A_n^{1/p^n} \cap K_\infty^*)$.

In the applications, we also pass to the limit on n for the Kummer pairing. We suppose that

$$A_n \subset A_{n+1}.$$

Let

$$\Omega_A = \bigcup K_\infty(A_n^{1/p^n}) \quad \text{and} \quad G_A = \mathrm{Gal}(\Omega_A/K_\infty).$$

We have a compatible system of pairings for $m \geq n$:

$$\begin{array}{ccccc} G_m & \times & A_m^{1/p^m}/(A_m^{1/p^m} \cap K_\infty{}^*) & \rightarrow & \mu_{p^m} \\ \downarrow & & \uparrow & & \uparrow \\ G_n & \times & A_n^{1/p^n}/(A_n^{1/p^n} \cap K_\infty{}^*) & \rightarrow & \mu_{p^n} \end{array}$$

The Galois groups on the left form a projective system, and the Kummer groups of field elements on the right of the pairing form an injective system. At each finite level, we have a compact-discrete duality. In the limit, we have a similar compact-discrete duality

$$G_A \times \varinjlim A_n^{1/p^n}/(A_n^{1/p^n} \cap K_\infty^*) \to \mu^{(p)}$$

with values in the p-primary roots of unity.

The action of Λ_n on G_n is compatible in the projective limit, so the limit group G_A is a topological compact Λ-module. We shall investigate its structure for various systems $\{A_n\}$ obtained from units and ideal classes in the next sections. It will also happen that we consider two groups, say

$$A \supset B,$$

in which case $\Omega_A \supset \Omega_B$. It is clear in each case that $\mathrm{Gal}(\Omega_A/\Omega_B)$ is a Λ-module, and that the Kummer pairings and involution described above also apply to this intermediate situation.

§2. The Maximal p-abelian p-ramified Extension of the Cyclotomic $\mathbf{Z}_p$-extension

A Galois extension is called ***p*-abelian** if its Galois group is a projective limit of finite p-abelian groups. It is called ***p*-ramified** if it is unramified at all primes (including infinity) not dividing p. We let:

$$M_p(K) = \text{the maximal } p\text{-abelian } p\text{-ramified extension of } K.$$
$$M_p^{\mathrm{nr}}(K) = \text{the maximal } p\text{-abelian unramified extension of } K.$$

We fix the prime number p and the field K, so we sometimes omit reference to them in the notation.

Even if K is infinite over $\mathbf{Q}$ we may define $M_p(K)$ and $M_p^{\mathrm{nr}}(K)$ as above. It is then immediate that

$$M_p(K) = \bigcup M_p(F),$$

where the union is taken over a family of subfields F of K finite over $\mathbf{Q}$, whose union is K, and which is cofinal with the family of all subfields of K finite over $\mathbf{Q}$. For instance, if K is finite over $\mathbf{Q}$, and K_∞ is a $\mathbf{Z}_p$-extension, then

$$M_p(K_\infty) = \bigcup_n M_p(K_n).$$

A similar remark applies for $M_p^{\mathrm{nr}}(K_\infty)$.

Throughout this section, we let:

K_∞ = cyclotomic $\mathbf{Z}_p$-extension of K_0, and we assume that K_0 contains the pth roots of unity if p is odd, contains i if $p = 2$.

Ω = maximal p-abelian p-ramified extension of K_∞.

E_n = units in K_n and $E = \bigcup E_n$.

$\Omega_E = K_\infty(E^{1/p^\infty})$

A_n = group of elements α in K_n^* such that $(\alpha) = \mathfrak{a}^{p^n}$ where $\mathfrak{a}$ is (fractional) ideal prime to p, and $A = \bigcup A_n$.

$\Omega_A = \bigcup K_\infty(A_n^{1/p^n})$

B_n = p-units in K_n = group of elements whose ideal factorization contains only ideals dividing p.

$\Omega_B = K_\infty(B^{1/p^\infty})$

We have the following diagram of fields.

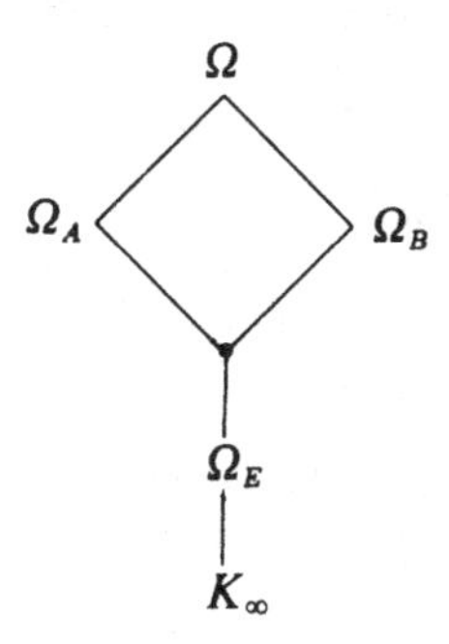

It is clear that Ω_A and Ω_B both contain Ω_E. In fact, both A and B contain E.

Lemma 1. $\Omega = \Omega_A \Omega_B$.

Proof. By Kummer theory, Ω is a composite of cyclic extensions. Let $K_\infty(\alpha^{1/p^m}) \subset \Omega$ for some $\alpha \in K_\infty$. Then $\alpha \in K_n$ for some n. We take $n \geq m$ and also such that

$$K_n(\alpha^{1/p^m}) \text{ is } p\text{-ramified over } K_n.$$

Then α necessarily has an ideal factorization

$$(\alpha) = \mathfrak{a}^{p^n}\mathfrak{b},$$

where $\mathfrak{b}$ is p-primary and $\mathfrak{a}$ is prime to p. Let h be the class number of K_n, and write $h = p^r d$ with d prime to p. Then

$$(\alpha^h) = (\alpha_1)^{p^n}(\beta)$$

where $(\alpha_1) = \mathfrak{a}^h$ and $(\beta) = \mathfrak{b}^h$. Furthermore,

$$K_{n+r}(\alpha^{h/p^{n+r}}) = K_{n+r}(\alpha^{d/p^n}) = K_{n+r}(\alpha^{1/p^n})$$

and also

$$K_{n+r}(\alpha^{h/p^{n+r}}) \subset K_{n+r}(\alpha_1^{1/p^{n+r}}, \beta^{1/p^{n+r}}, E_{n+r}^{1/p^{n+r}}).$$

This proves the lemma.

Theorem 2.1. *The Galois groups* $\mathrm{Gal}(\Omega_A/\Omega_E)$ *and* $\mathrm{Gal}(\Omega_B/\Omega_E)$ *are* Λ*-torsion modules. So* $\mathrm{Gal}(\Omega/\Omega_E)$ *is a* Λ*-torsion module.*

Proof. We shall analyze each Galois group separately, and get a closer view of its structure.

The extension Ω_A/Ω_E.

Let $G_{A/E} = \mathrm{Gal}(\Omega_A/\Omega_E)$. For now abbreviate $G_{A/E} = G$, and let

$$G_n = \mathrm{Gal}(\Omega_E(A_n^{1/p^n})/\Omega_E).$$

The field diagram is as follows.

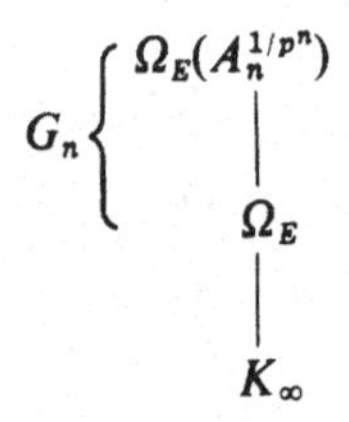

It is clear that G = projective limit of the groups G_n, and that G_n is a $\mathbf{Z}(p^n)[\Gamma_n]$-module, so in the limit, G is a Λ-module.

As in Chapter 5, let:

$$C_n = Cl^{(p)}(K_n) = p\text{-primary subgroup of ideal class group of } K_n.$$

Then we have a homomorphism

$$A_n^{1/p^n} \to C_n$$

given by

$$\alpha^{1/p^n} \to \mathfrak{a}$$

if $(\alpha) = \mathfrak{a}^{p^n}$. If u is a unit in Ω such that $u^{p^n} \in A_n$, then $u^{p^n} \in E_n$. The kernel of our homomorphism is therefore precisely E_n^{1/p^n}, so we have an injective homomorphism

$$A_n^{1/p^n}/E_n^{1/p^n} \to C_n,$$

which is also a Λ-homomorphism. Let $\mathscr{A}_n$ be its image. Then the Kummer pairing is isomorphic to a pairing with A_n, namely:

$$\begin{array}{ccccc} G_n & \times & A_n^{1/p^n}/E_n^{1/p^n} & \to & \mu_{p^n} \\ \updownarrow & & \updownarrow & & \\ G_n & \times & \mathscr{A}_n & \longrightarrow & \mu_{p^n} \end{array}$$

In addition, this isomorphism is compatible with the limiting process:

$$\begin{array}{ccccc} G_{n+1} & \times & \mathscr{A}_{n+1} & \to & \mu_{p^{n+1}} \\ \downarrow & & \uparrow & & \uparrow \\ G_n & \times & \mathscr{A}_n & \longrightarrow & \mu_{p^n} \end{array}$$

Hence we get a compact-discrete duality

$$G \times \mathscr{A} \to \mu^{(p)}$$

where $\mathscr{A}$ = direct limit of $\mathscr{A}_n$. By Chapter 5, Theorem 4.4, there exists $\lambda \in \Lambda$ such that $C^\lambda = 1$, so $C_n^\lambda = 1$ for all n, and $A_n^\lambda = 1$ for all n. By Kummer duality, for $\sigma \in G$ we get

$$\langle \sigma^{\lambda^*}, \alpha \rangle = \langle \sigma, \alpha^\lambda \rangle = 1 \quad \text{for all } \alpha \in A_n^{1/p^n} \text{ and all } n.$$

Hence $\sigma^{\lambda^*} = 1$, so λ^* annihilates G, which is therefore a torsion module over Λ as desired.

In addition, we note that the direct limits

$$\varinjlim A_n^{1/p^n}/E_n^{1/p^n} = \varinjlim \mathscr{A}_n \quad \text{and} \quad \varinjlim C_n = C_\infty$$

are equal since any element in C_∞ has a representative ideal prime to p. Consequently we get the additional information:

Theorem 2.2. *The Kummer pairing gives rise to a compact-discrete duality*

$$\mathrm{Gal}(\Omega_A/\Omega_E) \times C_\infty \to \mu^{(p)}.$$

Remark. Iwasawa has also shown that $C = \varprojlim C_n$ is quasi-isomorphic to $\mathrm{Hom}(\varinjlim C_n, \mathbf{Q}_p/\mathbf{Z}_p)$ (see Theorem 11, p. 266 of [Iw 12]).

The extension Ω_B/Ω_E.

Let $G_{B/E} = \mathrm{Gal}(\Omega_B/\Omega_E)$. For now abbreviate $G_{B/E} = G$. By the Lemma of Chapter 5, §1 we know that there is only a finite number of primes $\mathfrak{p}_1, \ldots, \mathfrak{p}_s$ dividing p in some finite extension K_d, such that $\mathfrak{p}_1, \ldots, \mathfrak{p}_s$ are totally ramified in K_∞. Let h be the class number of K_d. Let

$$\mathfrak{p}_1^h = (\pi_1), \ldots, \mathfrak{p}_s^h = (\pi_s).$$

Then

$$\Omega_B = \Omega_E(\pi_1^{1/p^\infty}, \ldots, \pi_s^{1/p^\infty}).$$

It is immediate that

$$\mathrm{Gal}(\Omega_B/\Omega_E) \approx \mathbf{Z}_p^{s'} \text{ with } s' \leq s.$$

In particular, the structure theorem for finitely generated Λ-modules implies that G cannot have any free part, so is a Λ-torsion module. This proves Theorem 2.1.

For additional information concerning the fixed field of

$$\mathrm{Gal}(\Omega/K_\infty)_{\mathrm{tor}}$$

(for referring to Λ-torsion), cf. for instance Coates [Co 1], Theorem 5. Iwasawa has an example showing that there are cases when the fixed field is not necessarily Ω_E.

Theorem 2.3. *Assume that there is only one prime in K_∞ lying above p. Then*

$$\Omega_E = \Omega_{E_p} = \Omega_B$$

where E_p is the group of p-units in K_∞, and

$$\Omega_{E_p} = \Omega(E_p^{1/p^\infty}).$$

Proof. We consider the diagram of fields:

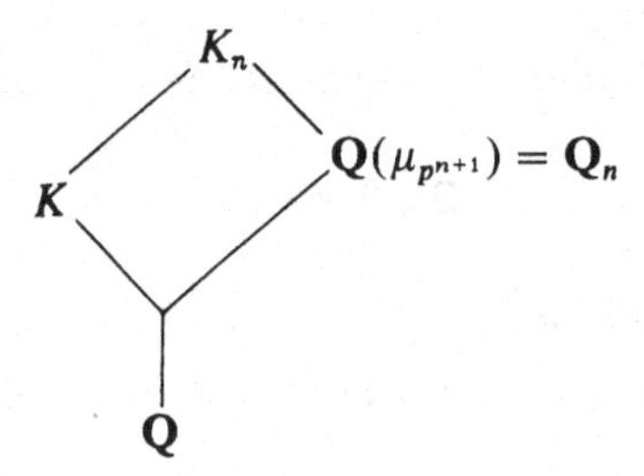

The ideal above p in $\mathbf{Q}_n$ is principal, say generated by the element $\lambda_n = 1 - \zeta_n$. The degree $[K_n : \mathbf{Q}_p]$ is bounded independently of n, and we have

$$(\lambda_n) = \mathfrak{p}_n^{e_n}$$

where e_n is the ramification index, bounded by this degree. Taking e to be the least common multiple of the integers e_n shows that $\mathfrak{p}_n^e$ is principal for all n. We apply this to the previous discussion of the extension $\Omega_B = \Omega_E(\pi_1^{1/p^\infty})$. As ideals we have

$$(\pi_1) = \mathfrak{p}_n^{j_n}$$

for n sufficiently large, and j_n is divisible by arbitrary large powers of p as $n \to \infty$. Furthermore

$$\Omega_B = \Omega_E(\pi_1^{e/p^\infty}).$$

It is then clear that $\Omega_B = \Omega_E$.

§3. Cyclotomic Units as a Universal Distribution

Let p be a prime number.

Let $\mathscr{E}_p$ be the group generated by $\pm\mu^{(p)}$ (p-power roots of unity) and by the elements

$$\zeta - 1, \quad \text{with } \zeta \in \mu^{(p)} \text{ and } \zeta \neq 1.$$

We call $\mathscr{E}_p$ **the cyclotomic p-units**. They satisfy the following relations:

CU 1. $$\sigma_{-1}(\zeta - 1) = -\zeta^{-1}(\zeta - 1)$$

CU 2. $$\prod_{\eta^p = 1} (\zeta\eta - 1) = \zeta^p - 1 \quad \text{if } \zeta^p \neq 1.$$

CU 3. $$\prod_{\substack{\zeta^{p^n} = 1 \\ \zeta \text{ primitive}}} (\zeta - 1) = p.$$

For this last one, note that the pth roots of unity satisfy

$$X^{p-1} + \cdots + 1 = 0.$$

Replacing X by $X + 1$ yields the equation for $\zeta - 1$, where ζ is a pth root of unity. The constant term is then p. Replacing X by $X^{p^{n-1}}$ yields the equation for the general case, proving **CU 3**. The other properties are obvious.

We may rewrite these relations to fit the formalism of distributions as follows. Let $a \in (\mathbf{Q}/\mathbf{Z})^{(p)}$ and $a \neq 0$. Define

$$g_a = e^{2\pi i a} - 1.$$

Let $V = \mathscr{E}_p/\pm\mu^{(p)}$ be the factor group of cyclotomic p-units by roots of unity.

Theorem 3.1. *The association of* $(\mathbf{Q}/\mathbf{Z})^{(p)} \to V$ *given by*

$$a \mapsto g_a \text{ (mod roots of unity)}$$

satisfies the distribution relations except at 0.

The theorem means that for $a \neq 0$ we have

$$\prod_{pb = a} g_b = g_a,$$

and is obvious in the light of **CU 2**.

Let $V_n = \mathscr{E}_{p,n}/\pm\mu_{p^n}$ be the factor group by roots of unity of p-units at level $\leq n$, i.e., generated by the roots of unity, and the elements $\zeta - 1$ where ζ is a p^nth root of unity $\neq 1$. Whether we take ζ to be primitive or not to generate V_n is immaterial since the distribution relation shows that we get all of them from the primitive ones.

Let $\mathcal{G}_n^+ = \text{Gal}(\mathbf{Q}(\mu_{p^n})/\mathbf{Q}) \bmod \sigma_{-1}$. The next theorem is due to Bass [Ba].

Theorem 3.2. *The group $\mathcal{G}_n^+$ operates simply transitively on the primitive elements of V_n, and the induced homomorphism*

$$\mathbf{Z}[\mathcal{G}_n^+] \to V_n \text{ such that } \sigma_c \mapsto g_{c/p^n}$$

is an isomorphism.

Proof. The homomorphism is obviously surjective. It is injective because $\mathbf{Z}[\mathcal{G}_n^+]$ is torsion free, and the ranks of the two groups are equal. This proves the theorem.

Theorem 3.3. *The factor group V_m/V_n for $m \geq n$ has no torsion.*

Proof. The embedding of V_n into V_m corresponds to the embedding of group rings

$$\mathbf{Z}[\mathcal{G}_n^+] \to \mathbf{Z}[\mathcal{G}_m^+]$$

which sends an element σ_c on the element $\sum \sigma_b$, where the sum is taken over $\sigma_b \in \mathcal{G}_m^+(c)$, the set of elements in $\mathcal{G}_m^+$ which project on σ_c under the canonical map

$$\mathcal{G}_m^+ \to \mathcal{G}_n^+.$$

If an element

$$\sum_c \sum_{b \in \mathcal{G}_m^+(c)} k(b)\sigma_b, \quad k(b) \in \mathbf{Z},$$

is a torsion element with respect to $\mathbf{Z}[\mathcal{G}_n^+]$, then all the coefficients $k(b)$ for $b \in \mathcal{G}_m^+(c)$ must be equal to each other, and hence the element already lies in $\mathbf{Z}[\mathcal{G}_n^+]$, as was to be shown.

Analogues of Theorem 3.2 and 3.3 in the modular case are proved in the Kubert–Lang series [KL 2, 3, 4, 5]. In that case, it is also shown that there are no units except the modular ones. Here in the cyclotomic case, say for the p-primary component, it is the Kummer–Vandiver conjecture whether the factor group $E/\mathcal{E}$ is without p-torsion.

For the rest of this section, it is convenient to use $\mathcal{E}_n$ to denote the proper group of cyclotomic units, i.e., the group of units of the form

$$\pi^\alpha,$$

where $\pi = \zeta - 1$, ζ is a primitive p^nth root of unity, and $\alpha = \sum_{b \in \mathbf{Z}(p^n)^*} k(b)\sigma_b$ is an element of $\mathbf{Z}[\mathcal{G}_n]_0$ of degree 0, i.e.,

$$\sum k(b) = 0.$$

Theorem 3.4. *Let p be odd, and let c be a primitive root* mod p^2. *Then $\mathcal{E}_n$ is generated over $\mathbf{Z}[\mathcal{G}_n]_0$ by the element*

$$v = \sigma_c \pi / \pi = \frac{\zeta^c - 1}{\zeta - 1},$$

Proof. We write an element α of degree 0 in the form

$$\alpha = \sum k(b)(\sigma_b - 1),$$

and observe that $\sigma_b - 1$ is divisible in the integral group ring by $\sigma_c - 1$ because σ_c is a generator of the cyclic group $\mathcal{G}_n$. This proves the theorem.

For $p = 2$ one has an analogous result using for c an element $\equiv 1 \bmod 4$ such that c generates $1 + 4\mathbf{Z}_2$. The group $\mathcal{G}_n = \mathrm{Gal}(\mathbf{Q}(\mu_{p^n})/\mathbf{Q})$ is not cyclic but a product of a cyclic group of order 2 and $\mathcal{G}_n^+$, which is cyclic.

It is convenient to reformulate the above theorem by passing to $\mathcal{G}_n^+$.

Theorem 3.5. *Let c be a generator of* $1 + 4\mathbf{Z}_2$ *if $p = 2$, and a primitive root* mod p^2 *if $p > 2$. Let*

$$v_n = \textit{image of } \frac{\zeta^c - 1}{\zeta - 1} \textit{ in } V_n,$$

and let V_n^0 be the subgroup of V_n represented by units (not just p-units). Then we have an isomorphism

$$V_n^0 \approx \mathbf{Z}[\mathcal{G}_n^+]_0 v_n,$$

so V_u^0 is free of rank 1 *over* $\mathbf{Z}[\mathcal{G}_n^+]_0$.

Proof. Clear.

The composite case.

Let V be the group of cyclotomic units of all levels, modulo the group of roots of unity. Then V is torsion free.

Let $c = (\ldots, c_p, \ldots)$ be a vector with a component $c_p \in \mathbf{Z}_p^*$ for each p such that c_p is a primitive root mod p^2 if p is odd, and $c_2 \equiv 1 \bmod 4$ and generates $1 + 4\mathbf{Z}_2$ if $p = 2$. We have an associated automorphism σ_c on the full cyclotomic extension of $\mathbf{Q}$.

Let $a \in \mathbf{Q}/\mathbf{Z}$ and $a \notin \mathbf{Z}$. We define

$$h(a) = \text{image in } V \text{ of } \frac{\sigma_c(e^{2\pi i a} - 1)}{e^{2\pi i a} - 1}.$$

Then h is an ordinary distribution in the sense of Chapter 2, §8 if we define $h(0) = 0$.

Theorem 3.6. *This distribution is the universal even ordinary distribution with value* 0 *at* 0, *and values into abelian groups on which multiplication by* 2 *is invertible.*

Proof. On $(1/N)\mathbf{Z}/\mathbf{Z}$ the group generated by the image of h has rank

$$\tfrac{1}{2}|\mathbf{Z}(N)^*| - 1,$$

which according to Kubert's Theorem 9.1(iii) of Chapter 2 is the maximal possible rank (the value 0 at 0 gives rise to the -1). The Kubert generators in $T_N/\pm 1$ must therefore be free generators, and the canonical map from the universal distribution to h must be an isomorphism.

The above is more or less Bass' theorem in a different formulation. (Also, as Bass states it, there is some difficulty with 2-torsion.) The idea of interpreting it in terms of the cyclotomic units forming a universal distribution is due to Kubert–Lang [KL 3], where a similar result is proved for the modular units. The essential step in the proof here is of course Kubert's theorem cited above, combined with the independence of the units. In the cyclotomic case, this comes back to the non-vanishing of the regulator, i.e., $L(1, \chi) \neq 0$. In the modular case, see [KL 2] and [KL 5].

§4. The Iwasawa–Leopoldt Theorem and the Kummer–Vandiver Conjecture

For simplicity throughout this section we assume that p is an odd prime. Also throughout this section, we let:

$$\begin{aligned}
K_\infty &= \mathbf{Q}(\mu^{(p)}) \text{ and } K_0 = \mathbf{Q}(\mu_p),\\
\mathcal{G} &= \mathrm{Gal}(K_\infty/\mathbf{Q}),\\
\mathcal{G}_n &= \mathrm{Gal}(K_n/\mathbf{Q}),\\
\mathcal{G}_n^+ &= \mathcal{G}_n \bmod \sigma_{-1} \text{ as in the preceding section.}\\
R_n &= \mathbf{Z}(p^n)[\mathcal{G}_n], \text{ and } R = \text{projective limit of } R_n.
\end{aligned}$$

Remark. It is easy to see that

$$R \approx \Lambda[\mathcal{G}_0]$$

where Λ is the usual Iwasawa algebra.

$h_n^+ =$ class number of K_n^+. *We assume the Kummer–Vandiver conjecture that $h_n^+ = (E_n : \mathcal{E}_n)$ is prime to p.*

$\Omega_n = K_\infty(V_n^{1/p^n}) = K_\infty(\mathcal{E}_{p,n}^{1/p^n})$. By the Kummer–Vandiver conjecture,

$$\Omega_n = K_\infty(E_{p,n}^{1/p^n}).$$

$$\Omega = \bigcup \Omega_n = K_\infty(E_p^{1/p^\infty}) = K_\infty(\mathcal{E}_p^{1/p^\infty}).$$

$$G_n = \mathrm{Gal}(\Omega_n/K_\infty) \text{ and } G = \mathrm{Gal}(\Omega/K_\infty).$$

We shall now develop the theory of G as R-module following the exposition of [KL 9]. We use an upper minus sign to denote, as usual, the (-1)-eigenspace. This applies for instance to R^-, G^-, etc.

Theorem 4.1. *Assuming the Kummer–Vandiver conjecture, we have $G = G^-$, and G is a 1-dimensional free module over R^-.*

Proof. By the Kummer–Vandiver conjecture and Theorem 3.3 we have

$$E_{p,n} \cap K_\infty^{*p^n} = E_{p,n}^{p^n} \mu^{(p)}.$$

Let us abbreviate for simplicity

$$V_n^{1/p^n} = E_{p,n}^{1/p^n} / E_{p,n} \cap K_\infty^*.$$

Then the Kummer theory pairing discussed in §1 can be described more explicitly as follows. From Theorem 3.2 we write an isomorphism

$$\frac{1}{p^n} \mathbf{Z}_p/\mathbf{Z}_p[\mathcal{G}_n]^+ \to V_n^{1/p^n},$$

using formal linear combinations with coefficients in $(1/p^n)\mathbf{Z}_p/\mathbf{Z}_p$. We have a model for Kummer duality, through the pairing

$$\mathbf{Z}(p^n)[\mathcal{G}_n] \times \frac{1}{p^n} \mathbf{Z}_p/\mathbf{Z}_p[\mathcal{G}_n] \to \mu_{p^n}$$

such that

$$\sum x(c)\sigma_c \times \sum y(c)\sigma_c \mapsto e^{2\pi i \sum x(c)y(c)c}.$$

This pairing induces a perfect duality

$$\mathbf{Z}(p^n)[\mathcal{G}_n]^- \times \frac{1}{p^n} \mathbf{Z}_p/\mathbf{Z}_p[\mathcal{G}_n]^+ \to \mu_{p^n},$$

as follows immediately from the formula

$$\langle \lambda^\rho, \xi^\rho \rangle = \langle \lambda, \xi \rangle^\rho$$

where ρ is complex conjugation, or for that matter any element of $\mathcal{G}_n$. Therefore we have an isomorphism of the Kummer pairing in terms of the group rings,

$$\begin{array}{ccccc} G_n & \times & V_n^{1/p^n} & \to & \mu_{p^n} \\ \updownarrow & & \updownarrow & & \updownarrow \\ R_n^- & \times & \frac{1}{p} R_n^+ & \to & \mu_{p^n} \end{array}$$

and this isomorphism is compatible with the limiting process, which can be represented by a diagram in terms of the group rings for $m \geq n$:

$$\begin{array}{ccccc} R_m^- & \times & \frac{1}{p^m} R_m^+ & \rightarrow & \mu_{p^m} \\ \downarrow & & \uparrow & & \uparrow \\ R_n^- & \times & \frac{1}{p^n} R_n^+ & \rightarrow & \mu_{p^n} \end{array}$$

It is then clear that

$$G = \lim G_n \approx \lim R_n^- = R^-,$$

as desired.

One would expect the units whose existence is conjectured by Stark [St] to play a similar role over totally real fields.

Theorem 4.2. *Let $C_n = Cl^{(p)}(K_n)$ be the p-primary part of the ideal class group of K_n and let C = projective limit of the C_n under the norm map. Under the Kummer–Vandiver conjecture, we have $C = C^-$, and C^- is cyclic as a Λ-module. In fact, the maximal unramified p-abelian extension of K_∞ is contained in Ω, so we have a natural surjective map*

$$G \rightarrow G_C \approx C,$$

the first map by restriction and the second by class field theory.

Proof. The field diagram (once the theorem is proved) is as follows.

$$\begin{array}{c} \Omega \\ | \\ \Omega^{\text{nr}} \\ | \\ K_\infty \end{array}$$

(G spans Ω over K_∞; G_C spans Ω^{nr} over K_∞.)

What we have to do is to show that the maximal p-abelian unramified extension of K_∞ is in fact contained in Ω. The rest of the theorem is then obvious from Theorem 4.1. It will suffice to prove that a finite cyclic unramified p-abelian extension of K_∞ is contained in Ω.

Let $K_\infty(\alpha)$ be unramified, with some element α such that α^{p^t} lies in K_∞. We first show that we may select α to be real. By Kummer–Vandiver, we have $G_C^+ = 1$ so $G_C = G_C^-$. Let σ be a generator for $\mathrm{Gal}(K_\infty(\alpha)/K_\infty)$. Then

$$\sigma\alpha = \zeta\alpha \quad \text{for some } p\text{th root of unity } \zeta.$$

Let ρ be complex conjugation. Then $\rho\sigma\rho^{-1} = \rho\sigma\rho = \sigma^{-1}$ since $G = G^-$. Hence $\rho\sigma\rho\alpha = \zeta^{-1}\alpha$, and therefore

$$\sigma\bar{\alpha} = \zeta\bar{\alpha},$$

so that $\sigma(\bar{\alpha}/\alpha) = \bar{\alpha}/\alpha$. Thus $\bar{\alpha}/\alpha = b$ lies in K_∞. But the norm of b from K_∞ to K_∞^+ is 1 (obvious), so by Hilbert's Theorem 90, there exists $\beta \in K_\infty$ such that $\beta/\bar{\beta} = b$. Then $\alpha\beta$ is real, and $K_\infty(\alpha) = K_\infty(\alpha\beta)$. This shows that we may assume α real.

For n sufficiently large, α^{p^t} lies in K_n^+, and $K_n^+(\alpha)$ is unramified over K_n^+ because p is odd. Hence we have an ideal factorization

$$(\alpha) = \mathfrak{a}^{p^t}$$

for some fractional ideal $\mathfrak{a}$ in K_n^+. The class of this ideal is principal by Kummer–Vandiver's conjecture. It is then immediate that

$$K_n(\alpha) = K_n(u^{1/p^t})$$

for some unit u, thereby concluding the proof.

In the next theorem we let

$$\Gamma_{m,n} = \mathrm{Gal}(K_m/K_n).$$

Theorem 4.3. (i) *For $m \geq n$ we have an injection*

$$\mathrm{Ker}(C_n \to C_m) \to H^1(\Gamma_{m,n}, E_m).$$

(ii) *The Kummer–Vandiver conjecture implies $H^1(\Gamma_{m,n}, E_m) = 0$, so*

$$C_n \to C_m$$

is injective.

Proof. Let $\mathfrak{a}$ be an ideal representing an element of C_n, becoming principal in K_m, say $\mathfrak{a} = (\alpha)$ with $\alpha \in K_m$. For any element $\sigma \in \Gamma_{m,n}$ we have $\sigma\mathfrak{a} = \mathfrak{a}$. Hence $\sigma\alpha$ is equal to α times some unit. The association

$$\mathfrak{a} \mapsto \text{cocycle class of } (\sigma\alpha/\alpha)$$

is a homomorphism of $\mathrm{Ker}(C_n \to C_m)$ into H^1 of the units, which is immediately verified to be injective.

Assume now the Kummer–Vandiver conjecture. Let $\mathscr{E}_m$ be the group of cyclotomic units. Then $E_m/\mathscr{E}_m$ has order prime to p, so

$$H^1(\mathscr{E}_m) \cong H^1(E_m).$$

For simplicity let $W_m = \mu_{p^{m+1}}$. Then we have exact sequences

$$0 \to W_m \to \mathscr{E}_{p,m} \to V_m \to 0$$

and

$$0 \to \mathscr{E}_m \to \mathscr{E}_{p,m} \to \mathbf{Z} \to 0$$

whence exact cohomology sequences

$$0 \to H^1(W_m) \to H^1(\mathscr{E}_{p,m}) \to H^1(V_m)$$

and

$$0 \to H^1(\mathscr{E}_m) \to H^1(\mathscr{E}_{p,m}) \to H^1(\mathbf{Z}).$$

Since $H^1(\mathbf{Z})$ is trivial and $H^1(V_m)$ is trivial (by elementary facts of cohomology of finite groups, and Theorem 3.2), it will now suffice to prove that $H^1(W_m)$ is trivial. By the theory of the Herbrand quotient (cf. for instance Chapter IX, §1 of my *Algebraic Number Theory*), the orders of $H^1(W_m)$ and $H^0(W_m)$ are equal. However,

$$H^0(W_m) = W_m^{\Gamma_{m,n}}/N_{m,n}W_m = W_n/N_{m,n}W_m.$$

where $N_{m,n}$ is the norm. Thus finally it suffices to prove that every p-power root of unity in K_n is the norm of an element in W_m. Let ζ be a generator of W_m. The elements of the Galois group are represented by p-adic integers of the form

$$1 + xp^{n+1} \quad \text{with } x \in \mathbf{Z}/p^{m-n}\mathbf{Z}.$$

Taking the norm yields

$$N_{m,n}\zeta = \prod_x \zeta^{1+xp^n} = \zeta^{p^{m-n}}$$

which is a primitive element in W_n and thus shows that $H^0(W_m)$ is trivial. This concludes the proof of the theorem.

Let K be a number field and let K' be an abelian extension with Galois group G. We assume that K and K' are stable under complex conjugation. We say that the extension K' of K is **odd** (resp. **even**) if its Galois group is in the (-1)-eigenspace (resp. the 1-eigenspace) for complex conjugation.

Lemma. *If $\alpha^{p^{n+1}} \in K_n^+$, then $K_n(\alpha)/K_n$ is an odd extension.*

Proof. Clear.

From the lemma, it follows that Ω_E/K_∞ is an odd extension, because the units are generated by real cyclotomic units and roots of unity.

Theorem 4.4. *Under the Kummer–Vandiver conjecture, the Kummer duality gives rise to a compact discrete duality*

$$\mathrm{Gal}(\Omega/\Omega_E)^+ \textit{ dual to } C_\infty^-$$

and also

$$\mathrm{Gal}(\Omega/\Omega^{\mathrm{nr}})^+ \textit{ dual to } \mathfrak{C}_\infty^-.$$

Proof. By Theorems 2.2, 2.3 and Lemma 1 of §2 we know that $\Omega = \Omega_A$ and that

$$\mathrm{Gal}(\Omega/\Omega_E) \text{ is dual to } C_\infty.$$

Taking eigenspaces for complex conjugation yields the first assertion. As to the second, we know from Theorem 4.2 that

$$\Omega^{\mathrm{nr}} \subset \Omega_E$$

and Ω_E is an odd extension of Ω^{nr}. Again considering the eigenspaces yields the second assertion.

7 Iwasawa Theory of Local Units

Iwasawa [Iw 8], [Iw 10] developed a theory of local units analogous to the global theory, taking projective limits, especially in the cyclotomic tower, and getting the structure of this projective limit modulo the closure of the cyclotomic units. He considers eigenspaces for the characters of $\mathrm{Gal}(K_0/\mathbf{Q}_p)$ where $K_0 = \mathbf{Q}_p(\zeta)$ with a primitive pth root of unity ζ. Since the cyclotomic units are essentially real, we consider only even non-trivial characters. Then the eigenspace is isomorphic to $\Lambda/(g)$, where g is a power series which is essentially the p-adic L-function.

The first section deals with the classical Kummer–Takagi exponents at the first level $\mathbf{Q}_p(\zeta)$, where ζ is a primitive pth root of unity, *p odd*. This is used in combination with Nakayama's lemma afterwards to get corresponding results in the cyclotomic tower. Throughout this chapter we assume that p is odd.

Coates–Wiles [C–W 4] have extended this theory to the case of elliptic curves with complex multiplication. In the process they have found substantial simplifications for Iwasawa's proofs, and the exposition of this chapter is essentially due to them. Note especially their generalization of the Kummer homomorphism to all levels—a key to the whole theory. Such a homomorphism extends to other formal groups besides the multiplicative group, and a quite general statement has also been given by Coleman [Col].

On the whole, this chapter may be viewed as giving a good introduction to the theories of Coates–Wiles. I am much indebted to them for keeping me up on their work.

§1. The Kummer–Takagi Exponents

Let ζ be a primitive pth root of unity, where p is an odd prime. Let $K_0 = \mathbf{Q}_p(\zeta)$. We let $\mathfrak{o}$, $\mathfrak{p}$ be the integers and prime ideal of K_0 respectively, and

let

$$\pi = \zeta - 1.$$

Let U_0 be the group of units $\equiv 1 \bmod \mathfrak{p}$ in K_0. We let $G_0 = \mathrm{Gal}(K_0/\mathbf{Q}_p)$, and

$$\varkappa_0 \colon G_0 \to \mu_{p-1} \subset \mathbf{Z}_p^*$$

be the homomorphism such that

$$\zeta^\sigma = \zeta^{\varkappa_0(\sigma)}, \quad \text{for } \sigma \in G_0.$$

For simplicity of typography in this section we shall write $\varkappa$ instead of $\varkappa_0$.

Let $f \in \mathbf{Z}_p[[X]]$. We recall the variables

$$T = 1 + X = e^Z,$$

and the differential operator

$$D = (1 + X)D_X = D_Z = TD_T.$$

This last equality holds only for rational functions of X (or T).

Let $u \in U_0$ so $u \equiv 1 \bmod \mathfrak{p}$. Let f be a power series $\equiv 1 \bmod (p, X)$ such that

$$u = f(\pi).$$

We then say that f is a **power series associated** with u. Such a power series is well defined up to a multiple of the irreducible polynomial $h(X)$ of π over $\mathbf{Z}_p$. Let f, f_1 be associated with u, so $f \equiv f_1 \bmod h$. Since f_1 is a unit power series, there exists a power series g such that

$$f = f_1(1 + gh) \quad \text{with } g \in \mathbf{Z}_p[[X]].$$

Then

$$f'/f = f_1'/f_1 + \frac{gh' + hg'}{1 + gh}$$

and

$$Df/f = Df_1/f_1 + \text{multiples of } h \text{ and } h'.$$

Since h is an Eisenstein polynomial, it follows that

$$D^{k-1}(Df/f)(0) \ [= D^k \log f(0)] \text{ is well defined mod } p \text{ for } 1 \leq k \leq p - 2.$$

We define the **Kummer homomorphism** for these values of k by

$$\varphi_k(u) = D^{k-1}(Df/f)(0) \bmod p.$$

It is indeed clear that

$$\varphi_k: U_0 \to \mathbf{Z}(p)$$

is a homomorphism. By the change of variables $X = e^Z - 1$, the formula

$$D^{k-1}(Df/f)(0)$$

is also valid for f as function of Z, i.e., if we set

$$f(X) = f_{G_a}(Z)$$

then

$$D^{k-1}(Df/f)(0) = D_Z^{k-1}(D_Z f_{G_a}/f_{G_a})(0).$$

We now develop systematically certain properties of the Kummer homomorphism. These will be extended in the Coates–Wiles manner later to all levels.

K 1. *If f_1, f_2 are associated with units u_1, u_2, then $f_1 f_2$ is associated with $u_1 u_2$. If f is associated with u and $a \in \mathbf{Z}_p$, then $f(X)^a$ is associated with u^a.*

Proof. The homomorphic property is clear, and has already been mentioned. The statement for $a \in \mathbf{Z}_p$ follows from positive integers by continuity.

K 2. *If f is associated with u, then a power series associated with u^σ is*

$$f((1 + X)^{\varkappa(\sigma)} - 1).$$

Proof. If $u = f(\pi)$ and $f = 1 + \cdots$, then

$$u^\sigma = f((1 + \pi)^{\varkappa(\sigma)} - 1).$$

So the property is obvious. Furthermore,

$$f((1 + X)^{\varkappa(\sigma)} - 1) = f(e^{\varkappa(\sigma)Z} - 1).$$

The next property then follows from the chain rule in terms of the variable Z.

K 3. $$\varphi_k(u^\sigma) = \varkappa(\sigma)^k \varphi_k(u).$$

Let χ be a character of G_0. Let

$$e(\chi) = \frac{1}{p-1} \sum_{\sigma \in G_0} \bar{\chi}(\sigma)\sigma$$

be the corresponding idempotent in the group algebra $\mathbf{Z}_p[G_0]$. Write $\chi = \varkappa^\alpha$ for some residue class $\alpha \bmod p-1$. Let $u \equiv 1 \bmod \mathfrak{p}$. Put

$$u^{e(\chi)} = u(\chi).$$

K 4(i). *If* $k \equiv \alpha \bmod p - 1$ *then* $\varphi_k(u(\chi)) = \varphi_k(u)$.

K 4(ii). *If* $k \not\equiv \alpha \bmod p - 1$ *then* $\varphi_k(u(\chi)) = 0$.

Proof. By **K 2** and **K 3** we find

$$\varphi_k(u^{e(\chi)}) = \varkappa^k(e(\chi))\varphi_k(u).$$

The property follows by orthogonality of characters.

The units

$$1 - \pi^k \quad \text{for } k = 1, 2, \ldots$$

generate U_0 topologically. We shall be especially interested in the values of k satisfying $1 \leq k \leq p-2$, and we shall orthogonalize these units with respect to the characters of G_0. We let

$$\eta_k = (1 - \pi^k)^{e(k)}$$

where we abbreviate

$$e(k) = e(\varkappa^k) = \frac{1}{p-1} \sum_{\sigma \in G_0} \varkappa^{-k}(\sigma)\sigma.$$

Lemma. *We have* $\eta_k \equiv 1 - \pi^k \bmod \pi^{k+1}$, *for* $1 \leq k \leq p-2$.

Proof. We have

$$\begin{aligned} \eta_k &\equiv \prod (1 - (\zeta^\sigma - 1)^k)^{-\varkappa^{-k}(\sigma)} \bmod \pi^{p-1} \\ &\equiv \prod (1 + \varkappa^{-k}(\sigma)(\zeta^\sigma - 1)^k). \end{aligned}$$

Say $\zeta^\sigma = \zeta^a$. Then

$$\varkappa^{-k}(\sigma)(\zeta^\sigma - 1)^k \equiv a^{-k}(\zeta^a - 1)^k \equiv (\zeta - 1)^k \bmod \pi^{p-1},$$

as was to be shown.

Theorem 1.1. *Let* $1 \leq k, j \leq p - 2$.

(i) $\varphi_k(\eta_k) = -k \bmod p$.

(ii) $\varphi_k(\eta_j) = 0$ *if* $k \neq j$.

Proof. The second assertion is a special case of **K 4(ii)**. As to the first,

$$\varphi_k(\eta_k) = \varphi_k(1 - \pi^k).$$

An associated power series of $1 - \pi^k$ is $f(X) = 1 - X^k$, and

$$f'/f(X) = \frac{-k}{1 - X^k}.$$

Then

$$\begin{aligned} Df/f(X) &= -k(1 + X) \sum_{\nu=0}^{\infty} X^{\nu k} \\ &= -k\, e^Z \sum_{\nu=0}^{\infty} (e^Z - 1)^{\nu k} \\ &\equiv -k\, e^Z \bmod Z^k. \end{aligned}$$

Hence

$$D^{k-1}(Df/f)(0) = -k$$

as desired.

By the lemma, and a trivial recursion procedure, any unit $\equiv 1 \bmod \mathfrak{p}$ has a product expression

$$u = \eta_1^{t_1} \cdots \eta_{p-2}^{t_{p-2}} \bmod \pi^{p-1},$$

and the exponents t_k are called the **Kummer–Takagi exponents**. They are well defined mod p.

Theorem 1.2. *Let u be as above. Then*

$$t_k \equiv -\frac{1}{k} \varphi_k(u) \bmod p.$$

Proof. Immediate, from **K 4** and the fact that φ_k is a $\mathbf{Z}_p$-morphism.

The Kummer Generators

The rest of this section will not be needed, but is included for completeness of reference, and as an introduction to [C–W 2], [C–W 4]. It is convenient to phrase the results in terms of the Lubin–Tate formal groups, so for the rest of this section, we assume that the reader is acquainted with the basic facts of these groups as explained in §1 and §2 of the next chapter, as well as the existence of the logarithm on such groups, as explained in §6 of the next chapter.

Let A be a Lubin–Tate formal group over the $\mathfrak{p}$-adic field K, and associated prime π. We let B be the basic Lubin–Tate group associated with the Frobenius polynomial

$$X^q + \pi X.$$

Let W be the local parameter on B, and Z the parameter on the additive group, so we have

$$Z = \lambda_B(W) \equiv W \bmod W^{q-1}$$

by Lemma 2 of §6 in the next chapter.

Let

$$g_B(W) = b_0 + b_1 W + \cdots$$

be a power series with coefficients in K, and let

$$g_{\mathbf{G}_a}(Z) = d_0 + d_1 Z + \cdots$$

be the power series obtained by putting $g_B(W) = g_{\mathbf{G}_a}(\lambda_B(W))$. Then it is clear that

$$b_k = d_k \quad \text{for } k = 0, \ldots, q - 1.$$

Taking the logarithmic derivative, i.e., the operator

$$g \mapsto g'/g$$

for any power series g, we then obtain:

Lemma. *If*

$$g_B'/g_B(W) = \sum_{k=1}^{\infty} c_{k,B} W^{k-1}$$

$$g_{\mathbf{G}_a}'/g_{\mathbf{G}_a}(Z) = \sum_{k=1}^{\infty} c_{k,\mathbf{G}_a} Z^{k-1}$$

then

$$c_{k,B} = c_{k,\mathbf{G}_a} \quad \textit{for } k = 1, \ldots, q - 2.$$

Proof. This is trivial from the chain rule, writing

$$g_B(W) = g_{\mathbf{G}_a}(W + O(W^{q-1})).$$

Let w_0 be an element of B_π such that

$$w_0^{q-1} + \pi = 0.$$

The field

$$K_0 = K(w_0)$$

is tamely ramified, with different $\mathfrak{p}_0^{q-2}$.

Let u be a unit in K_0, and let $g \in \mathfrak{o}[[W]]$ be a power series such that

$$u = g(w_0).$$

If g_1 is another such power series, then

$$g(W) = g_1(W)h(W)$$

where $h(W)$ is the irreducible polynomial of w_0 over K. From this it is immediate that

$g'/g(w_0)$ *is well defined modulo* $\mathfrak{p}_0^{q-2}$*, and lies in* $\mathfrak{o}$.

In particular, if we write

$$g'/g(w_0) = \sum_{k=1}^{\infty} c_k w_0^{k-1}, \quad \text{with } c_k \in \mathfrak{o},$$

then c_k is well defined mod $\mathfrak{p}$ for $1 \leq k \leq q - 2$. We define

$$\varphi_k(u) = c_k \quad \text{for } 1 \leq k \leq q - 2.$$

Then it is clear that

$$\varphi_k \colon \mathfrak{o}_0^* \to \mathfrak{o}/\mathfrak{p}$$

is a homomorphism, which we shall call the **Kummer homomorphism** of degree k. We shall determine the value of this homomorphism in special interesting cases.

There is a character χ of G_0 into μ_{q-1} such that

$$\sigma w_0 = \chi(\sigma) w_0.$$

To avoid technical complications, we now assume that $K = \mathbf{Q}_p$ so that

$\mathfrak{o} = \mathbf{Z}_p$. We let

$$e_k = \frac{1}{p-1} \sum_{\sigma \in G_0} \chi^{-k}(\sigma)\sigma$$

be the idempotent in the group ring $\mathbf{Z}_p[G_0]$ for the character χ^k. If u is a unit $\equiv 1 \bmod w_0$, we can **define**

$$u^t \quad \text{with} \quad t \in \mathbf{Z}_p$$

in the obvious manner. We pick a sequence of integers $m \in \mathbf{Z}$ approaching t p-adically, and the ordinary powers u^m approach a limit, which is by definition u^t.

The units

$$1 - w_0^k, \quad \text{with } k = 1, 2, \ldots$$

form a topological system of generators for the units $\equiv 1 \bmod \mathfrak{p}_0$ in $\mathfrak{o}_0$. Let

$$G_0 = \mathrm{Gal}(K_0/K), \quad \text{so } G_0 \approx \mu_{p-1}.$$

We orthogonalize a basis for the units. We let

$$\eta_k = (1 - w_0^k)^{e_k}, \quad \text{for } k = 1, \ldots, p-1.$$

Then:

(i) $\eta_k \equiv 1 - w_0^k \bmod \mathfrak{p}_0^{k+1}$.

(ii) $\sigma\eta_k = \eta_k^{\chi^k(\sigma)}$, i.e., $\eta_k \in U_0(k)$, where $U_0(k)$ is the χ^k-eigenspace of U_0, and U_0 are the units $\equiv 1 \bmod \mathfrak{p}_0$ in K_0.

The second statement is obvious by the standard properties of the idempotent e_k. For the first, we simply expand the product

$$\begin{aligned} \eta_k &\equiv \prod (1 - w_0^k)^{-\chi^{-k}(\sigma)\sigma} \bmod w_0^{k+1} \\ &\equiv \prod (1 - \chi(\sigma)^k w_0^k)^{-\chi^{-k}(\sigma)} \\ &\equiv (1 + w_0^k)^{p-1} \\ &\equiv 1 - w_0^k \end{aligned}$$

as was to be shown.

Theorem 1.3. *Let* $j, k = 1, \ldots, p-2$. *Then:*

(i) $\varphi_k(\eta_j) = 0$ *if* $k \neq j$.

(ii) $\varphi_k(\eta_k) = -k \bmod p$.

If u *is a unit* $\equiv 1 \bmod \mathfrak{p}_0$ *and*

$$u \equiv \eta_1^{t_1} \cdots \eta_{p-2}^{t_{p-2}} \bmod w_0^{p-1},$$

then

$$t_k \equiv -\frac{1}{k}\varphi_k(u) \bmod p.$$

Proof. Taking the logarithmic derivative formally, we have:

$$\begin{aligned}\frac{d\log}{dw_0}\eta_k &= \frac{d\log}{dw_0}\prod_\sigma (1-\chi^k(\sigma)w_0^k)^{-\chi^k(\sigma)}\\ &= \sum_\sigma -\chi^{-k}(\sigma)\frac{-k\chi^k(\sigma)w_0^{k-1}}{1-\chi^k(\sigma)w_0^k}\\ &= k\sum_{j=0}^{\infty}\sum_\sigma \chi^{kj}(\sigma)w_0^{k(j+1)-1}.\end{aligned}$$

For $j = 0$ we get a term $k(p-1)w_0^{k-1} \equiv -kw_0^{k-1} \bmod w_0^{p-2}$. This shows that $\varphi_k(\eta_k) = -k$. On the other hand, if $k(j+1) - 1 \leq p - 3$, we have

$$k(j+1) \leq p-2 \quad \text{so} \quad kj \leq p-3.$$

Then

$$\chi^{kj} \text{ is not trivial,}$$

so the orthogonality relations show that the coefficient of the corresponding power of w_0 is 0. This proves (i) and (ii). The last assertion then follows from the homomorphic property of the map φ_k, thus proving the theorem.

Let $\mathscr{A} = \{a_i\}$ be a finite family of integers prime to p, and let $\mathscr{N} = \{n_i\}$ be a finite family of integers satisfying

$$\prod a_i^{n_i} \equiv 1 \bmod p \quad \text{and} \quad \sum n_i = 0.$$

Let

$$u = u(\mathscr{A}, \mathscr{N}) = \prod (\zeta^{a_i} - 1)^{n_i}.$$

Then u is a cyclotomic unit, and $u \equiv 1 \bmod \mathfrak{p}_0$.

Theorem 1.4. *Let*

$$u(\mathscr{A}, \mathscr{N}) = \eta_1^{t_1}\cdots\eta_{p-2}^{t_{p-2}} \bmod w_0^{p-1}.$$

Then

$$t_k = \frac{(-1)^k}{k\cdot k!}B_k\sum n_i a_i^k \bmod p.$$

where B_k is the Bernoulli number.

Proof. Let A be the formal multiplicative group, B the special Lubin–Tate group associated with it. The power series

$$g_{\mathbf{G}_a}(Z) = e^{aZ} - 1$$

corresponds to the power series $g_B(W)$ such that

$$g_B(w_0) = \zeta^a - 1$$

where ζ is a pth root of unity. Directly from the definition of the Bernoulli numbers,

$$\frac{Z}{e^Z - 1} = \sum_{k=0}^{\infty} B_k \frac{Z^k}{k!}$$

it follows trivially that

$$g'_{\mathbf{G}_a}/g_{\mathbf{G}_a}(Z) = a + \sum_{k=0}^{\infty} \frac{1}{k!} B_k\, a^k\, Z^{k-1}.$$

Since the operation $g \mapsto g'/g$ sends multiplication to addition, the theorem follows from Lemma 1 of §6 in the next chapter, and Theorem 1.3.

§2. Projective Limit of the Unit Groups

Let:

$K_n = \mathbf{Q}_p(W_n)$, $W_n = \mu_{p^{n+1}}$.

$\mathfrak{o}_n, \mathfrak{p}_n$ = integers and maximal ideal in K_n respectively.

U_n = units $\equiv 1 \bmod \mathfrak{p}_n$ in K_n.

U'_n = units whose norms to $\mathbf{Q}_p$ are equal to 1

= units which are infinitely divisible in the projective system of units under the norm maps $N_{m,n}$ with $m \geq n$.

We have given two conditions describing U'_n, and it is easy to prove that they are equivalent. Indeed, we have the formula for the norm residue symbol:

$$(u, K_m/K_n) = (N_n u, K_m/\mathbf{Q}_p)$$

where N_n is the norm from K_n to $\mathbf{Q}_p$. If $N_n u = 1$ then u is a norm from K_m for every m. Conversely, if the left-hand side is 1 for all m, then

$$N_n u \equiv 1 \bmod p^m$$

for all positive integers m, so $N_n u = 1$. This proves the equivalence.

Let:

$G_n = \mathrm{Gal}(K_n/\mathbf{Q}_p)$ and $G_\infty = \mathrm{Gal}(K_\infty/\mathbf{Q}_p)$.

$\Gamma_n = \mathrm{Gal}(K_n/K_0)$ and $\Gamma = \lim \Gamma_n = \mathrm{Gal}(K_\infty/K_0)$.

Note that G_0 operates on K_n, U_n, U_n' and

$$G_n \approx \Gamma_n \times G_0, \quad \text{while} \quad G_\infty \approx \Gamma \times G_0.$$

Let γ be a topological generator of Γ. Then $\Gamma^{p^n} = \mathrm{Gal}(K_\infty/K_n)$ is generated by γ^{p^n}.

We have an exact sequence of Galois modules

$$1 \to U_n' \to U_n \xrightarrow{N_n} \text{subgroup of } \mathbf{Z}_p^* \to 1.$$

From this sequence we conclude that for each character χ of G_0 and $\chi \neq 1$,

$$U_n(\chi) \approx U_n'(\chi).$$

In the next lemma, by $\mathrm{rank}_{\mathbf{Z}_p}$ we mean (as usual) the rank of a module modulo torsion over $\mathbf{Z}_p$.

Lemma 1. (i) $\mathrm{rank}_{\mathbf{Z}_p} U_n(\chi) = p^n$.

(ii) *If* $\chi \neq 1, \varkappa_0$ *then*

$$U_n(\chi) \approx \mathbf{Z}_p^{(p^n)}.$$

Proof. The integers $\mathfrak{o}_n$ contain a free submodule over the group ring

$$\mathbf{Z}_p[\mathrm{Gal}(K_n/\mathbf{Q}_p)],$$

and for large r, $1 + p^r\mathfrak{o}_n$ is Galois-isomorphic to the above submodule under the exponential map, and is contained in U_n with finite index, so the first part of the lemma is clear.

For the second part, the only torsion in U_n consists of the roots of unity W_n, which is a $\varkappa_0$-eigenspace. Hence for $\chi \neq 1, \varkappa_0$ we have an isomorphism

$$U_n(\chi) \approx \mathbf{Z}_p^{(p^n)}$$

as desired.

We consider the groups U_n as forming a projective system under the norm maps, and we let

$$U = \varprojlim U_n$$

be the projective limit. Then from the definition of U_n' we see that also

$$U = \varprojlim U_n'.$$

Note that U is a topological, compact $\mathbf{Z}_p$-module, and also a Λ-module, where

$$\Lambda = \varprojlim \mathbf{Z}_p[\Gamma_n].$$

If u is an element of U, then we view u as a vector

$$u = (\ldots, u_n, \ldots)$$

with components $u_n \in U_n$ such that $N_{m,n}u_m = u_n$ for $m \geq n$, and we also write

$$u = \lim u_n.$$

Lemma 2. *U has no $\mathbf{Z}_p$-torsion.*

Proof. Otherwise there exists a fixed power p^r and an element $u = \lim u_n$ such that $u_n^{p^r} = 1$ for all n. Then u_n is a root of unity, and if $u_n \neq 1$ for some n, then the order of u_m becomes arbitrarily large as m becomes large, which is impossible.

Theorem 2.1. *For each character $\chi \neq 1$, $\varkappa_0$ of G_0 there is a Λ-isomorphism*

$$U(\chi) \approx \Lambda.$$

In other words, $U(\chi)$ is free of dimension 1 over Λ.

The proof will occupy the rest of this section, and will result from a sequence of lemmas. A "natural" basis element for $U(\chi)$ over Λ will be given in the next section.

We shall apply Galois and class field theory in a manner similar to the global case. For simplicity of notation, if X is a Γ-module, we let:

$$X_{(n)} = X/(\gamma^{p^n} - 1)X \quad \text{and} \quad X^{(n)} = \text{fixed elements under } \gamma^{p^n}.$$

For simplicity of notation, throughout this section, denote by K_n^{ab} the maximal p-abelian extension of K_n, and similarly let K_∞^{ab} be the maximal p-abelian extension of K_∞. We have a tower of fields

$$K_n \subset K_\infty \subset K_n^{ab} \subset K_\infty^{ab}.$$

Recall that Γ operates by conjugation on $\text{Gal}(K_\infty^{ab}/K_\infty)$.

Lemma 3. *If $\chi \neq 1$ then we have isomorphisms*

$$\begin{aligned} \text{Gal}(K_n^{ab}/K_\infty)(\chi) &\approx \text{Gal}(K_\infty^{ab}/K_\infty)_{(n)}(\chi) \\ &\approx \text{Gal}(K_n^{ab}/K_n)(\chi). \end{aligned}$$

Proof. This is clear from the fact that K_n^{ab} is the maximal abelian extension of K_n contained in K_∞^{ab}, together with the exact sequence

$$0 \to \mathrm{Gal}(K_n^{ab}/K_\infty) \to \mathrm{Gal}(K^{ab}/K_n) \to \mathrm{Gal}(K_\infty/K_n) \to 0$$

together with the fact that the last term is $\approx \mathbf{Z}_p$.

The Galois group $\mathrm{Gal}(K_n^{ab}/K_n)$ is isomorphic by class field theory with the completion of K_n^* under the topology of subgroups of finite index. There is a topological isomorphism as abelian groups

$$K_n^* \approx \mathbf{Z} \times \mathfrak{o}_n^*.$$

Given a choice of prime element π in K_n, the isomorphism has the form

$$K_n^* = \pi^{\mathbf{Z}} \times \mathfrak{o}_n^*.$$

The completion of K_n^* in the topology of subgroups of finite index is therefore

$$\overline{K_n^*} \approx \pi^{\overline{\mathbf{Z}}} \times \mathfrak{o}_n^*$$

as abelian groups (not Galois modules), where

$$\overline{\mathbf{Z}} = \prod_l \mathbf{Z}_l \quad \text{(product taken over all primes } l\text{)}.$$

On the other hand we have an exact sequence of Galois modules

$$1 \to U_n \to \overline{K_n^*}/\mu_{p-1} \to \overline{\mathbf{Z}} \to 0.$$

Since G_0 operates trivially on $\overline{\mathbf{Z}}$, for each $\chi \neq 1$ of G_0 we have an isomorphism

$$U_n(\chi) \approx (\overline{K_n^*}/\mu_{p-1})(\chi).$$

The isomorphism of local class field theory

$$\overline{K_n^*}/\mu_{p-1} \approx \mathrm{Gal}(K_n^{ab}/K_n)$$

preserves the Γ and G_0 structures of both groups. Passing to the projective limit over n, it follows from the previous isomorphism that

$$U(\chi) = \varprojlim U_n(\chi) \approx \mathrm{Gal}(K_\infty^{ab}/K_\infty)(\chi),$$

whence we obtain the next theorem from Lemma 3.

Theorem 2.2. *For $\chi \neq 1$ we have an isomorphism*

$$U_n(\chi) \approx U(\chi)/(\gamma^{p^n} - 1)U(\chi) = U(\chi)_{(n)}.$$

Lemma 4. *Let M be a finitely generated Λ-module such that*

$$M/(\gamma^{p^n} - 1)M$$

is free over $\mathbf{Z}_p$ of rank p^n for all n. Then M is quasi-isomorphic to Λ.

Proof. Obvious from the structure theorem in Chapter 5.

The lemma is applied to the unit groups, using Lemma 1 and Theorem 2.2. We therefore conclude that there is an exact sequence of Λ-modules

$$0 \to A \to U(\chi) \to \Lambda \to B \to 0$$

where A, B are finite. Since U has no $\mathbf{Z}_p$-torsion by Lemma 2, it follows that $A = 0$. The next lemma will conclude the proof.

Lemma 5. *In the exact sequence, we have $B = 0$, for $\chi \neq 1, \varkappa_0$.*

Proof. From the exact sequence

$$0 \to U(\chi) \to \Lambda \to B \to 0$$

we get the exact (cohomology) sequence

$$0 \to U(\chi)^{(n)} \to \Lambda^{(n)} \to B^{(n)} \to U(\chi)_{(n)}.$$

[This is no big deal in the present instance. The last map is obtained by taking an element $b \in B^{(n)}$, lifting back to any $c \in \Lambda$, and sending $c \mapsto (\gamma^{p^n} - 1)c$. This is well defined in $U(\chi)/(\gamma^{p^n} - 1)U(\chi)$, and the sequence is trivially verified to be exact.]

Trivially $\Lambda^{(n)} = 0$. Hence we obtain an injection

$$0 \to B^{(n)} \to U(\chi)_{(n)} \approx U_n(\chi) \quad \text{by Theorem 2.2.}$$

But $U_n(\chi)$ has no torsion by Lemma 1(ii). Hence

$$B^{(n)} = 0 \quad \text{for all } n.$$

Since B is finite, this implies that $B = 0$, and proves Theorem 2.1.

§3. A Basis for $U(\chi)$ over Λ

For each $n \geq 0$ we let $W_n = \mu_{p^{n+1}}$, and we fix a family of primitive p^{n+1}th roots of unity $\zeta_n \in W_n$ such that

$$\zeta_{n+1}^p = \zeta_n.$$

We let

$$x_n = \zeta_n - 1.$$

The notation remains that of the preceding section. If $\sigma \in G_\infty$ then there is an isomorphism

$$\varkappa : G_\infty \to \mathbf{Z}_p^*$$

such that for all n,

$$\zeta_n^\sigma = \zeta_n^{\varkappa(\sigma)}.$$

As before, we let

$$\varkappa_0 : G_0 \to \mu_{p-1}$$

be the corresponding isomorphism at the first level. If χ is a character of G_0, with values in μ_{p-1}, then

$$\chi = \varkappa_0^k$$

for some k determined mod $p - 1$.

Given $\chi \neq 1, \varkappa_0$ we shall construct an element $\xi \in U$ such that the element

$$\xi(\chi) = e(\chi)\xi$$

is a basis of $U(\chi)$ over Λ. It is natural to construct ξ_n of the form

$$\boxed{\xi_n = \omega(b)^{-1}(b - x_n)}$$

where $b \in \mathbf{Z}_p^*$, and ω is the Teichmuller character. We have divided by $\omega(b)$ so that $\xi_n \equiv 1 \bmod \mathfrak{p}_n$. For each $n \geq 1$ we want that

$$N_{n,n-1}(b - x_n) = b - x_{n-1}.$$

But x_n is a root of $(1 + X)^p = \zeta_{n-1}$ so the equation for $b - x_n$ over K_{n-1} is

$$(1 + b - Y)^p - \zeta_{n-1} = 0,$$

and from the constant term we see that

$$(1 + b)^p - \zeta_{n-1} = N_{n,n-1}(b - x_n) = b - x_{n-1} = b - (\zeta_{n-1} - 1).$$

Thus

$$(1 + b)^p = 1 + b, \quad \text{so} \quad (1 + b)^{p-1} = 1.$$

Therefore we select any

$$\boxed{b = \lambda - 1, \quad \text{with any} \quad \lambda \neq 1 \quad \text{and} \quad \lambda \in \mu_{p-1}}$$

to get the desired b. A choice of λ determines such ξ_n and we write

$$\xi_n^{(\lambda)} \quad \text{instead of} \quad \xi_n$$

if we wish to emphasize the dependence on λ. Since

$$\omega(b)^p = \omega(b),$$

it follows that for $n \geq 1$,

$$N_{n,n-1}\omega(b) = \omega(b)^p = \omega(b),$$

and so the elements ξ_n form a projective system in U.

Lemma 1. *Given $\chi = \varkappa_0^k \neq 1$, $\varkappa_0$ there exists $\lambda \in \mu_{p-1}$ such that if we let $b = \lambda - 1$, and*

$$\xi_0 = \xi_0^{(\lambda)} = \omega(b)^{-1}(b - x_0),$$

then:

(i) $\varphi_k(\xi_0) \not\equiv 0 \bmod p$;

(ii) $\xi_0(\chi) = e(\chi)\cdot\xi_0$ *generates* $U_0(\chi)$ *over* $\mathbf{Z}_p$.

Proof. We shall check below that for a suitable choice of λ (depending on k) the Kummer–Takagi exponent given by Theorem 1.2 is $\not\equiv 0 \bmod p$. Then $\xi_0(\chi)$ generates $U_0(\chi)/U_0(\chi)^p$, and hence generates $U_0(\chi)$ over $\mathbf{Z}_p$ by Nakayama's lemma.

Now for the computation of the Kummer–Takagi exponents, we need only compute $\varphi_k(\xi_0)$ by **K 4**. We have

$$\xi_0 = \frac{b}{\omega(b)}(1 - x_0/b).$$

The associated power series is

$$f(X) = \frac{b}{\omega(b)}(1 - X/b).$$

Then

$$f'/f(X) = -\frac{1}{b}\frac{1}{1 - X/b} = -\frac{1}{b}\sum_{\nu=0}^{\infty}(X/b)^\nu,$$

We want to prove that

$$D^{k-1}(Df/f) \not\equiv 0 \bmod p.$$

We have

$$Df/f = (1 + X)f'/f(X) = \frac{X+1}{X-b} = 1 + \frac{\lambda}{T-\lambda}.$$

But $D = TD_T$. Hence it will suffice to prove that for $2 \leq k \leq p-2$ we have

$$(TD_T)^{k-1}\left(\frac{1}{T-\lambda}\right)\bigg|_{T=1} \not\equiv 0 \bmod p.$$

By induction, it is immediately shown that

$$(TD_T)^m\left(\frac{1}{T-\lambda}\right) = \pm\frac{T\lambda^{m-1} + P_m(T, \lambda)}{(T-\lambda)^{m+1}}$$

where $P_m(T, \lambda)$ is a polynomial in λ of degree $\leq m-2$, with coefficients in $\mathbf{Z}[T]$. Hence

$$(TD_T)^{k-1}\left(\frac{1}{T-\lambda}\right)\bigg|_{T=1} = \pm\frac{\lambda^{k-2} + P_{k-1}(1, \lambda)}{(1-\lambda)^k}.$$

The numerator $\lambda^{k-2} + P_{k-1}(1, \lambda)$ is a polynomial in λ of degree $\leq p-4$. It is clearly not identically zero mod p, and so it has at most $p-4$ roots mod p. We can therefore choose $\lambda \neq 1$ in μ_{p-1} such that λ is not a root of the polynomial mod p. This completes the proof.

Theorem 3.1. *Let $\chi \neq 1, \varkappa_0$. We can choose $\lambda \in \mu_{p-1}$ such that the element*

$$\xi(\chi) = \xi^{(\lambda)}(\xi)$$

generates $U(\chi)$ over Λ, i.e.,

$$U(\chi) = \Lambda \cdot \xi(\chi),$$

and such an element is a free basis for $U(\chi)$ over Λ.

Proof. We know from Theorem 2.2 that

$$U_0(\chi) = U(\chi)/(\gamma - 1)U(\chi),$$

and so by Lemma 1, $e(\chi)\cdot\xi$ generates $U(\chi) \bmod \mathfrak{m}_\Lambda \cdot U(\chi)$. By Nakayama's lemma, it follows that $\xi(\chi)$ generates $U(\chi)$ over Λ. Since $U(\chi) \approx \Lambda$ by Theorem 2.1, such a generator is also a basis, thereby proving the theorem.

§4. The Coates–Wiles Homomorphism

In this section we give the extension of the Kummer homomorphism to all levels, based on a refinement of the associated power series.

Theorem 4.1. *To every element $u \in U$ there is a unique power series f_u in $\mathbf{Z}_p[[X]]$ such that*

$$f_u(x_n) = u_n.$$

This power series satisfies $f_u(X) \equiv 1 \bmod (p, X)$, and the map

$$u \mapsto f_u$$

is a homomorphism of U into the multiplicative group of power series $\equiv 1 \bmod (p, X)$.

We first note that uniqueness is obvious since a power series has only a finite number of zeros (Weierstrass preparation theorem).

The proof of existence will proceed via several steps, which also develop systematically other properties of these series. First:

CW 0.
$$f_\xi(X) = \frac{1}{\omega(b)}(b - X)$$

is the power series associated with our element ξ. Indeed,

$$f_\xi(x_n) = \frac{1}{\omega(b)}(b - x_n) = \xi_n.$$

Next we note two formal properties of the power series f_u which is called the **associated power series** to u.

CW 1. *If $a \in \mathbf{Z}_p$ then the power series associated with u^a is $f_u(X)^a$.*

Proof. This is first obvious when a is a positive integer, and is then true for all $a \in \mathbf{Z}_p$ by continuity.

CW 2. *If f_u is associated with u, and $\sigma \in G_\infty$, then there is a power series associated with u^σ, namely*

$$f_{u^\sigma}(X) = f_u((1 + X)^{\kappa(\sigma)} - 1).$$

Proof. If $u_n = \sum a_i x_n^i$ with $a_i \in \mathbf{Z}_p$, then

$$u_n^\sigma = \sum a_i(\zeta_n^\sigma - 1)^i,$$

so the property is obvious from the definitions.

We are now ready to prove the theorem, i.e., we must show that every u has an associated power series. The two properties **CW 1** and **CW 2** show that the set of elements in U having an associated power series is a $\Lambda[G_0]$-submodule of U, and contains ξ. So it contains $\Lambda[G_0]\xi$. In particular, taking

Theorem 3.1 into account, we have already shown that it contains $U(\chi)$ for all $\chi \neq 1, \varkappa_0$. This is enough for the applications we have in mind.

For the remaining eigenspaces of $\chi = 1$ or $\varkappa_0$, the element ξ does not suffice to generate these spaces, and one must show how to find an associated power series for additional elements generating these spaces. This is not too difficult and will be left to the reader, especially since a generalization of the associated power series to all Lubin–Tate formal groups has been given by Coleman [Col].

Let as usual

$$D = (1 + X)D_X = D_Z.$$

Since f_u'/f_u has coefficients in $\mathbf{Z}_p$, it is clear that

$$Df_u/f_u(X) \in \mathbf{Z}_p[[X]].$$

For each integer $k \geq 1$ we define the **Coates–Wiles homomorphism** φ_k on U by

$$\varphi_k(u) = D^k \log f_u(0) = D^{k-1}(Df_u/f_u)(0).$$

We see that $\varphi_k \colon U \to \mathbf{Z}_p$ maps U into $\mathbf{Z}_p$ by the preceding remark.

CW 3. *For $\sigma \in G_\infty$, $\varphi_k(u^\sigma) = \varkappa(\sigma)^k \varphi_k(u)$.*

Proof. If $u \mapsto u^\sigma$ then **CW 2** gives the power series associated with u^σ, and the assertion is then obvious by the chain rule applied to

$$D^k \log f_{u^\sigma}(0) = D_Z^k \log f_u(e^{\varkappa(\sigma)Z} - 1)|_{Z=0}.$$

CW 4. *Let $\chi = \varkappa_0^\alpha$ where α is a residue class* mod $p - 1$.

(i) *If $k \equiv \alpha$* mod $p - 1$, *then $\varphi_k(u(\chi)) = \varphi_k(u)$.*

(ii) *If $k \not\equiv \alpha$* mod $p - 1$, *then $\varphi_k(u(\chi)) = 0$.*

Proof. Let $e(\chi)$ be the idempotent associated with χ. Then by **CW 1** and **CW 3** we find

$$\varphi_k(u^{e(\chi)}) = \varkappa_0^k(e(\chi))\varphi_k(u).$$

The property follows by orthogonality of characters.

CW 5. *For $u \in U$ and $g \in \Lambda = \mathbf{Z}_p[[X]]$, we have*

$$\varphi_k(g \cdot u) = g(\varkappa(\gamma)^k - 1)\varphi_k(u).$$

Proof. The assertion is true when $g(X) = 1$, and when $g(X) = 1 + X = \gamma$ by **CW 3**. Thus it is true when $g(X) = X$, i.e.,

$$\varphi_k(X \cdot u) = (\varkappa(\gamma)^k - 1)\varphi_k(u).$$

The property follows for arbitrary polynomials by induction, and arbitrary g by continuity.

Theorem 4.2. *Given a congruence class $\alpha \bmod p-1$, there exists a power series h_α such that for any $k \equiv \alpha \bmod p-1$, we have*

$$(1 - p^{k-1})\varphi_k(\xi) = h_\alpha(\varkappa(\gamma)^k - 1).$$

If α is even $\not\equiv 0 \bmod p-1$, then we can choose λ such that

$$\varphi_k(\xi^{(\lambda)})$$

is a unit, and h_α is a unit in $\mathbf{Z}_p[[X]]$.

Proof. Let

$$f_1(X) = D \log f_\xi(X) = (1 + X)f_\xi'/f_\xi(X).$$

Then by **Meas 6** of Chapter 4,

$$\varphi_k(\xi) = D^{k-1}f_1(0) = \int_{\mathbf{Z}_p} x^{k-1}\, d\mu_{f_1}(x).$$

Then a computation shows that

$$\begin{aligned}(1 - p^{k-1})\varphi_k(\xi) = D^{k-1}(\mathrm{U}f_1)(0) &= \int_{\mathbf{Z}_p^*} a^{k-1}\, d\mu_{\mathrm{U}f_1}(a) \\ &= \int_{\mathbf{Z}_p^*} a^k\, d\mu(a)\end{aligned}$$

for some measure μ. By decomposing the integral over cosets of μ_{p-1} in $\mathbf{Z}_p^*$ we can write

$$\int_{\mathbf{Z}_p^*} \omega(a)^\alpha \langle a \rangle^s\, d\mu(a) = \sum_{r \in \mu_{p-1}} \int_{1+p\mathbf{Z}_p} a^s\, d\mu_r(a)$$

where μ_r is a measure with support in $1 + p\mathbf{Z}_p$. By Example 2 of Chapter 4, §1, we conclude that for each r there is a power series f_r such that

$$\int_{1+p\mathbf{Z}_p} a^s\, d\mu_r(a) = f_r(\varkappa(\gamma)^s - 1),$$

and we let

$$h = \sum_{r \in \mu_{p-1}} f_r.$$

Then for $k \equiv \alpha \bmod p - 1$, we get

$$h(\varkappa(\gamma)^k - 1) = \int_{\mathbf{Z}_p^*} \omega(a)^\alpha \langle a \rangle^k \, d\mu(a)$$

$$= \int_{\mathbf{Z}_p^*} a^k \, d\mu(a).$$

This concludes the proof of the existence of h. There remains to show that $\varphi_k(\xi^{(\lambda)})$ is a unit—it is then trivial that h is a unit power series. But this last property is clear from Lemma 1 of §3, as was to be shown.

§5. The Closure of the Cyclotomic Units

Let $\mathscr{E}_n$ be the group of cyclotomic units, i.e., the group generated by

$$W_n = \pm \mu_{p^{n+1}} \quad \text{and elements} \quad \frac{\zeta^a - 1}{\zeta - 1}$$

where ζ is primitive p^{n+1}th root of unity. Let:

$$V_n = \text{closure of } \mathscr{E}_n \cap U_n \text{ in } U_n = \bar{\mathscr{E}}_{n,1}$$

$$V = \varprojlim V_n.$$

If we wish to preserve the $\mathscr{E}$-notation, then we may write

$$V = \varprojlim \bar{\mathscr{E}}_{n,1} = \bar{\mathscr{E}}_{\infty,1}.$$

The group V_n is a $\mathbf{Z}_p[G_0]$-module, and V is a Λ-module. Since $\mathscr{E}_n/W_n$ comes from the real subfield, χ-eigenspaces occur only for even characters χ of G_0.

Before analyzing the projective limit of V_n, we recall in the p-adic context some facts about finite levels. In the global fields, we have an isomorphism

$$\mathscr{E}_n/W_n \approx \mathbf{Z}[G_n^+]_0,$$

where the index on the right indicates the augmentation ideal. Cf. Theorem 3.2 of Chapter 6. Since the cyclotomic units are independent over $\mathbf{Z}_p$ by the non-vanishing of the p-adic regulator, we obtain a G_n-isomorphism

$$\bar{\mathscr{E}}_n/W_n \approx \mathbf{Z}_p[G_n^+]_0.$$

Hence for each even $\chi \neq 1$,

$$(\bar{\mathscr{E}}_n/W_n)(\chi) \approx \mathbf{Z}_p[G_n^+]_0(\chi) = \mathbf{Z}_p[G_n^+]_0 \, e(\chi).$$

Let $c \in \mathbf{Z}_p^*$ be a primitive root mod p^2, and let

$$v_n = \omega(c)\frac{\zeta_n - 1}{\zeta_n^c - 1}.$$

Then $v_n \equiv 1 \bmod \mathfrak{p}_n$. Furthermore for any even character $\chi \neq 1$ the element

$$v_n(\chi) = v_n^{e(\chi)}$$

lies in $\mathscr{E}_n$, so in $V_n(\chi)$. (The root of unity $\omega(c)$ disappears when we project on the eigenspace for χ.) Note that the elements v_n form a compatible system in the cyclotomic tower, that is

$$N_{m,n}v_m = v_n \quad \text{for } m \geq n.$$

We let

$$v = \varprojlim v_n.$$

Then

$$v(\chi) = \lim v_n(\chi) = \lim v_n^{e(\chi)}.$$

Theorem 5.1. *For each even character $\chi \neq 1$ we have*

$$V_n(\chi) = \mathbf{Z}_p[G_n]v_n(\chi),$$

and hence

$$V(\chi) = \Lambda \cdot v(\chi).$$

Proof. Immediate from Theorem 3.2 of Chapter 6.

We note that the power series associated with the element v is

$$\boxed{f_v(X) = \omega(c)\frac{X}{(1 + X)^c - 1}.}$$

Let χ be an even character of G_0, and $\chi \neq 1$. Let $\xi = \xi^{(\lambda)}$ be the constructed element of U such that $\xi(\chi)$ is a basis for $U(\chi)$ over Λ. We have

$$U(\chi) = \Lambda \cdot \xi(\chi) \quad \text{and} \quad V(\chi) = \Lambda \cdot v(\chi).$$

Let us write

$$v(\chi) = g_\chi \cdot \xi(\chi) \quad \text{with some } g_\chi \in \Lambda.$$

Theorem 5.2. *Let* $\chi = \varkappa_0^\alpha$ *be an even character* $\neq 1$. *We have an isomorphism*

$$U(\chi)/V(\chi) \approx \Lambda/g_\chi\Lambda.$$

The power series g_χ *(determined up to a unit in* Λ*) can be selected such that it is equal to the power series* g *satisfying*

$$g(\varkappa(\gamma)^k - 1) = (1 - p^{k-1})(1 - c^k)\frac{1}{k}B_k$$

for any even positive integer k *such that* $k \equiv \alpha \bmod p - 1$.

Proof. We have

$$\begin{aligned} \varphi_k(v) &= \varphi_k(v(\chi)) && \text{by } \mathbf{CW\,4} \\ &= g_\chi(\varkappa(\gamma)^k - 1)\varphi_k(\xi(\chi)) && \text{by } \mathbf{CW\,5} \\ &= g_\chi(\varkappa(\gamma)^k - 1)\varphi_k(\xi) && \text{by } \mathbf{CW\,4} \\ &= g(\varkappa(\gamma)^k - 1)(1 - p^{k-1})^{-1}, \end{aligned}$$

where $g = g_\chi h_\alpha$, and h_α is the power series of Theorem 4.2. Since h_α is a unit power series, g and g_χ generate the same $\Lambda e(\chi)$-ideal. There remains to prove

$$\varphi_k(v) = (1 - c^k)\frac{1}{k}B_k.$$

But

$$\begin{aligned} (1 + X)f_v'/f_v(X) &= \frac{1 + X}{X} - \frac{c(1 + X)^c}{(1 + X)^c - 1} \\ &= \frac{e^Z}{e^Z - 1} - \frac{ce^{cZ}}{e^{cZ} - 1} \\ &= Df_v/f_v. \end{aligned}$$

So for $k \geq 2$,

$$\begin{aligned} \varphi_k(v) &= D^{k-1}(Df_v/f_v)(0) \\ &= (1 - c^k)\frac{1}{k}B_k \end{aligned}$$

by the definition of Bernoulli numbers. This concludes the proof.

The values of the power series g show that it is essentially the p-adic L-function.

Remark. In this chapter we have proved a local statement which would be immediate if one had the Kummer–Vandiver conjecture, as explained in Chapter 6, §4. The corresponding *global conjecture* can also be formulated as follows.

Let $K_\infty = \mathbf{Q}(\mu^{(p)})$, *and let* M_p *be the maximal p-abelian p-ramified extension of* K_∞. *Let* χ *be any even character* $\neq 1$. *Then there is a quasi-isomorphism*

$$\mathrm{Gal}(M_p/K_\infty)(\chi) \sim \Lambda/g_\chi\Lambda$$

where g_χ *is as in Theorem* 5.2.

8 Lubin–Tate Theory

This chapter reproduces with little change the approach to local class field theory given by Lubin–Tate [L–T]. Using special power series associated with prime elements in a p-adic field, they construct maximal abelian totally ramified extensions by means of torsion points on formal groups, thus obtaining a merging of class field theory and Kummer theory by means of these groups.

The theory applies in particular to the cyclotomic case. The p^nth torsion point on a suitable group will be seen to be the classical cyclotomic numbers

$$\zeta - 1$$

where ζ is a p^nth root of unity.

§1. Lubin–Tate Groups

Let $\mathfrak{o}$ be a ring. By a **formal group** over $\mathfrak{o}$ we mean a power series

$$F(X, Y) \in \mathfrak{o}[[X, Y]]$$

in two variables satisfying the three conditions:

FG 1. $F(X, Y) \equiv X + Y \pmod{\text{degree } 2}$.

FG 2. $F(X, F(Y, Z)) = F(F(X, Y), Z)$.

FG 3. $F(X, Y) = F(Y, X)$.

Strictly speaking, our formal groups should be called commutative one-parameter formal groups, but we won't deal with any others. The expression mod degree 2 means modulo the power series of degree ≥ 2. Using the associativity with $Y = Z = 0$ it follows at once that

$$F(X, Y) \equiv X + Y \bmod XY,$$

i.e., $F(X, 0) = X$ and $F(0, Y) = Y$.

It is an easy matter to show recursively that given a formal group as above, there exists a unique power series $\lambda(X)$ such that

$$\lambda(X) \equiv -X \pmod{\text{degree } 2}$$

and

$$F(X, \lambda(X)) = F(\lambda(X), X) = 0.$$

If this could not be proved, we would assume it as an axiom. We leave the proof as an exercise. For the more extensive foundations of formal groups in any number of variables, cf. Fröhlich [Fr].

Example. The **formal multiplicative group $\mathbf{G}_m$** is defined by

$$F(X, Y) = X + Y + XY = (1 + X)(1 + Y) - 1.$$

If a is a positive integer, and $[a]$ denotes "addition" on $\mathbf{G}_m$ a times, then

$$[a](X) = (1 + X)^a - 1.$$

If M is an algebra over $\mathfrak{o}$ (always assumed commutative) and M is nilpotent (in the sense that every element of M is nilpotent—some positive power of the element is 0) then the formal group F defines an additive group law on the set of elements of M, by the association

$$(x, y) \mapsto F(x, y)$$

for x, y in M. Instead of $F(x, y)$ we would also write

$$F(x, y) = x +_F y, \quad \text{or} \quad x\,[+]\,y, \quad \text{or} \quad x\,[+]_F\,y.$$

The set of elements of M with this group law could be denoted by M_F. On the other hand, it is useful to use a slightly different notation. We view F as defining a functor

$$M \mapsto M_F.$$

We may also denote this functor by a letter like A (or A_F if we wish to make the reference to F explicit), and then denote

$$M_F = A(M)$$

to be the set of points of A in M. As a set, it consists of the elements of M, and it is also an additive group with the group law determined as above.

Suppose that $\mathfrak{o}$ is a **complete** valuation ring, with quotient field K and maximal ideal $\mathfrak{m}_K$. We also write $\mathfrak{o} = \mathfrak{o}_K$. Then $\mathfrak{m}_K = \mathfrak{m}$ is topologically nilpotent, in the sense that arbitrarily large powers of an element tend to 0. For any positive integer k, $\mathfrak{m}/\mathfrak{m}^k$ is a nilpotent $\mathfrak{o}$-algebra, and $A(\mathfrak{m}/\mathfrak{m}^k)$ is a group, as we saw. By continuity, it follows that $A(\mathfrak{m})$ is also a group. Addition between elements x, y in $\mathfrak{m}$ is again given by

$$(x, y) \mapsto F(x, y).$$

Let L be any algebraic extension with valuation ring $\mathfrak{o}_L$ and maximal ideal $\mathfrak{m}_L$. Then we also have the completion $\hat{\mathfrak{o}}_L$ if L is infinite over K, with maximal ideal $\hat{\mathfrak{m}}_L$, and it is clear that $A(\hat{\mathfrak{m}}_L)$ can again be defined as group with the group law given by the same formula as above.

By an **endomorphism** of the formal group F (or A_F), we mean a power series $f(X)$ such that

$$f(F(X, Y)) = F(f(X), f(Y)).$$

We say that f is **defined over** $\mathfrak{o}$ if the coefficients of f lie in $\mathfrak{o}$. It is then clear that such an endomorphism defines an endomorphism of $A(\mathfrak{m})$ by the association

$$x \mapsto f(x), \quad \text{for } x \text{ in } \mathfrak{m}.$$

Similarly, a **homomorphism** f of a formal group F into a formal group F' is a power series such that

$$f(F(X, Y)) = F'(f(X), f(Y)).$$

This relation could also be written

$$f(X +_F Y) = f(X) +_{F'} f(Y).$$

Such homomorphism induces a group homomorphism

$$A(\mathfrak{m}) \to A'(\mathfrak{m}),$$

where $A'(\mathfrak{m})$ is the group whose underlying set is $\mathfrak{m}$, and whose group law is that determined by F'.

We shall be interested in a special kind of formal group. From now on, we assume that $\mathfrak{o}_K$ is a discrete valuation ring, and we let π be a prime element in $\mathfrak{m}_K$. We assume that $\mathfrak{o}_K/\mathfrak{m}_K$ is finite with q elements. We let:

$\mathscr{F}_\pi$ = set of power series $f \in \mathfrak{o}[[X]]$ such that

$$f(X) \equiv \pi X \text{ mod degree } 2$$
$$f(X) \equiv X^q \text{ mod } \pi.$$

Example. The power series (polynomial) $f(X) = X^q + \pi X$ is an element of $\mathscr{F}_\pi$, actually its simplest element, which will be called the **special** or **basic Frobenius polynomial** associated with π.

Example. Let

$$f(X) = (1 + X)^p - 1 = X^p + \cdots + pX.$$

Then $f(X)$ is an element of $\mathscr{F}_p$.

The elements of $\mathscr{F}_\pi$ will be called the **Frobenius power series** determined by π.

Theorem 1.1. *To each Frobenius power series f in $\mathscr{F}_\pi$ there exists a unique formal group F_f (defined over $\mathfrak{o}$) such that f is an endomorphism of F_f.*

The formal group associated with $f(X) = X^q + \pi X$ in Theorem 1.1 will be called the **special** or **basic Lubin–Tate group** associated with the prime π.

The proof of this theorem will follow from a general lemma, as will the fact that the formal group F_f then admits $\mathfrak{o}$ in a natural way as a ring of endomorphisms commuting with f.

Lemma. *Let f and g be Frobenius power series in $\mathscr{F}_\pi$. Let*

$$L(X_1, \ldots, X_n) = a_1 X_1 + \cdots + a_n X_n$$

be a linear form with coefficients a_i in $\mathfrak{o}$. There exists a unique series $F(X_1, \ldots, X_n) \in \mathfrak{o}[[X_1, \ldots, X_n]]$ such that

$$F(X_1, \ldots, X_n) \equiv L(X_1, \ldots, X_n) \text{ mod degree } 2$$

and

$$f(F(X_1, \ldots, X_n)) = F(g(X_1), \ldots, g(X_n)).$$

Proof. We abbreviate $X = (X_1, \ldots, X_n)$ and $g(X) = (g(X_1), \ldots, g(X_n))$. We show by induction on r that the congruences

$$F_r(X) \equiv L(X) \text{ mod deg } 2 \quad \text{and} \quad f(F_r(X)) \equiv F_r(g(X)) \text{ mod deg } r + 1$$

have a solution $F_r(X)$ in $\mathfrak{o}[X]$ which is unique mod deg $r + 1$. This is true for $r = 1$ with $F_1(X) = L(X)$. Suppose it true for $r \geq 1$. We let

$$F_{r+1} = F_r + H_{r+1}$$

where H_{r+1} is a homogeneous polynomial of degree $r+1$ with coefficients in $\mathfrak{o}$. We have:

$$f(F_{r+1}(X)) \equiv f(F_r(X)) + \pi H_{r+1}(X) \bmod \deg r+2$$
$$F_{r+1}(g(X)) \equiv F_r(g(X)) + \pi^{r+1}H_{r+1}(X) \bmod \deg r+2.$$

To satisfy the desired relation up to degree $r+1$, we must take

$$H_{r+1}(X) = \frac{f(F_r(X)) - F_r(g(X))}{\pi^{r+1} - \pi} \bmod \deg r+2.$$

The coefficients are in $\mathfrak{o}$ because

$$f(F_r(X)) - F_r(g(X)) \equiv (F_r(X))^q - F_r(X^q) \equiv 0 \pmod{\pi}.$$

It is then clear that

$$F(X) = \lim F_r(X) \in \mathfrak{o}[[X]]$$

is the desired unique solution satisfying the conditions of the lemma.

Addendum to the lemma. *The completeness of $\mathfrak{o}$ was not assumed or used in the proof. Furthermore, the proof shows that F_f is the only power series with coefficients in an extension field of K satisfying the conditions of the lemma.*

Theorem 1.1 is immediate from the lemma. Indeed, F_f is the unique power series $F(X, Y)$ such that

$$F(X, Y) \equiv X + Y \bmod \deg 2,$$

and

$$f(F(X, Y)) = F(f(X), f(Y)).$$

The other two formal group properties are seen to be satisfied by showing that the left-hand side and right-hand side of **FG 2** (resp. **FG 3**) are each the unique solution of a system of conditions as in the lemma.

We call F_f the **Lubin–Tate formal group** associated with f. If we want to use the other notation, we also write it $A(f)$, or simply A if the reference to f is clear from the context.

We shall now see that F_f admits $\mathfrak{o}$ as a ring of endomorphisms in a natural way. We prove slightly more. For each pair of elements $f, g \in \mathscr{F}_\pi$ and $a \in \mathfrak{o}$, we let $a_{f,g}$ or $[a]_{f,g}$ denote the unique solution of

$$a_{f,g}(X) \equiv aX \bmod \deg 2$$
$$f \circ a_{f,g} = a_{f,g} \circ g.$$

We write a_f or $[a]_f$ instead of $a_{f,f}$ for simplicity.

Theorem 1.2. *The association $a \mapsto a_f$ is an injective ring homomorphism of* $\mathfrak{o}$ *into* End(F_f), *such that*

$$\pi_f = f.$$

More generally, the association $a \mapsto a_{f,g}$ is an injective additive homomorphism of $\mathfrak{o}$ *into* Hom(F_g, F_f), *satisfying the composition rule*

$$a_{f,g} \circ b_{g,h} = [ab]_{f,h}$$

and

$$\begin{aligned}[a + b]_{f,g}(X) &= F_f(a_{f,g}(X), b_{f,g}(X)) \\ &= a_{f,g}(X) +_{F_f} b_{f,g}(X).\end{aligned}$$

Proof. In each case, one checks immediately that both the left-hand side and right-hand side of the desired identity are solutions of the type given in the Lemma, whose solution is unique.

It is clear that if $f, g \in \mathscr{F}_\pi$ then the element $1_{f,g}$ is an isomorphism between F_g and F_f. Thus the isomorphism class of F_f is uniquely determined by π.

Furthermore, from Theorem 1.2, we may also view F_f as an $\mathfrak{o}$-module via the operation a_f for $a \in \mathfrak{o}$, and the above isomorphism is obviously an $\mathfrak{o}$-isomorphism.

As a matter of notation, we shall use the three notations

$$\boxed{a_f,\ a_A \quad \text{or} \quad [a]}$$

to denote the same power series. After a while, the polynomial f in $\mathscr{F}_\pi$ becomes mostly irrelevant, and we think in terms of the group law A. Thus a_A or $[a]$ when A is fixed become more satisfying to work with.

Let L be the completion of an algebraic extension of K. Then we may view $A(\mathfrak{m}_L)$ as an $\mathfrak{o}$-module in the obvious way. The operation of $\mathfrak{o}$ on $A(\mathfrak{m}_L)$ is given by

$$x \mapsto a_f(x) \quad \text{for } x \in \mathfrak{m}_L.$$

Of course, if L is finite over K, then L is equal to its own completion. By functoriality, we also see that the formal isomorphisms $1_{f,g}$ induce isomorphisms

$$A(g)(\mathfrak{m}_L) \xrightarrow{\approx} A(f)(\mathfrak{m}_L).$$

In view of this isomorphism, it is often convenient to omit f or g from the notation and write $[a](x)$ for the operation of $\mathfrak{o}$ on $A(\mathfrak{m}_L)$ for x in $\mathfrak{m}_L$.

Let L be a Galois extension, with Galois group G over K. The operations of elements of G on L extend to the completion by continuity, so we may replace L by its completion. Since the power series $a_{f,g}$, $a_{f,f}$, F_f have coefficients in $\mathfrak{o}$, it is then clear that the operations which they define on $\mathfrak{m}_L$ commute with the action of G on $\mathfrak{m}_L$.

§2. Formal p-adic Multiplication

Again we let $\mathfrak{o}_K$ be a discrete valuation ring with quotient field K, which we assume complete. We let $\mathfrak{m}_K$ be the maximal ideal, and let $q = \text{card } \mathfrak{o}_K/\mathfrak{m}_K$ be finite, a power of the prime p. We let f be a Frobenius power series over $\mathfrak{o}$, associated with the prime element π in $\mathfrak{m}_K$, and we let F_f or $A = A(f)$ be the corresponding Lubin–Tate group.

For each $a \in \mathfrak{o}$, we let $A_a(f)$ be the set of elements x in the maximal ideal $\mathfrak{m}_{K^a}$ of the algebraic closure of K such that

$$a_f(x) = 0.$$

In other words, A_a is the kernel of $[a]$. If a is a unit, then $\text{Ker } a_f = 0$, so we are really concerned with A_{π^n} for positive integers n. Of course, A_a depends on f so we should write $A_a(f)$. However, if g is another Frobenius power series in $\mathfrak{o}$ associated with the same π, then the isomorphism $1_{f,g}$ induces an isomorphism

$$A_a(g) \xrightarrow{\approx} A_a(f),$$

which commutes with Galois isomorphisms. Further, if σ is an automorphism of K^a over K, then

$$a_f(\sigma x) = \sigma(a_f(x)),$$

so $A_a(f)$ is a Galois module, and the extension

$$K(A_a(f)) \text{ over } K$$

is independent of the choice of f in $\mathscr{F}_\pi$. We shall see in a moment that it is a separable extension, whence it is a Galois extension, and is finite for $a \neq 0$ because a non-zero power series has only a finite number of zeros (Weierstrass preparation, or more naively, use the power series $X^q + \pi X$ in $\mathscr{F}_\pi$).

Consider the case $n = 1$, so consider $K(A_\pi)$. Then

$$K(A_\pi) = K(x)$$

where x is a root of $X^q + \pi X = 0$, $x \neq 0$, or in other words, x is a root of

$$X^{q-1} + \pi = 0.$$

Thus $K(A_\pi)$ is a Kummer extension (since the $(q-1)$th roots of unity are in K), with abelian Galois group, cyclic of order $q-1$, and totally ramified over π.

Let $x \in A_\pi$ and $x \neq 0$. The map

$$a \mapsto a_A(x)$$

gives a homomorphism of $\mathfrak{o}$ into A_π, whose kernel is obviously $\pi\mathfrak{o}$. Since A_π has q elements, it follows that

$$A_\pi \approx \mathfrak{o}/\pi\mathfrak{o}$$

as $\mathfrak{o}$-module. In particular, $\mathrm{End}_\mathfrak{o}\, A_\pi \approx \mathfrak{o}/\pi\mathfrak{o}$, and

$$\mathrm{Aut}_\mathfrak{o}\, A_\pi \approx (\mathfrak{o}/\pi\mathfrak{o})^*.$$

We have a representation

$$\varkappa\colon G_0 = \mathrm{Gal}(K(A_\pi)/K) \to \mathrm{Aut}_\mathfrak{o}\, A_\pi \approx (\mathfrak{o}/\pi\mathfrak{o})^*.$$

Since G_0 and $(\mathfrak{o}/\pi\mathfrak{o})^*$ have the same cardinality, namely $q-1$, it follows that this representation is an isomorphism.

We have similar results in the π^n-tower.

Theorem 2.1. (i) *The group A_{π^n} is a free* 1-*dimensional module over $\mathfrak{o}/\pi^n\mathfrak{o}$.*

(ii) *$K(A_{\pi^n})$ is abelian over K, totally ramified, and we have a natural isomorphism*

$$\varkappa\colon \mathrm{Gal}(K(A_{\pi^n})/K) \approx (\mathfrak{o}/\pi^n\mathfrak{o})^*.$$

Proof. Let $(x_1, x_2, \ldots, x_n)$ be a sequence with $x_k \in A_{\pi^k}$, such that $x_1 \neq 0$ and $\pi_f(x_k) = x_{k-1}$. Without loss of generality we may assume

$$f(X) = X^q + \pi X.$$

For $k > 1$ we see that x_k is a root of

$$X^q + \pi X - x_{k-1} = 0.$$

Relatively to the field $K(A_{\pi^{n-1}})$ this is an Eisenstein equation, and so we have shown inductively that $K(A_{\pi^n})$ is totally ramified. Since A_{π^n} is stable under the Galois action, and since the equation

$$X^q + \pi X - x_{k-1} = 0$$

is separable, it follows that $K(A_{\pi^n})/K$ is Galois. As before, we get a representation of the Galois group in $\mathrm{Aut}_\mathfrak{o}\, A_{\pi^n}$. The map

$$a \mapsto a_f(x_n)$$

induces an injection of $\mathfrak{o}/\pi^n\mathfrak{o}$ into A_{π^n}, whence an $\mathfrak{o}$-isomorphism by counting cardinalities, and it follows as for $n = 1$ that we have an isomorphism as in (ii), thus proving the theorem.

Passing to the limit, we may form the projective limit $T_\pi(A)$, consisting of all infinite vectors

$$(x_0, x_1, \dots)$$

such that $\pi_f(x_n) = x_{n-1}$ and $\pi_f(x_0) = 0$. It is then immediate that $T_\pi(A)$ is a free 1-dimensional module over $\mathfrak{o}$.

Let

$$K_n = K(A_{\pi^{n+1}}), \qquad K_\infty = \bigcup K_n.$$

Then K_∞ is an abelian, totally ramified extension of K, and

$$\varkappa\colon \mathrm{Gal}(K_\infty/K) \approx \mathfrak{o}^*$$

in the natural way. If u is a unit in $\mathfrak{o}^*$, then we have a corresponding element of the Galois group, denoted by σ_u, which is such that

$$\sigma_u = [u]_f^{-1}$$

in the representation on $T_\pi(A(f))$. If we wish to omit the reference to f, we simply write $[u]$. Thus on a vector as above, we have

$$\sigma_u^{-1}(x_0, x_1, \dots) = ([u](x_0), [u](x_1), \dots).$$

It is also convenient to have a notation for the representation of the Galois group in $\mathfrak{o}^*$. We let

$$\varkappa\colon G_K \to \mathfrak{o}^*$$

be this representation, where $G_K = \mathrm{Gal}(K^a/K)$, such that

$$\boxed{\sigma x = [\varkappa(\sigma)]_f(x) \quad \text{for} \quad x \in A^{(\pi)}.}$$

Example. We shall now give the standard example with the **formal multiplicative group**

$$F(X, Y) = X + Y + XY.$$

Over the p-adic integers $\mathfrak{o} = \mathbf{Z}_p$ we have the Frobenius series given by

$$f(X) = (1 + X)^p - 1 = X^p + \cdots + pX$$

associated with the prime p. Let A be the corresponding Lubin–Tate formal group. Then in fact A is defined by the power series

$$F_f(X, Y) = F(X, Y),$$

i.e., A is the formal multiplicative group. Then A_{p^n} consists of those elements in the maximal ideal of the algebraic closure satisfying the equation

$$(1 + X)^{p^n} - 1 = 0$$

and these elements are none other than

$$\zeta - 1,$$

where ζ is a p^nth root of unity.

Theorem 2.2. *The prime π is a norm from every extension $K(A_{\pi^n})$.*

Proof. Consider first the bottom level of the tower $K(A_\pi)$ over K, obtained from the equation

$$X^{q-1} + \pi = 0.$$

Let α be a root. Then

$$(-1)^{q-1}N(\alpha) = \pi.$$

If q is odd then π is the norm of α. If q is even then $q - 1$ is odd, the degree $[K(A_\pi) : K]$ is odd, and -1 itself is a norm. Hence π is the norm of $-\alpha$. This proves the theorem in case $n = 1$. For the proof in general, let $\pi = x_0$, and let

$$(x_0, x_1, \ldots)$$

be such that

$$\pi_f(x_n) = x_{n-1}.$$

Thus x_1 is an element of A_π, $x_1 \neq 0$, and x_{n-1} is a norm of $\pm x_n$ from the field $K(x_n)$ over $K(x_{n-1})$. The argument is similar and equally trivial, as desired.

Theorem 2.3. *Let B be the special Lubin–Tate group associated with the prime π and the Frobenius polynomial $X^q + \pi X$. Let $\zeta \in \mu_{q-1}$. Then:*

(i) $$[\zeta](X) = \zeta X.$$

(ii) *If $F(X, Y)$ is the group law for B, and*

$$F(X, Y) = X + Y + \sum a_{ij}X^iY^j,$$

then $a_{ij} = 0$ unless $i + j \equiv 1 \pmod{p-1}$.

Proof. Let x_n be a generator of $B_{\pi^{n+1}}$ such that

$$[\pi](x_n) = x_{n-1}.$$

Since x_0 is a root of $X^{q-1} + \pi = 0$ it follows by a trivial recursion that the irreducible polynomial for x_n over K is a polynomial in X^{q-1}. Therefore we can find an automorphism

$$\sigma_n \in \mathrm{Gal}(K_n/K)$$

such that $\sigma_n x_n = \zeta x_n$. Since elements of the Galois group commute with $[\pi]$, there exists an element $\sigma \in \mathrm{Gal}(K_\infty/K)$ such that

$$\sigma x_n = \zeta x_n \quad \text{for all } n.$$

By Theorem 2.1 there exists $a \in \mathfrak{o}_K^*$ such that $\sigma x_n = [a](x_n)$ for all n. Since $\sigma^{q-1} = 1$, it follows that a is a $(q-1)$th root of unity. But also

$$\zeta x_n = [a]x_n \equiv a x_n \bmod x_n^2$$

so $\zeta - a \equiv 0 \bmod x_n$. This is impossible since both ζ, a are in μ_{q-1}, unless $\zeta = a$, thereby proving (i).

Secondly, for every $\zeta \in \mu_{q-1}$ we have

$$\zeta F(X, Y) = F(\zeta X, \zeta Y) = \zeta X + \zeta Y + \sum a_{ij}\zeta^{i+j}X^iY^j.$$

Then (ii) follows immediately.

§3. Changing the Prime

We shall now analyze what happens when going from one prime π to another prime $\pi' = \pi u$ where u is a unit of $\mathfrak{o}_K$. Since we have to refer to the primes, we let

$$K^{(\pi)} = K_\infty^{(\pi)} = \bigcup K(A_{\pi^n}).$$

We let A' be the Lubin–Tate group associated with π', so

$$K^{(\pi')} = \bigcup K(A'_{\pi'^n}).$$

We let L be the completion of the maximal unramified extension of K, with ring of integers $\mathfrak{o}_L$. We let K_{nr} be the maximal unramified extension of K, and $\mathfrak{o}_{\mathrm{nr}}$ its valuation ring, with maximal ideal $\mathfrak{m}_{\mathrm{nr}}$. We let φ be the Frobenius automorphism of K_{nr}, extended by continuity to L.

Theorem 3.1. *Let f, f' be Frobenius power series over $\mathfrak{o}$, associated with the primes π, π' respectively. Let ε be a unit of $\mathfrak{o}_L$ such that $\varepsilon^\varphi/\varepsilon = u$. (Such units exist.) Then there exists a unique isomorphism*

$$\theta\colon F_f \to F_{f'}$$

defined over $\mathfrak{o}_L$ which commutes with the operation of $\mathfrak{o}$, that is for all a in $\mathfrak{o}$.

$$\theta \circ a_f = a_{f'} \circ \theta,$$

and such that

$$\theta(X) \equiv \varepsilon X \bmod \deg 2.$$

This power series θ satisfies

$$\theta^\varphi = \theta \circ u_f.$$

Proof. The existence of the unit ε such that $\varepsilon^\varphi/\varepsilon = u$ is easily obtained by a recursive procedure, and is left to the reader. We then construct a power series $\theta(X) = \varepsilon X + \cdots$ to satisfy $\theta^\varphi = \theta \circ u_f$ as follows. Let $\theta_1(X) = \varepsilon X$. Suppose we have found $\theta_r(X)$ of degree r such that

$$\theta_r^\varphi(X) \equiv \theta_r(u_f(X)) \bmod \deg r + 1.$$

We wish to find some element $b \in \mathfrak{o}_L$ such that the series

$$\theta_{r+1}(X) = \theta_r(X) + bX^{r+1}$$

satisfies the same congruence to one higher degree. We have:

$$\theta_{r+1}^\varphi(X) = \theta_r^\varphi(X) + b^\varphi X^{r+1}$$
$$\theta_{r+1} \circ u_f(X) = \theta_r \circ u_f(X) + bu_f(X)^{r+1}.$$

The condition on b is therefore that

$$b^\varphi - bu^{r+1} = c,$$

where c is the $(r+1)$th coefficient of $\theta_r \circ u_f(X) - \theta_r^\varphi(X)$. Write $b = a\varepsilon^{r+1}$. The condition on b is equivalent with the condition

$$a^\varphi - a = c/\varepsilon^{\varphi(r+1)}.$$

A recursive procedure, letting $a_{n+1} = a_n + x\pi^{n+1}$ shows that we can solve for x at each step to make the equation valid mod π^n, whence solve the equation for a in $\mathfrak{o}_L$. This concludes the construction of θ.

We shall now see that θ almost satisfies the other conditions of the theorem, and that it is easy to adjust it to get these other conditions exactly. Let

$$g = \theta \circ \pi_f' \circ \theta^{-1}.$$

It is obvious that $\theta \circ F_f \circ \theta^{-1}$ commutes with $\theta \circ \pi_f' \circ \theta^{-1} = g$. We contend that g is a Frobenius power series associated with π', and has coefficients in $\mathfrak{o} = \mathfrak{o}_K$. Once we have proved this contention, we then conclude that the power series

$$\theta(F_f(\theta^{-1}(X), \theta^{-1}(Y))$$

has the properties characterizing $F_g(X, Y)$ (it is obvious that this power series is congruent to $X + Y$ mod deg 2). The Lemma and addendum to Theorem 1.1 show that the power series is equal to $F_g(X, Y)$. Similarly, we verify that $\theta \circ a_f \circ \theta^{-1}$ has the properties which characterize a_g, so is equal to a_g. In this manner, we have proved the theorem except for the fact that

$$\theta\colon F_f \to F_g$$

is an isomorphism from F_f to F_g. Replacing θ by $1_{f',g} \circ \theta$ then concludes the proof.

All that remains to be done is settle the contention. We have:

$$\theta\pi_{f'}\theta^{-1}(X) \equiv \varepsilon\pi'\varepsilon^{-1}X \equiv \pi'X \text{ mod deg } 2.$$

Also,

$$\begin{aligned}\theta\pi_f'\theta^{-1}(X) = \theta u_f\pi_f\theta^{-1}(X) &= \theta^{\varphi}(f(\theta^{-1}(X))) \\ &\equiv \theta^{\varphi}(\theta^{-1}(X)^q) \text{ mod } \pi \\ &\equiv \theta^{\varphi}(\theta^{-\varphi}(X^q)) \text{ mod } \pi \\ &\equiv X^q \qquad\qquad \text{mod } \pi.\end{aligned}$$

There remains only to prove that the coefficients of $\theta\pi_f'\theta^{-1}$ lie in $\mathfrak{o}$, and it suffices to show that they are fixed under the Frobenius automorphism φ. We have:

$$(\theta\pi_f'\theta^{-1})^{\varphi} = \theta^{\varphi}fu_f\theta^{-\varphi} = \theta^{\varphi}f\theta^{-1} = \theta u_f f\theta^{-1} = \theta\pi_f'\theta^{-1},$$

which concludes the proof of the theorem.

§4. The Reciprocity Law

Let K_{nr} as before be the maximal unramified abelian extension of K. Local class field theory would immediately show that the composite field

$$K^{(\pi)}K_{\text{nr}}$$

is the maximal abelian extension of K. On $K^{(\pi)}$ we have a good model for the Galois group given by the association

$$a \mapsto [a], \quad a \in \mathfrak{o}^*,$$

and on K_{nr} we have the Frobenius automorphism, which generates the Galois group. We wish now to give an independent proof that the field $K^{(\pi)}K_{\text{nr}}$ is independent of the choice of π, and that the structure of the Galois group is in fact determined independently of that choice also.

Theorem 4.1. *The field $K^{(\pi)}K_{\text{nr}}$ is independent of π. Let $a \in K^*$. Write*

$$a = u\pi^m$$

for some unit u, and some integer m. Let $r_\pi(a)$ be the automorphism of $K^{(\pi)}K_{\text{nr}}$ such that:

$$r_\pi(a) = \sigma_u \text{ on } K^{(\pi)}$$

$$r_\pi(a) = \varphi^m \ (\varphi = \textit{Frobenius}) \text{ on } K_{\text{nr}}.$$

Then the association $a \mapsto r_\pi(a)$ is independent of the choice of π.

Proof. Let L be the completion of K_{nr} as in the preceding section. Let A be the Lubin–Tate formal groups associated with the prime π, and let A' be associated with the prime π'. Since A and A' are isomorphic over L by Theorem 3.1, it is clear that

$$LK^{(\pi)} = LK^{(\pi')}.$$

However, K_{nr} is algebraically closed in L. The two totally ramified extensions

$$K^{(\pi)}K_{\text{nr}} \quad \text{and} \quad K^{(\pi')}K_{\text{nr}}$$

become equal when lifted to L. By elementary field theory, they must be equal as extensions of K_{nr}, thus proving the first assertion in the theorem.

The set of prime elements π' generates the multiplicative group K^*. To prove the independence of $r_\pi(a)$ from the choice of π, it will therefore suffice to prove that for all π',

$$r_\pi(\pi') = r_{\pi'}(\pi').$$

These automorphisms coincide on K_{nr} since they both give rise to the Frobenius element. It will therefore suffice to prove that they coincide on $K^{(\pi')}$. We keep the notation of Theorem 3.1. Write $\pi' = u\pi$ with some unit u. Since $r_{\pi'}(\pi') =$ identity on $K^{(\pi')}$, we are reduced to showing this same property for $r_\pi(\pi')$.

Let f be a Frobenius power series associated with π. The field $K^{(\pi')}$ is generated by the elements $\theta(x)$ with $x \in A^{(\pi)}(f)$. Hence we are reduced to showing that such elements are fixed by $r_\pi(\pi')$. Indeed:

$$\begin{aligned} r_\pi(\pi')\theta(x) &= r_\pi(u)r_\pi(\pi)\theta(x) \\ &= r_\pi(u)\theta^{\varphi}(x) \\ &= \theta^{\varphi}(u_f{}^{-1}(x)) \\ &= \theta(x). \end{aligned}$$

This concludes the proof of the theorem.

One may now use local class field theory to guarantee that $K^{(\pi)}K_{\mathrm{nr}}$ is the maximal abelian extension of K. Let

$$(a, K_{\mathrm{ab}}/K)$$

be the norm residue symbol mapping K^* into $\mathrm{Gal}(K_{\mathrm{ab}}/K)$ from local class field theory. Then we find

$$r(a) = (a, K_{\mathrm{ab}}/K).$$

Indeed, both automorphisms induce the Frobenius automorphism on K_{nr}, and for any prime element π, both automorphisms induce the identity on $K^{(\pi)}$, since π is a norm from every finite subextension of $K^{(\pi)}$ by Theorem 2.2. Since $r(a)$ and $(a, K_{\mathrm{ab}}/K)$ coincide on all prime elements, they coincide on K^*.

§5. The Kummer Pairing

It should be clear that the formalism of formal groups is completely analogous to the classical formalism on the multiplicative group or the group of Witt vectors. In a similar way, one can develop "Kummer theory" completely analogously to the standard way (cf. for instance *Algebra*, Chapter VIII, §8), or the way Witt did it in characteristic p for his vectors (Crelle 1935–1936). The possibility of doing this in the context of Lubin–Tate groups was first noted by Frohlich [Fr]. Of course, some new phenomena arise. Applications to explicit reciprocity laws as in Coates–Wiles [C–W 2] and Wiles [Wi] will be postponed to a later chapter.

Let A be a Lubin–Tate group associated with the prime π. We let

$$K_n = K(A_{\pi^{n+1}})$$

so $K_0 = K(A_\pi)$. We let $\mathfrak{o}_n$ be the ring of integral elements in K_n and let $\mathfrak{p}_n$ be its maximal ideal. We write $A(\mathfrak{p}_n)$ as usual for the set of elements $\mathfrak{p}_n$ with the group law defined by A. We define a pairing

$$A(\mathfrak{p}_n) \times K_n^* \to A_{\pi^{n+1}}.$$

Let $x \in A(\mathfrak{p}_n)$, let t be an element of $A(\mathfrak{p}_\infty)$ such that $[\pi^{n+1}]t = x$, and let $\alpha \in K_n^*$. Note that actually $t \in A(\mathfrak{p}^{2n+1})$. Define the symbol

$$\langle x, \alpha \rangle_n = \sigma_\alpha t -_A t,$$

where $\sigma_\alpha = (\alpha, K_n^{ab}/K_n)$ is the automorphism of K_n^{ab} over K_n arising from local class field theory. Then it is clear that $\langle x, \alpha \rangle$ lies in $A_{\pi^{n+1}}$ and is independent of the choice of t such that $[\pi^{n+1}]t = x$. We call it the **local (Kummer) symbol** (relative to the formal group A and the multiplicative group). If we want to specify A in the notation we write

$$\langle x, \alpha \rangle_n^A.$$

Example. The formal multiplicative group. For $\beta \equiv 1 \bmod \mathfrak{p}_n$ and $\alpha \in K_n^*$ we define the classical **norm residue symbol**

$$(\beta, \alpha)_n = \langle \beta - 1, \alpha \rangle_n^A + 1,$$

where A is the formal multiplicative group.

The local symbol trivially satisfies the following properties.

LS 1. *It is $\mathfrak{o}_K$-linear in x and multiplicative in α.*

In particular, the symbol induces a pairing

$$A(\mathfrak{p}_n)/[\pi^{n+1}]A(\mathfrak{p}_n) \times K_n^*/K_n^{*p^{n+1}} \to A_{\pi^{n+1}},$$

and it is clear that in the pairing

$$A(\mathfrak{p}_n) \times K_n^* \to A_{\pi^{n+1}}$$

the kernel on the left is exactly $[\pi^{n+1}]A(\mathfrak{p}_n)$, because if x is not a $[\pi^{n+1}]$-multiple in $A(\mathfrak{p}_n)$, then its $[\pi^{n+1}]$ root t generates a proper extension of K_n, so the Galois group operates non-trivially.

LS 2. *If $\theta: A \to A'$ is an isomorphism over $\mathfrak{o}$ between two Lubin–Tate groups associated with the same prime π, then*

$$\langle x, \alpha \rangle_n^{A'} = \theta(\langle \theta^{-1}(x), \alpha \rangle_n^A).$$

LS 3. *If σ is an automorphism of K^{ab} over K then*

$$\langle \sigma x, \sigma\alpha \rangle_n = \sigma \langle x, \alpha \rangle_n.$$

LS 4. $\langle x, \alpha \rangle_n = 0$ *if and only if α is a norm from $K_n(t)$, where $[\pi^{n+1}]t = x$.*

This last property uses a fact from local class field theory which could be proved from the Lubin–Tate formalism, but which we shall take for granted. Otherwise, the other properties are obvious.

As an application of **LS 4**, let $N_{m,n}$ denote the norm from K_m to K_n for $m \geq n$. Then the orthogonal complement of $A_{\pi^{n+1}}$ under the pairing is given by

$$A_{\pi^{n+1}}^{\perp} = N_{2n+1,n} K_{2n+1}^{*}.$$

LS 5. *Let $m \geq n$ and let $N_{m,n}$ denote the norm from K_m to K_n. Then for $x \in A(\mathfrak{p}_n)$ and $\alpha \in K_m^*$ we have*

$$[\pi^{m-n}]\langle x, \alpha \rangle_m = \langle x, N_{m,n}(\alpha) \rangle_n.$$

In other words, $N_{m,n}$ is the transpose of $[\pi^{m-n}]$.

Again this is clear from the functorial properties of the norm residue symbol which we assume. We can then define the symbol in a limit situation as follows. We let

$T(K_\infty^*)$ = group of sequences $(\alpha_0, \alpha_1, \ldots)$ with $\alpha_n \in K_n^*$ such that for all $n \geq 0$,

$$N_{n+1,n}\alpha_{n+1} = \alpha_n.$$

Thus $T(K_\infty^*)$ is just the projective limit of the groups K_∞^n under the norm mappings. We may then define a pairing

$$A(\mathfrak{p}_\infty) \times T(K_\infty^*) \to T_\pi(A)$$

by letting

$$x \times (\alpha_0, \alpha_1, \ldots) \mapsto (\langle x, \alpha_0 \rangle_0, \langle x, \alpha_1 \rangle_1, \ldots).$$

On the right-hand side, the components $\langle x, \alpha_m \rangle_m$ are defined for all m sufficiently large, i.e., $m \geq n$ such that $x \in A(\mathfrak{p}_n)$. The components $\langle x, \alpha_k \rangle_k$ for $k < n$ may then be defined by applying the appropriate power of $[\pi]$ to the nth component. Property **LS 5** shows that this is well defined. The pairing is $\mathfrak{o}_K$-linear in x and multiplicative in $\alpha \in T(K_\infty^*)$.

Of course it is a considerable restriction on an element α_n in K_n^* to be liftable to an infinite vector of consecutive norms. In fact, let N_n be the norm from K_n to K, and let

$$N_n^{-1}(\pi^{\mathbf{Z}}) = \text{group of elements } \alpha \in K_n^* \text{ whose norm } N_n\alpha \text{ is a power of } \pi.$$

Lemma 1. *We have* $N_n^{-1}(\pi^{\mathbf{Z}}) = \bigcap_{m \geq n} N_{m,n}K_m^*$.

Proof. Suppose $N_n\alpha \in \pi^{\mathbf{Z}}$. Then

$$(\alpha, K_m/K_n) = (N_n\alpha, K_m/K) \in (\pi^{\mathbf{Z}}, K_m/K) = 1$$

because π is a norm from each extension K_m by Theorem 2.2. Hence α is a norm from K_m, thus proving one inclusion.

Conversely, suppose that α is a norm from each K_m for $m \geq n$. Then

$$1 = (\alpha, K_m/K_n) = (N_n\alpha, K_m/K).$$

Let $N_n\alpha = \pi^r u$ where u is a unit in K. Since π is a norm from K_m by Theorem 2.2, we conclude that

$$(u, K_m/K) = 1 \quad \text{for all } m.$$

Hence $u \equiv 1 \bmod \mathfrak{p}^m$ for all m, so $u = 1$. This proves the reverse inclusion, and proves the lemma.

LS 6. *Assume that* $[\pi](X)$ *is a polynomial. If p is odd then*

(i) $$\langle \alpha, \alpha \rangle_n = 0.$$

Whether p is odd or even, we have

(ii) $$\langle \alpha, -\alpha \rangle_n = 0,$$

(iii) $$\langle x, -1 \rangle_n = 0 \quad \textit{if } \operatorname{ord}_\pi x \geq e + 1$$

and so

(iv) $$\langle \alpha, \alpha \rangle_n = 0 \quad \textit{if } \operatorname{ord}_\pi \alpha \geq e + 1.$$

Proof. Let

$$[\pi^{n+1}](X) = f_n(X).$$

Let t be a root of $f_n(X) = \alpha$. The extension $K_n(t)$ is independent of the choice of t. Thus, if we factor $f_n(X) - \alpha$ into irreducible polynomials over K_n, say

$$f_n(X) - \alpha = \prod_{j=1}^{s} f_{n,j}(X),$$

then for each j we obtain $K_n(t)$ by adjoining a single root of $f_{n,j}(X)$ to K_n. Therefore, if $c_{n,j}$ denotes the constant term of $f_{n,j}(X)$, and d is the common degree of the irreducible polynomials $f_{n,j}$ then we conclude that $(-1)^d c_{n,j}$ is a norm. But

$$-\alpha = \prod_{j=1}^{s} c_{n,j}.$$

Hence $-\alpha = (-1)^{ds}$ times a norm, and $(-1)^{ds} = (-1)^{q^n+1}$ is a norm. This proves the first two assertions. For the last two (relevant only for $p = 2$), it is clear from bilinearity that

$$[2]\langle x, -1\rangle = 0.$$

If $\operatorname{ord}_\pi \alpha \geq e + 1$, then $\alpha = [2]y$ for some y, and so $\langle \alpha, -1\rangle = 0$, thus proving the last two assertions.

The following lemma gives information on the factor group

$$A(\mathfrak{p}_n)/[\pi^{n+1}]A(\mathfrak{p}_n),$$

by showing that near the origin, the operator $[\pi]$ operates very regularly. We let π_n denote a prime element in K_n.

Lemma 2. *Assume $k > q^n$. Then $A(\pi\pi_n^k\mathfrak{o}_n) = [\pi]A(\pi_n^k\mathfrak{o}_n)$.*

Proof. The inclusion $\supset$ is obvious. We prove the reverse inclusion. Let $z = \pi\pi_n^k t$ with $t \in \mathfrak{o}_n$. We must solve

$$x^q + \pi x = z \quad \text{with} \quad x = \pi_n^k y \quad \text{and} \quad y \in \mathfrak{o}_n.$$

This is equivalent to

$$\pi_n^{qk} y^q + \pi_n^{q^n(q-1)+k} y = \pi\pi_n^k t.$$

But $k > q^n$ implies that $qk > q^n(q - 1) + k$, so we are reduced to solving

$$f(y) = ay^q + y - t = 0$$

with a divisible by π_n. Since $f(t) \equiv 0 \bmod \pi_n$, and $f'(t) \equiv 1$, Hensel's lemma does it.

As consequences of the lemma, we find for instance:

$$A(\pi^{n+2}\mathfrak{o}_n) \subset [\pi^{n+1}]A(\mathfrak{p}_n)$$
$$A(\pi_n^{q^{n+1}+1}\mathfrak{o}_n) \subset [\pi]A(\mathfrak{p}_n).$$

This second inclusion comes from using $k = q^n + 1$.

Theorem 5.1. *Let w_i be elements of $\mathfrak{p}_n$ for $i = 1, \ldots, q^{n+1}$, such that $\operatorname{ord}_{\mathfrak{p}_n} w_i = i$. Then these elements generate $A(\mathfrak{p}_n)$ mod $[\pi]A(\mathfrak{p}_n)$, and therefore generate $A(\mathfrak{p}_n)$ mod$[\pi^{n+1}]A(\mathfrak{p}_n)$ over $\mathfrak{o}$.*

Proof. Since $X +_A Y \equiv X + Y \bmod \deg 2$, given $x \in \mathfrak{p}_n$ we can find $a_1 \in \mathfrak{o}$ such that

$$x -_A [a_1]w_1 \equiv 0 \bmod \mathfrak{p}_n^2,$$

because $[a_1]w_1 \equiv a_1 w_1 \bmod \mathfrak{p}_n^2$. We may then proceed recursively to find $a_2, \ldots, a_{q^{n+1}}$ such that

$$x \equiv [a_1]w_1 +_A [a_2]w_2 +_A \cdots +_A [a_{q^{n+1}}]w_{q^{n+1}} \bmod \mathfrak{p}_n^{q^{n+1}+1}.$$

By the lemma, this congruence also holds mod$[\pi]A(\mathfrak{p}_n)$. Hence the w_i generate $A(\mathfrak{p}_n)$ mod$[\pi]A(\mathfrak{p}_n)$, whence by Nakayama's lemma, they also generate $A(\mathfrak{p}_n)$ mod$[\pi^{n+1}]A(\mathfrak{p}_n)$. This proves the theorem.

The special case when $K = \mathbf{Q}_p$ is of importance in the cyclotomic theory (and elsewhere), and some refined statements can be given as in the next two theorems due to Coates–Wiles [C–W 1], [C–W 2].

Theorem 5.2. *Assume $K = \mathbf{Q}_p$. Then $A(\mathfrak{p}_n)/[\pi^{n+1}]A(\mathfrak{p}_n)$ is free over $\mathfrak{o}/\pi^{n+1}\mathfrak{o}$. Suppose that A is the basic Lubin–Tate group. Let I be the set of integers i satisfying*

$$1 \le i < p^{n+1} \quad \text{with} \quad (i, p) = 1, \quad \text{or} \quad i = p^{n+1}.$$

Let x_0 be a non-zero element of A_π and let

$$(x_0, x_1, \ldots, x_n, \ldots)$$

be an element of $T_\pi(A)$, that is $[\pi]x_{k+1} = x_k$. Then the elements

$$\{x_n^i\} \quad \text{with } i \in I$$

form a basis.

Proof. For $i \in I$ we let $w_i = x_n^i$. On the other hand, if p^r divides i exactly, we take

$$w_i = [\pi^r](x_n^{i/p^r}).$$

This shows that the elements x_n^i with $i \in I$ generate $A(\mathfrak{p}_n) \bmod [\pi^{n+1}]A(\mathfrak{p}_n)$, over $\mathfrak{o}$.

There remains to prove that they are linearly independent mod $\pi^{n+1}\mathfrak{o}$. We first show that they are linearly independent in $A(\mathfrak{p}_n)/[\pi]A(\mathfrak{p}_n)$ over $\mathfrak{o}/\pi\mathfrak{o}$. Suppose we have a relation

$$(*) \qquad \sum\nolimits_A [a_i]x_n^i \equiv 0 \bmod [\pi]A(\mathfrak{p}_n),$$

where $\sum_A$ indicates the sum with respect to the group law of A, and some coefficient a_i is a unit, say a_k. We may assume $a_k = 1$.

Case 1. $k < p^{n+1}$. For any $x \in \mathfrak{p}_n$ we know that

$$[\pi]x = x^p + \cdots + px.$$

Either the term x^p dominates this expression, in which case $\operatorname{ord}_{\mathfrak{p}_n} x$ is divisible by p, or some other term dominates, which means that

$$p \cdot \operatorname{ord} x > p^n(p-1) + \operatorname{ord} x,$$

so $\operatorname{ord} x > p^n$, and $\operatorname{ord} [\pi]x > p^{n+1}$. This implies that we cannot have a relation of congruence (*) because as with ordinary addition, if $y, y' \in A(\mathfrak{p}_n)$ and $\operatorname{ord} y \neq \operatorname{ord} y'$ then

$$\operatorname{ord}(y +_A y') = \min(\operatorname{ord} y, \operatorname{ord} y').$$

Hence there cannot be any cancellation in the sum of the left-hand side of (*), thus concluding the proof in Case 1.

Case 2. $k = p^{n+1}$. Then by Case 1 we may suppose that a_i is divisible by π for all $i \neq k$, and therefore $x_n^{p^{n+1}}$ lies in $[\pi]A(\mathfrak{p}_n)$. We use the hypothesis that A is the basic Lubin–Tate group, and then there exists $y \in \mathfrak{p}_n$ such that

$$y^p + \pi y = x_n^{p^{n+1}}.$$

But $x_n^{p^n} \sim x_0$ and $x_0^{p-1} = -\pi$. The above equation is clearly impossible if y is not divisible by x_0 because the orders on the left-hand side cannot match the order on the right-hand side. Then we divide by $x_0{}^p$ to find

$$(y/x_0)^p - (y/x_0) = x_n^{p^{n+1}}/x_p^0 = \text{unit}.$$

Reading this equation mod $\mathfrak{p}_n$ yields a solution of

$$Y^p - Y \equiv \text{unit} \bmod p,$$

in the residue class field $\mathbf{Z}/p\mathbf{Z}$, which is impossible. This proves the theorem.

Theorem 5.3. *Assume that $K = \mathbf{Q}_p$, and that $K(A_\pi)$ does not contain the* pth *roots of unity. Then the local pairing*

$$A(\mathfrak{p}_n)/[\pi^{n+1}]A(\mathfrak{p}_n) \times K_n^*/K_n^{*p^{n+1}} \to A_{\pi^{n+1}}$$

is exact on both sides, i.e., the kernels are 0 *on both sides.*

Proof. In Theorem 5.2 we have determined the order of

$$A(\mathfrak{p}_n)/[\pi^{n+1}]A(\mathfrak{p}_n).$$

It is a standard exercise of local algebraic number theory [L 1], Chapter II, §3 to determine that

$$\text{order of} \quad K_n^*/K_n^{*p^{n+1}} = (A(\mathfrak{p}_n) : [\pi^{n+1}]A(\mathfrak{p}_n))p^r$$

where p^r is the order of the group of p-power roots of unity in K_n. If $K(A_\pi)$ does not contain μ_p then neither does K_n. Hence

$$K_n^*/K_n^{*p^{n+1}}$$

has the same order as $A(\mathfrak{p}_n)/[\pi^{n+1}]A(\mathfrak{p}_n)$, and we know that the kernel on the left is trivial. Since $A_{\pi^{n+1}}$ is cyclic in the present case, it follows by the duality of finite abelian groups that the kernel on the right of the pairing must also be trivial, as desired.

Remark. When the pth roots of unity are in $K(A_\pi)$, in particular when $A = \mathbf{G}_m$, the above argument definitely shows that the kernel on the right is non-zero.

§6. The Logarithm

Let A be a formal group, defined by a power series $F(X, Y)$ over some ring $\mathfrak{o}$ with quotient field K of characteristic 0. It can be shown that there exists an isomorphism

$$\lambda: A \to \mathbf{G}_a$$

with the additive group, i.e., a power series with coefficients in some extension of K such that

$$\lambda(X +_F Y) = \lambda(X) + \lambda(Y),$$
$$\lambda(X) \equiv X \bmod \deg 2.$$

The $+$ sign on the right-hand side is the ordinary addition. That power series is then uniquely determined, and its coefficients lie in K. It is called the **logarithm** on A, and we write $\lambda = \lambda_A$ if we need to refer to A explicitly.

Example. Suppose $A = \mathbf{G}_m$ is the formal multiplicative group. Then the log is given by

$$\lambda(X) = \log(1 + X),$$

where the log here is the usual series from calculus.

It is easy to show that any endomorphism of $\mathbf{G}_a$ is given by multiplication with a scalar, i.e., if a power series h satisfies

$$h(X + Y) = h(X) + h(Y),$$

then $h(X) = aX$ for some constant a. Hence the uniqueness of the log λ_A follows at once. In this section we shall prove its existence for Lubin–Tate groups, and additional properties, following Wiles [W] in preparation for the explicit reciprocity laws.

Lemma 1. *The limit*

$$\lambda(X) = \lim \frac{1}{\pi^n} \pi_A^n(X)$$

exists, and gives a formal isomorphism of the Lubin–Tate formal group A with the additive group $\mathbf{G}_a$.

Remark. The limit is to be understood in the following sense. Each term

$$\frac{1}{\pi^n} \pi_A^n(X) = \sum c_k^{(n)} X^k$$

is a power series. By the existence of the limit, we mean that for each k,

$$\lim_n c_k^{(n)} = c_k$$

exists as $n \to \infty$, and then $\lambda(X)$ is defined to be $\sum c_k X^k$. The convergence will not be uniform in k.

Proof of the lemma. We look at the difference

$$\frac{1}{\pi^{n+r}} \pi_A^{n+r}(X) - \frac{1}{\pi^n} \pi_A^n(X) = \frac{1}{\pi^{n+r}} (\pi_A^r \circ \pi_A^n(X) - \pi^r \pi_A^n(X)).$$

Let

$$\pi_A^n(X) = \pi^n X + g_n(X)$$

where

$$g_n(X) = \sum a_{ij}^{(n)} \pi^i X^j \quad \text{and} \quad i + j \geq n + 1, \qquad j \geq 2,$$

and the coefficients $a_{ij}^{(n)}$ are in $\mathfrak{o}_K$. Then the right-hand side of the required difference is equal to

$$\frac{1}{\pi^{n+r}} g_r(\pi^n X + g_n(X)).$$

We are interested in the coefficients of monomials of degree $\leq k$ for a fixed k. Reading all expressions mod X^{k+1} we see that we may assume

$$i \geq n + 1 - (k + 1).$$

Hence $\pi^n X + g_n(X)$ is divisible by π^{n-k}. Similarly, the power series expression for g_r is divisible at least by π^{r-k}. Since g_r has degree ≥ 2 it follows that $g_r(\pi^n X + g_n(X))$ is divisible at least by

$$\pi^{r-k}\pi^{2(n-k)} = \pi^{2n+r-3k}.$$

Dividing by π^{n+r} shows that the required difference tends to 0 as $n \to \infty$. This proves that the desired limit exists.

It is clear that $\lambda(X) \equiv X \bmod \deg 2$.

There remains to prove that λ satisfies the homomorphism property. We have:

$$\frac{1}{\pi^n} \pi_A^n(X +_A Y) = \frac{1}{\pi^n} (\pi_A^n X +_A \pi_A^n Y)$$

$$= \frac{1}{\pi^n} (\pi_A^n X + \pi_A^n Y) + \frac{1}{\pi^n} \sum_{i+j \geq 2} c_{ij} \pi_A^n(X)^i \pi_A^n(Y)^j,$$

where c_{ij} are the coefficients of the formal group

$$X +_A Y = \sum c_{ij} X^i Y^j.$$

For each fixed k, m the coefficient of $X^k Y^m$ in the sum on the right-hand side tends to 0 as n tends to infinity, so the additivity follows.

Lemma 2. *The* log λ_A *commutes with the action of* $\mathfrak{o}$, *that is,*

$$\lambda_A(a_A(X)) = a\lambda_A(X) \quad \textit{for } a \in \mathfrak{o}_K.$$

For the basic Lubin–Tate group B, if

$$\lambda_B(X) = X + \sum_{i=2}^{\infty} a_i X^i$$

then $a_i = 0$ *unless* $i \equiv 1 \bmod q - 1$.

Proof. The function $X \mapsto \lambda_A(a_A(X))$ is an additive formal power series. such that

$$\lambda_A(a_A(X)) \equiv aX \bmod \deg 2.$$

The uniqueness of the logarithm shows that this function is $a\lambda_A$.

For the basic Lubin–Tate group, we take $a = \zeta$ where ζ is a primitive $(q-1)$th root of unity, and apply Theorem 2.3 to conclude the proof.

Lemma 3. (i) *Let λ' denote the formal derivative $d\lambda/dX$. Then $\lambda'_A(X)$ has coefficients in $\mathfrak{o}$.*

(ii) *The series $\lambda_A(X)$ can be written in the form*

$$\lambda_A(X) = \sum g_i(X) \frac{X^{q^i}}{\pi^i}$$

where $g_i(X) \in \mathfrak{o}[[X]]$. In particular, it converges on the maximal ideal.

(iii) *Suppose $q \geq 3$ and let $x \in K^a$ have* $\operatorname{ord}_\pi x \geq 1$. *Then*

$$\lambda_A(x) \equiv x \bmod x^2.$$

Proof. For (i) we differentiate with respect to Y the relation

$$\lambda_A(F(X, Y)) = \lambda_A(X) + \lambda_A(Y)$$

and get

$$\lambda'_A(F(X, Y))D_2F(X, Y) = \lambda'_A(Y).$$

We then put $Y = 0$, and find

$$\lambda'_A(X)D_2F(X, 0) = 1.$$

But from $F(X, Y) \equiv X + Y \bmod \deg 2$, it follows that $D_2F(X, 0)$ is a power series whose constant term is 1, and with coefficients in $\mathfrak{o}$. This proves the first assertion.

As for (ii), it suffices to prove the result for the basic Lubin–Tate group whose Frobenius power series is given by

$$X^q + \pi X,$$

because if $\psi: A \to B$ is an isomorphism such that $\psi(X) \equiv X \bmod \deg 2$, then $\lambda_A = \lambda_B \circ \psi$.

It also clearly suffices to prove the following statement:

The power series $[\pi^n](X)$ *lies in the module*

$$\mathfrak{o}[[X]]\pi^j X^k$$

with $j \geq 0$, $k \geq 1$ *and* $j + [\log_q k] \geq n$.

We prove this by induction. It is obvious for $n = 1$. Assume it for n. Let

$$[\pi^n](X) = f_n(X) = \sum g_{jk}(X)\pi^j X^k.$$

Then

$$[\pi^{n+1}](X) = f_n(X)^q + \pi f_n(X).$$

It is immediate that $\pi f_n(X)$ satisfies the induction hypothesis with respect to $n + 1$. For the term $f_n(X)^q$, it will consist of cross terms which binomial-type coefficients divisible by p, hence by π, thus satisfying the desired conditions on the exponents, or terms

$$g_{jk}(X)^q \pi^{jq} X^{kq}.$$

The $\log_q$ of the exponent of X is increased by one, and so the desired inequality is also satisfied. This proves (ii).

Part (iii) is obvious from (ii).

Observe that in the simplest case of the ordinary log,

$$\log(1 + X) = X - \frac{X^2}{2} + \cdots.$$

If $p = 2$, the first term after X gives trouble. If $p = 3$, the next term which might give trouble is

$$\frac{X^3}{3},$$

but in this case, the assumption $\operatorname{ord}_p x \geq 1$ shows that (iii) holds. After that, things only get better.

Lemma 4. *Let* $e_A(Z)$ *be the power series (with coefficients in* K*) which is the inverse of* $\lambda_A(X)$*. Let* D *be the disc in* $\mathfrak{m}_{K^a}$ *consisting of those elements* z *such that*

$$\operatorname{ord}_\pi z > \frac{1}{q - 1}.$$

Then $e_A(Z)$ converges on this disc, and induces the inverse isomorphism to λ_A, on the groups

$$A(D) \underset{e_A}{\overset{\lambda_A}{\rightleftarrows}} \mathbf{G}_a(D).$$

For z in this disc we have

$$\operatorname{ord} z = \operatorname{ord} \lambda_A(z) = \operatorname{ord} e_A(z).$$

Proof. Let $y \in D$, $y \neq 0$. Define

$$\lambda_y(X) = \frac{1}{y} \lambda_A(yX).$$

Then for $i > 0$,

$$\operatorname{ord} \frac{1}{y} \frac{y^{q^i}}{\pi^i} = (q^i - 1) \operatorname{ord} y - i > \frac{q^i - 1}{q - 1} - i.$$

By Lemma 3, it follows that $\lambda_y(X)$ has integral coefficients. Let E_y be the power series such that $E_y \circ \lambda_y(X) = X$. Replacing X with $y^{-1}X$ we see that $E_y(Z) = y^{-1}e_A(yZ)$. Since E_y has integral coefficients (because $\lambda_y(X) = X +$ higher terms), we conclude that

$$\frac{1}{y} e_A(yZ)$$

has integral coefficients. Let $e_A(Z) = \sum a_n Z^n$. Then $a_n y^{n-1}$ is integral for all n and all y in D. It follows that in fact, $a_n y^n$ tends to 0 (p-adically) as n tends to infinity for each y in D. Furthermore, we then conclude that

$$e_A(y) \in y\mathfrak{o}_{K^a}$$

for all y in D, and in particular,

$$\operatorname{ord} e_A(y) \geq \operatorname{ord} y.$$

On the other hand, again using Lemma 4, it is immediate that for x in D, we have

$$\operatorname{ord} \lambda_A(x) \geq \operatorname{ord} x.$$

Since e_A and λ_A give inverse mappings, we get

$$\operatorname{ord} e_A(y) = \operatorname{ord} y = \operatorname{ord} \lambda_A(y),$$

thus proving Lemma 4.

We may then recover immediately a lemma proved in connection with Theorem 5.1.

Corollary. (i) $A(\pi^{n+2}\mathfrak{o}_n) \subset [\pi^{n+1}]A(\mathfrak{p}_n)$.

(ii) $\lambda_A A(\mathfrak{p}_n) \supset \pi\mathfrak{o}_n$.

Proof. Clear.

Lemma 5. *The kernel of λ_A in the maximal ideal of the algebraic closure of K is precisely A_{tor}, the group of torsion points on A, or in other words, the group $A^{(\pi)}$.*

Proof. A point x is a torsion point if and only if $[\pi^n]x$ is a torsion point for some positive integer n, or for every large positive integer n. But $[\pi^n]x$ approaches 0, and for large n, lies in the neighborhood of 0 where the exponential and log on A give inverse mappings. Since on the additive group, there are no elements of finite order, it follows that the kernel of λ_A is precisely $A^{(\pi)}$.

§7. Application of the Logarithm to the Local Symbol

We recall that the finite extension K_n is self dual as a vector space over K by means of the trace. This means we have a non-degenerate K-linear pairing

$$K_n \times K_n \to K$$

given by

$$(x, y) \mapsto T_n(xy).$$

Let $\mathfrak{a}$ be an ideal in K_n. We denote by $\mathfrak{a}^\perp$ the set of elements $y \in K_n$ such that

$$T_n(xy) \in \mathfrak{o} = \mathfrak{o}_K.$$

Of course, we have the notion of perpendicularity with respect to any given pairing, and the context will always make clear which is meant. We have

$$\mathfrak{a}^\perp = \operatorname{Hom}_{\mathfrak{o}}(\mathfrak{a}, \mathfrak{o}).$$

Indeed, let $\psi: \mathfrak{a} \to \mathfrak{o}$ be a $\mathfrak{o}$-homomorphism. Then ψ can trivially be extended to a K-linear functional $K_n \to K$ denoted by the same letter. But then for some $\alpha \in K_n$ we have

$$\psi(x) = T_n(x\alpha) \quad \text{for all } x \in K_n,$$

and $\alpha \in \mathfrak{a}^{\perp}$ by assumption. Our identification of $\mathfrak{a}^{\perp}$ with $\mathrm{Hom}_{\mathfrak{o}}(\mathfrak{a}, \mathfrak{o})$ then follows at once. If $\mathfrak{D}_n = \mathfrak{D}_{K_n/K}$ is the different, then

$$\boxed{\mathfrak{a}^{\perp} = \mathfrak{a}^{-1}\mathfrak{D}_n^{-1}.}$$

We return to the pairing given by the local symbol

$$A(\mathfrak{p}_n)/[\pi^{n+1}]A(\mathfrak{p}_n) \times K_n^* \to A_{\pi^{n+1}}.$$

We had already noted as a consequence of **LS 4** that

$$\boxed{A_{\pi^{n+1}}^{\perp} = N_{2n+1,n}K_{2n+1}^*.}$$

Observe that we are dealing with two orthogonality signs: One referring to the local symbol, and one referring to the duality

$$K_n \times K_n \to K/\mathfrak{o}$$

(where the trace is viewed as having values in the factor group $K/\mathfrak{o}$), applied in particular to an ideal

$$\mathfrak{a} \times \mathfrak{a}^{\perp} \to \mathfrak{o}.$$

Then we have the pairing

$$A(\mathfrak{p}_n)/([\pi^{n+1}]A(\mathfrak{p}_n) +_A A_{\pi^{n+1}}) \times A_{\pi^{n+1}}^{\perp} \to A_{\pi^{n+1}}.$$

Since $\mathrm{Ker}\,\lambda_A = A_{\mathrm{tor}}$, we have $\mathrm{Ker}\,\lambda_A \cap K_n = A_{\pi^{n+1}}$. Applying the log map of A, we get a pairing

$$\lambda A(\mathfrak{p}_n)/\pi^{n+1}\lambda A(\mathfrak{p}_n) \times A_{\pi^{n+1}}^{\perp} \to A_{\pi^{n+1}}.$$

Let $\mathfrak{a} = \lambda A(\mathfrak{p}_n)$, so the factor group on the left is $\mathfrak{a}/\pi^{n+1}\mathfrak{a}$. In the light of the exact duality

$$\mathfrak{a}/\pi^{n+1}\mathfrak{a} \times \mathfrak{a}^{\perp}/\pi^{n+1}\mathfrak{a}^{\perp} \to \mathfrak{o}/\pi^{n+1}\mathfrak{o},$$

there exists a unique group homomorphism

$$\psi_n \colon A_{\pi^{n+1}}^{\perp} \to \lambda A(\mathfrak{p}_n)^{\perp}/\pi^{n+1}\lambda A(\mathfrak{p}_n)^{\perp}$$

such that for $x \in A(\mathfrak{p}_n)$ and $\alpha \in A_{\pi^{n+1}}^{\perp}$ we have

LS 7. $$\boxed{\langle x, \alpha\rangle_n^A = [T_n(\lambda_A(x)\psi_n(\alpha))]_A(x_n).}$$

This formula has been written without abbreviations, but of course in the future we frequently omit indices A, n, etc. If $\sigma \in \mathrm{Gal}(K^{a}/K)$ and $\sigma \mapsto \varkappa(\sigma)$ is its representation in $\mathfrak{o}^*$ on $T_\pi(A)$, then

LS 8. $$\psi_n(\sigma\alpha) = \varkappa(\sigma)\psi_n(\alpha)^{\sigma}.$$

Proof. We have

$$\sigma\langle\sigma^{-1}x, \alpha\rangle = \langle x, \sigma\alpha\rangle = [T_n(\lambda_A(x)\psi_n(\sigma\alpha))]x_n,$$

and also

$$\sigma\langle\sigma^{-1}x, \alpha\rangle = [T_n((\sigma^{-1}\lambda_A(x))\psi_n(\alpha))](\sigma x_n).$$

But $\sigma x_n = [\varkappa(\sigma)]x_n$ by the definition of $\varkappa$, with $\varkappa(\sigma) \in \mathfrak{o}^*$. Using $[ab] = [a][b]$ and $T_n(\sigma y) = T_n(y)$ concludes the proof of **LS 8**.

9 Explicit Reciprocity Laws

Iwasawa [Iw 8] proved general explicit reciprocity laws extending the classical results of Artin–Hasse, for applications to the study of units in cyclotomic fields. These were extended by Coates–Wiles [CW 1] and Wiles [Wi] to arbitrary Lubin–Tate groups. Although Wiles follows Iwasawa to a large extent, it turns out his proofs are simpler because of the formalism of the Lubin–Tate formal groups. We essentially reproduce his paper in the present chapter.

We assume that K is a finite extension of $\mathbf{Q}_p$ (i.e. has characteristic 0) because we want to use the logarithm.

We allow $p = 2$, and I am indebted to R. Coleman for showing me how Wiles' paper extends with essentially no change to that case, by using (ii), (iii), (iv) of **LS 6**, and the minus sign in **DL 6**.

We let:

$$A = \text{Lubin–Tate group associated with the prime element } \pi.$$

We let:

$$(x_0, x_1, \ldots) \in T_\pi(A) \quad \text{with } x_0 \neq 0.$$

$K_n = K(x_n) = K(A_{\pi^{n+1}})$.

N_n = norm from K_n to K, and $N_{m,n}$ = norm from K_m to K_n for $m \geq n$.

T_n = trace from K_n to K.

We abbreviate $\langle\ ,\ \rangle_n$ to $\langle\ ,\ \rangle$ unless we wish to specify the level at which the symbol is taken.

$N_n^{-1}(\pi^{\mathbf{Z}})$ = subgroup of K_n^* consisting of those elements whose norm to K lies in $\pi^{\mathbf{Z}}$ (i.e., is a power of π).

$= \bigcap_{m \geq n} N_{m,n} K_m^*$ by Lemma 1 of Chapter 8, §5.

$T(K_\infty^*)$ = group of vectors $(\alpha_0, \alpha_1, \ldots)$ with $\alpha_n \in K_n^*$ such that

$$N_{m,n}\alpha_m = \alpha_n.$$

§1. Statement of the Reciprocity Laws

Theorem 1.1. *Suppose* $\alpha \in \mathfrak{o}_n$ *and* $\alpha \equiv 1 \bmod \mathfrak{p}_n$. *Then*

$$N_n\alpha \equiv 1 \bmod \pi^{n+1}$$

and

$$\langle x_n, \alpha \rangle_n = \left[\frac{1}{\pi^{n+1}} (N_n\alpha^{-1} - 1)\right] x_n.$$

Proof. By the formalism of the norm residue symbol, we know that

$$1 = (\alpha, K_n/K_n) = (N_n\alpha, K_n/K).$$

Hence $[N_n\alpha]x_n = x_n$ by the Lubin–Tate theory, so the first assertion is clear.
We choose $t = x_{2n+1}$ so that $[\pi^{n+1}]t = x_n$. Then

$$\langle x_n, \alpha \rangle = \sigma_\alpha t -_A t.$$

Since $(\alpha, K_n(t)/K_n) = (N_n\alpha, K_n(t)/K)$ we obtain from Lubin–Tate theory

$$\begin{aligned} \langle x_n, \alpha \rangle &= [N_n\alpha^{-1}]t -_A t \\ &= [N_n\alpha^{-1} - 1]t. \end{aligned}$$

Using the first congruence and the fact that $[\pi^{n+1}]t = x_n$ yields the theorem.

Corollary 1. *Let* $\alpha \equiv 1 \bmod \mathfrak{p}_n$. *Assume that* K *is unramified over* $\mathbf{Q}_p$. *Then*

$$\langle x_n, \alpha \rangle = \left[-\frac{1}{\pi^{n+1}} T_n(\log \alpha)\right] x_n$$

where the log is the ordinary log on the multiplicative group.

Proof. Since π is unramified, we can write

$$N_n\alpha^{-1} = 1 + z$$

where $z \equiv 0 \bmod p^{n+1}$. Since $p \neq 2$ it follows that

$$\log N_n \alpha^{-1} \equiv z \bmod z^2.$$

Hence

$$-T_n(\log \alpha) = \log N_n \alpha^{-1} \equiv N_n \alpha^{-1} - 1 \bmod p^{2n+2}.$$

Since K is unramified over $\mathbf{Q}_p$, we have $\pi \sim p$, and the corollary follows.

Corollary 2. *Let $A = \mathbf{G}_m$ be the formal multiplicative group. Let ζ be a primitive p^{n+1}th root of unity, and let $\alpha \equiv 1 \bmod \mathfrak{p}_n$. Then*

$$(\zeta, \alpha) = \zeta^{-(1/p^{n+1})T_n(\log \alpha)}.$$

Proof. Special case of Corollary 1.

The law of Corollary 2 is one of Artin–Hasse's laws, obtained here by Wiles as a special case of the Lubin–Tate formalism. We have written the symbol with the usual parentheses, transfering its meaning to the multiplicative group.

We shall now state the main result of this chapter. Let $\alpha \in K_n^*$. Let $r = \operatorname{ord}_{\mathfrak{p}_n} \alpha$. Let $g(X) = c_r X^r + \cdots$ be a power series in $\mathfrak{o}[[X]]$ with a unit c_r such that

$$\alpha = g(x_n).$$

Of course, there exist infinitely many such power series. Let

$$D = d/dX$$

be the ordinary derivative of formal power series, so that

$$Dg(X) = g'(X).$$

Define

$$\boxed{D_n L(\alpha) = g'/g(x_n).}$$

The operator $D_n L$ depends on the choice of element

$$(x_0, x_1, \ldots) \in T_\pi(A),$$

and it depends on the choice of g. We shall see later to what extent it does not depend on g.

We define

$$\boxed{\delta_n(\alpha) = \frac{1}{\pi^n} \frac{1}{\lambda'(x_n)} D_n L(\alpha).}$$

Again this depends on the choice of g.

Let $x \in A(\mathfrak{p}_n)$ and let $\alpha_m \in K_m^*$. In Lemma 3.2 we shall give conditions under which the symbol

$$\boxed{[x, \alpha_m]_m = \frac{1}{\pi} T_m(\lambda_A(x)\delta_m(\alpha_m))}$$

is well defined mod π^{n+1} independently of m. These conditions involve either x being sufficiently divisible in $A(\mathfrak{p}_n)$, or m being sufficiently large. The value of the symbol lies *a priori* in $K/\pi^{n+1}\mathfrak{o}$, but it will turn out that under suitable conditions, its value lies in $\mathfrak{o}/\pi^{n+1}\mathfrak{o}$, so that it can be viewed as an operator on $A_{\pi^{n+1}}$. This was the reason for selecting the bracket in the notation. Precisely, the conditions are as follows.

Condition (i) $m \geq 2n + 1$ *and there exists an integer*

$$k \geq [n/2] + 2(n + 1) \qquad \textit{if } p \textit{ is odd}$$

$$k \geq \max\{[n/2] + 2(n + 1), e + 1\} \quad \textit{if } p = 2,$$

such that $\alpha_m = N_{k,m}\alpha_k$ *with* $\alpha_k \in K_k^*$.

Condition (ii) $m \geq n$ *and* $\operatorname{ord}_\pi x \geq [n/2] + 2$.

Theorem 1.2. *Let* $x \in A(\mathfrak{p}_n)$ *and* $\alpha \in K_n^*$. *Suppose* $\alpha = N_{m,n}\alpha_m$ *for some* $\alpha_m \in K_m^*$. *Under either one of the conditions* (i), (ii), *the symbol* $[x, \alpha_m]_m$ *has value in* $\mathfrak{o}/\pi^{n+1}\mathfrak{o}$ *and we have equality*

$$\langle x, \alpha\rangle_n^A = [x, \alpha_m]_m^A(x_n).$$

An important case is that when

$$\alpha = (\alpha_0, \alpha_1, \ldots) \in T(K_\infty^*).$$

Thus α_n satisfies an infinitely regressive norm-divisibility condition. In that case we may define the symbol

$$[x, \alpha] = [x, \alpha_m]_m$$

for arbitrarily large m, and its value will be the same for any $m \geq 2n + 1$. It gives the formula

$$\langle x, \alpha_n \rangle_n = [x, \alpha]_A(x_n).$$

Example. The formal multiplicative group. For $\beta \equiv 1 \bmod \mathfrak{p}_n$ and $\alpha \in K_n^*$ we defined the classical **norm residue symbol**

$$(\beta, \alpha)_n = \langle \beta - 1, \alpha \rangle_n^A + 1,$$

where $A = \mathbf{G}_m$. Consider the special case when $K = \mathbf{Q}_p$ and $\pi = p$. Let ζ be a primitive p^{n+1}th root of unity. Then

$$x_n = \zeta - 1.$$

We have $\lambda_A(X) = \log(1 + X)$, so

$$\lambda_A'(x_n) = \frac{1}{1 + x_n} = \zeta^{-1}.$$

Let $x \in \mathfrak{p}_n = \mathbf{G}_m(\mathfrak{p}_n)$. Then we find for $m \geq 2n + 1$:

$$(x, \alpha)_n = \zeta^\nu \quad \text{where } \nu = \frac{1}{p^{m+1}} \mathrm{T}_m\,(\zeta \log (1 + x)\, D_{x_m} \log \alpha).$$

This is Iwasawa's formula [Iw 10].

Finally there is another Artin–Hasse reciprocity law generalized by Coates–Wiles to the Lubin–Tate case for level 0.

Theorem 1.3. *Let $x \in A(\mathfrak{p}_0^2)$ and $\alpha \in K_0^*$. Then the symbol $[x, \alpha]_0$ has values in $\mathfrak{o}/\pi\mathfrak{o}$, and we have*

$$\langle x, \alpha \rangle_0 = [x, \alpha]_0^A(x_0).$$

The rest of the sections will be devoted to the proofs. By **LS 2**, we may assume without loss of generality that A is basic, so that we can apply **LS 6**.

§2. The Logarithmic Derivative

In this section we investigate systematically the logarithmic derivative, when it is well defined (modulo certain powers of the prime), and also to what extent the mapping $\delta_n(\alpha)$ is well defined. We let:

$$\mathfrak{D}_n = \text{different of } K_n \text{ over } K;$$

$$\mathfrak{D}_{m,n} = \text{different of } K_m \text{ over } K_n \text{ for } m \geq n.$$

Then

$$\mathfrak{D}_0 = \pi\mathfrak{p}_0^{-1}, \qquad \mathfrak{D}_{n+1,n} = \pi\mathfrak{o}_{n+1}, \qquad \mathfrak{D}_n = \pi^n \mathfrak{D}_0 \mathfrak{o}_n.$$

These are immediate by considering the basic Lubin–Tate generators w_n, satisfying the equations:

$$w_0^{q-1} + \pi = 0 \quad \text{and} \quad w_{n+1}^q + \pi w_{n+1} - w_n = 0.$$

The relative different is obtained by taking the derivative, and evaluating at w_0 and w_{n+1} respectively. The given values fall out.

Let $\alpha \in K_n$ and write α as a power series

$$\alpha = g(x_n) \quad \text{with} \quad g(X) = c_r X^r + \text{higher terms}$$

and c_r equal to a unit. Let $g(X) = X^r h(X)$. Then

$$g'/g(X) = \frac{r}{X} + h'/h(X).$$

Hence $g'/g(x_n)$ is integral if $r = 0$, and in any case lies in $\mathfrak{p}_n^{-1}$. If

$$g_1(X) = X^r h_1(X) \quad \text{and} \quad g_2(X) = X^r h_2(X)$$

are two power series whose values at x_n are equal to α, and $f(X)$ is the irreducible polynomial of x_n over K, then

$$g_1(X) - g_2(X) = X^r f(X)\varphi(X)$$

for some power series $\varphi(X) \in \mathfrak{o}[[X]]$. Hence

$$g_1'/g_1(x_n) - g_2'/g_2(x_n) = \frac{x_n^r}{\alpha} f'(x_n)\varphi(x_n) \equiv 0 \bmod \mathfrak{D}_n.$$

This shows that $g'/g(x_n)$ is well defined modulo the different.

DL 1. *The map $D_n L$ is a homomorphism*

$$D_n L\colon K_n^* \to \mathfrak{p}_n^{-1} \bmod \mathfrak{D}_n.$$

and the image of the units lies in $\mathfrak{o}_n \bmod \mathfrak{D}_n$.

This is obvious from the previous discussion. Since $\lambda'(X)$ is a power series starting with 1 and with integral coefficients, it follows that $\lambda'(x_n)$ is a unit. Hence from the definition of δ_n, we find:

DL 2. *The map δ_n is a homomorphism*

$$\delta_n\colon K_n^* \to K_n \bmod \mathfrak{D}_0\mathfrak{o}_n.$$

Its image lies in $\pi^{-n}\mathfrak{p}_n^{-1} \bmod \mathfrak{D}_0\mathfrak{o}_n$, *and the image of the units lies in* $\pi^{-n} \bmod \mathfrak{D}_0\mathfrak{o}_n$.

As the elements of K_n^* are generated by powers of x_n and units, the computation of δ_n is reduced to $\delta_n(x_n)$ and $\delta_n(\text{units})$. Note:

$$\boxed{\delta_n(x_n) = \frac{1}{\pi^n}\frac{1}{\lambda'(x_n)}\frac{1}{x_n}.}$$

DL 3. *For $\sigma \in \mathrm{Gal}(K^a/K)$, and $\alpha \in K_n^*$,*

$$\delta_n(\alpha^\sigma) = \varkappa(\sigma)\delta_n(\alpha)^\sigma.$$

Proof. Write $\alpha = g(x_n)$ as usual. Then

$$\sigma\alpha = g(\sigma x_n) = g([\varkappa(\sigma)]x_n).$$

Thus we let $g_\sigma(X) = g([\varkappa(\sigma)](X))$, so that

$$\sigma\alpha = g_\sigma(x_n).$$

On one hand, we have

$$\pi^n\,\delta_n(\alpha)^\sigma = \frac{1}{\lambda'(\sigma x_n)}\,g'/g(\sigma x_n) = \frac{1}{\lambda'(\sigma x_n)}\frac{g'([\varkappa(\sigma)]x_n)}{\sigma\alpha}. \tag{1}$$

On the other hand, since $\lambda \circ [\varkappa(\sigma)](X) = \varkappa(\sigma)\lambda(X)$, we find

$$\lambda'(\sigma x_n)[\varkappa(\sigma)]'(x_n) = \varkappa(\sigma)\lambda'(x_n). \tag{2}$$

Furthermore

$$\pi^n\,\delta_n(\sigma\alpha) = \frac{1}{\lambda'(\sigma x_n)}\frac{g_\sigma'(x_n)}{\sigma\alpha}. \tag{3}$$

and

$$g_\sigma'(X) = g'([\varkappa(\sigma)](X)[\varkappa(\sigma)]'(X).$$

Putting (1), (2), (3), (4) together yields the desired property.

DL 4. *Let $m \geq n$ and let α be a unit in K_n^*. Then*

$$\delta_m(\alpha) \equiv \delta_n(\alpha) \bmod \mathfrak{D}_0\mathfrak{o}_m.$$

Proof. The proof is similar to **DL 3**. We know that

$$\lambda \circ [\pi^{m-n}](X) = \pi^{m-n}\lambda(X),$$

so by the chain rule,

$$\lambda'(x_n)[\pi^{m-n}]'(x_m) = \pi^{m-n}\lambda'(x_m).$$

We have the representations

$$\alpha = g(x_n) = g([\pi^{m-n}]x_m) = g_m(x_m)$$

where

$$g_m(X) = g \circ [\pi^{m-n}](X).$$

Since α is assumed to be a unit, the power series g starts with a unit, and so does the power series g_m, so both these power series can be used to compute the logarithmic derivative. The rest of the proof then follows immediately from the chain rule and the definitions.

DL 5. *Let $m \geq n$, and $\alpha_m \in K_m^*$. Then*

$$\delta_n(N_{m,n}\alpha_m) \equiv T_{m,n}\delta_m(\alpha_m) \bmod \mathfrak{o}_n.$$

Proof. Without loss of generality we may assume that $m = n + 1$. We first deal with the case when α_m is a unit. We find:

$$\delta_n(N_{m,n}\alpha_m) = \delta_m\left(\prod_\sigma \alpha_m^\sigma\right) \equiv \sum_\sigma \delta_m(\alpha_u^\sigma) \qquad \text{by } \textbf{DL 4}$$

$$= \sum_\sigma \varkappa(\sigma)\delta_m(\alpha_m)^\sigma, \qquad \text{by } \textbf{DL 3}.$$

The sums are taken over $\sigma \in \mathrm{Gal}(K_m/K_n)$. For such σ we must have

$$\varkappa(\sigma) \equiv 1 \bmod \pi^{n+1}$$

because $[\varkappa(\sigma)]$ is the identity on $A_{\pi^{n+1}}$. Since $\delta_m(\alpha_m)^\sigma$ lies in $\pi^{-m}\mathfrak{o}_m = \pi^{-(n+1)}\mathfrak{o}_m$, the desired congruence follows.

It will then suffice to prove **DL 5** next for $\alpha_m = x_m$, because of the multiplicativity of the function δ. For simplicity, let us first suppose that the Frobenius power series associated with the Lubin–Tate group is in fact a polynomial,

$$[\pi](X) = f(X) = \pi X + \cdots + X^q,$$

and that the coefficient of X^q is exactly 1. For instance, the basic Lubin–Tate group and the formal multiplicative group are of this type. *Under this additional assumption, we have in fact the stronger property with equality instead of the congruence:*

DL 6. $\delta_n(N_{m,n}(-x_m)) = T_{m,n}\delta_m(-x_m)$, *where* $\delta_m(x_m) = \dfrac{1}{\pi^m}\dfrac{1}{\lambda'(x_m)x_m}$.

Remark. The minus signs are there to take care of the case $p = 2$. If $p \neq 2$, they can be omitted.

Proof. We may again suppose $m = n + 1$. We have (as in the proof of Theorem 2.2 of Chapter 8)

$$N_{n+1,n}(-x_{n+1}) = -x_n.$$

The formula to be proved amounts to

$$\frac{1}{\lambda'(x_n)x_n} = T_{n+1,n}\left(\frac{1}{\pi\lambda'(x_{n+1})x_{n+1}}\right).$$

We have $x_n = f(x_{n+1})$, and since $\lambda(f(X)) = \pi\lambda(X)$, we have

$$(\lambda \circ f)'(X) = \lambda'(f(X))f'(X) = \pi\lambda'(X).$$

We put $X = x_{n+1}$, and see that the formula amounts to

$$T_{n+1,n}\left(\frac{x_n}{f'(x_{n+1})x_{n+1}}\right) = 1.$$

We replace x_n by $f(x_{n+1}) = \pi x_{n+1} + \cdots + x_{n+1}^q$. Let $\alpha = x_{n+1}$. Standard orthogonality relations of elementary algebra (see for instance *Algebra*, Chapter VII, §6) yield

$$\begin{aligned} T_{n+1,n}\left(\frac{\alpha^{i-1}}{f'(\alpha)}\right) &= 1 \quad \text{if } i = q \\ &= 0 \quad \text{if } i < q. \end{aligned}$$

This proves what we wanted.

The proof of **DL 5** in general when $\alpha_m = x_m$ follows exactly the same pattern, but we end up only with the asserted congruence. We give the details.

By the Weierstrass theorem, we may factor in $\mathfrak{o}_n$,

$$f(X) - x_n = g(X)h(X)$$

where

$$\begin{aligned} g(X) &= b_0 + \cdots + b_{q-1}X^{q-1} + X^q, \qquad && b_i \equiv 0 \bmod \mathfrak{p}_n \\ h(X) &= c_0 + c_1X + \cdots \text{ is a unit power series,} && c_0 \in \mathfrak{o}_n^*. \end{aligned}$$

Then $f'(x_{n+1}) = g'(x_{n+1})h(x_{n+1})$. Proceeding as before, we are reduced to proving the congruence

$$T_{n+1,n}\left(\frac{x_n}{g'(x_{n+1})h(x_{n+1})x_{n+1}}\right) \equiv 1 \bmod \mathfrak{p}_n.$$

Again we replace x_n by $f(x_{n+1})$. From the factorization we have

$$c_0 \equiv 1 \pmod{x_n}.$$

Hence

$$h(x_{n+1})^{-1} \equiv 1 \bmod x_{n+1}.$$

From the orthogonality relation, we obtain a contribution of 1 from the trace of one term. From the definition of the different (which is precisely $g'(x_{n+1})$) it is then clear that the traces of all the other terms are $\equiv 0 \bmod \mathfrak{p}_n$, as desired.

Property **DL 5** can be expressed in the projective limit as usual. Let

$T(K_\infty/\mathfrak{o}_\infty)$ = projective limit of the additive groups $K_n/\mathfrak{o}_n$ under the trace maps,
= group of vectors $(z_0, z_1, \ldots)$ with $z_n \in K_n/\mathfrak{o}_n$ such that

$$T_{n+1,n}z_{n+1} = z_n.$$

Then the map

$$\delta: T(K_\infty^*) \to T(K_\infty/\mathfrak{o}_\infty)$$

given by

$$(\ldots, \alpha_n, \ldots) \mapsto (\ldots, \delta_n(\alpha_n), \ldots)$$

is well defined, and is a homomorphism.

§3. A Local Pairing with the Logarithmic Derivative

Having derived the necessary formalism for the values of $\delta_n(\alpha)$, we may now combine this with the logarithm on A to define the symbol

$$[x, \alpha]_m = \frac{1}{\pi} T_m(\lambda(x)\delta_m(\alpha)), \qquad \text{for } \alpha \in K_m^*.$$

Lemma 3.1. *The symbol* $[x, \alpha]_m$ *is well defined* mod π^{n+1} *in each of the following cases:*

(i) $x \in A(\mathfrak{p}_n)$ *and* $m \geq 2n + 1$;

(ii) $x \in A(\mathfrak{p}_n^{2q^n})$ *and* $m \geq n$.

Proof. By **DL 2** we know that $\delta_m(\alpha)$ is well defined mod $\mathfrak{D}_0\mathfrak{o}_m$. Hence the symbol is defined mod π^{n+1} if

$$T_m\left(\frac{1}{\pi}\lambda(x)\mathfrak{D}_0\mathfrak{o}_m\right) \subset \pi^{n+1}\mathfrak{o}.$$

By the definition of the different $\mathfrak{D}_m^{-1} = \pi^{-m}\mathfrak{D}_0^{-1}\mathfrak{o}_m$, this is equivalent with:

$$\pi^{m-n-2}\lambda(x)\mathfrak{D}_0^2 \text{ is integral.}$$

It will even be shown that in case (i), $\pi^{m-n-2}\lambda(x)\mathfrak{D}_0$ is integral.

For future reference, we prove congruences which imply the above, and list them systematically.

C 1. If $x \in A(\mathfrak{p}_n)$ and $m \geq 2n + 1$ then

$$T_m\left(\frac{1}{\pi}\lambda(x)\mathfrak{o}_m\right) \subset \pi^{n+1}\mathfrak{o}.$$

C 2. If $x \in A(\mathfrak{p}_n^{2q^n})$ and $m \geq n$ then

$$T_m\left(\frac{1}{\pi}\lambda(x)\mathfrak{D}_0\mathfrak{o}_m\right) \subset \pi^{n+1}\mathfrak{o}.$$

C 3. If $x \in A(\mathfrak{p}_n^{2q^n}\mathfrak{D}_0)$ and $m \geq n$ then

$$T_m\left(\frac{1}{\pi}\lambda(x)\mathfrak{o}_m\right) \subset \pi^{n+1}\mathfrak{o}.$$

Observe that the $\mathfrak{D}_0$ does not occur inside the trace in **C 1** and **C 3**. Only **C 1** and **C 2** are needed for Lemma 3.1 but **C 3** will be needed for Lemma 3.2. We now give the proofs.

Suppose first that $m = 2n + 1$ (the worst case of (i)). We have to verify that

$$\pi^{n-1}\lambda(x)\mathfrak{D}_0 \text{ is integral.}$$

Recall that $\operatorname{ord}_\pi \mathfrak{D}_0 = (q - 2)/(q - 1)$.

By Chapter 8, §6, Lemma 3 we know that $\lambda(x)$ is a power series in x whose terms are either integral, or at worst with a factor

$$\frac{x^{q^i}}{\pi^i}, \quad \text{and} \quad i \geq 1.$$

Suppose $x \in A(\mathfrak{p}_n)$. Then

$$\operatorname{ord} \pi^{n-1}\lambda(x)\mathfrak{D}_0 \geq n - 1 + q^i\frac{1}{q^n(q - 1)} - i + \frac{q - 2}{q - 1}.$$

We need the right-hand side ≥ 0. For $i \leq n - 1$ this is obvious, because $n - 1 - i \geq 0$ and the other terms are positive. For $i \geq n$ the estimate is equally easy.

Suppose next that $x \in A(\mathfrak{p}_n^{2q^n})$ and only that $m \geq n$, say $m = n$ which is the worst case. Then x lies in the disc of "good" convergence for the log and exponent, and thus

$$\operatorname{ord} \lambda(x) = \operatorname{ord} x \geq \frac{2}{q-1}.$$

For **C 2**, it suffices to verify that $\pi^{-2}\lambda(x)\mathfrak{D}_0^2$ is integral, or equivalently

$$-2 + \frac{2}{q-2} + 2\frac{q-2}{q-1} \geq 0,$$

which is obviously the case. The proof for **C 3** is the same.

We remark that in the range where the symbol is well defined, it is $\mathfrak{o}_K$-linear in x and multiplicative in α. In any case, within the ranges of Lemma 1, we view the symbol as having values in

$$K \bmod \pi^{n+1}\mathfrak{o}_K.$$

The next lemma will show that the value $[x, \alpha_m]_m$ is independent of m when α_m is the component of an infinite vector

$$\alpha = (\alpha_0, \alpha_1, \ldots) \in T(K_\infty^*),$$

and m is sufficiently large. We define $[x, \alpha]$ to be this value.

Lemma 3.2. *Let $k \geq m \geq n$. Let $\alpha_m \in K_m^*$ and $\alpha_k \in K_k^*$ be such that*

$$\alpha_m = N_{k,m}\alpha_k.$$

Then

$$[x, \alpha_k]_k \equiv [x, \alpha_m]_m \bmod \pi^{n+1}$$

in either case:

(i) $x \in A(\mathfrak{p}_n)$ *and* $m \geq 2n + 1$;

(ii) $x \in A(\mathfrak{p}_n^{2q^n}\mathfrak{D}_0)$ *and* $m \geq n$.

Proof. We have:

$$\begin{aligned}
[x, \alpha_m]_m &= [x, N_{k,m}\alpha_k]_m \\
&\equiv \frac{1}{\pi} T_m(\lambda(x)\delta_m(N_{k,m}\alpha_k)) \bmod \pi^{n+1} \\
&\equiv \frac{1}{\pi} T_m(\lambda(x)T_{k,m}\delta_k(\alpha_k)) \qquad \text{by } \mathbf{DL\ 5} \text{ and } \mathbf{C\ 1} \text{ or } \mathbf{C\ 3} \\
&= \frac{1}{\pi} T_k(\lambda(x)\delta_k(\alpha_k)) \\
&= [x, \alpha_k]_k,
\end{aligned}$$

as was to be shown.

§4. The Main Lemma for Highly Divisible x and $\alpha = x_n$

Only the statement of the main lemma will be used later, and we recommend to the reader to read the next sections before reading the proof of the main lemma.

Lemma 4.1. *Let $x \in A(\mathfrak{p}_n)$. Suppose that*

$$\operatorname{ord}_\pi x \geq [n/2] + 1 + 2e,$$

where e is the ramification index of K over $\mathbf{Q}_p$. Then

$$\langle x, x_n \rangle_n = [x, x_n]_n^A(x_n).$$

Proof. First we remark that for the applications, the exact nature of the order condition on x is irrelevant, and the reader will find it easier just to assume that the lemma is being proved under the assumption

$$\operatorname{ord}_\pi x \geq [n/2] + 50{,}000e,$$

or any other high multiple of e, which would do just as well. Also, instead of $[n/2]$, any expression like $[n^{1-\varepsilon}]$ would do just as well. In the next section it will be shown how to use a duality to lower back such expressions to the precise ones which we ultimately want.

We let $\operatorname{ord}_\pi x \geq \tau(n)$, and derive sufficient conditions (also more or less necessary) for the method of proof to yield the lemma.

During the course of the proof we shall constantly be interchanging logarithms with the first term in the expansion. If $\operatorname{ord}_\pi x \geq 1$ then

$$\lambda(x) \equiv x \bmod \frac{x^2}{\pi}$$

If p is odd, we even have

$$\lambda(x) \equiv x \bmod x^2.$$

Furthermore, if $\operatorname{ord}_\pi y \geq e$ then

$$\log(1 + y) \equiv y \bmod \frac{y^2}{2}.$$

The formula to be proved is

$$\langle x, x_n \rangle = \left[\frac{1}{\pi^{n+1}} T_n\left(\lambda(x) \frac{1}{\lambda'(x_n)x_n}\right)\right]_A(x_n).$$

We first want to replace $\lambda(x)$ by x on the right-hand side. It suffices for this that

$$T_n\left(\frac{1}{\pi^{n+1}}\frac{x^2}{\lambda'(x_n)x_n}\right) \equiv 0 \bmod \pi^{n+1}.$$

Since $\lambda'(x_n)$ is a unit, this is equivalent with

$$(1) \qquad 2\tau(n) - n - 3 + \frac{q-2}{q-1} - \frac{1}{q^n(q-1)} \geq 0.$$

Certainly $\tau(n) \geq (n/2) + 2$ suffices. Again since $\lambda'(x_n)$ is a unit, proving the formula for all x is equivalent to proving

$$\langle \lambda'(x_n)x, x_n\rangle = \left[\frac{1}{\pi^{n+1}} T_n\left(\frac{x}{x_n}\right)\right]_A (x_n).$$

This will be done by the sequence of following steps.

Step 1. $\langle \lambda'(x_n)x, x_n\rangle = \langle x + x_n, x_n\rangle$

Step 2. $\langle x_n + x, x_n\rangle = [-1]\langle x_n, x_n + x\rangle$

Step 3. $\langle x_n, x_n + x\rangle = \left\langle x_n, 1 + \frac{x}{x_n}\right\rangle$. We then apply the basic reciprocity law of Theorem 1.1 to show that this is equal to

$$\left[\frac{1}{\pi^{n+1}} T_n(x/x_n)\right]_A (x_n)$$

to conclude the proof.

Step 1. We shall use the expression involving the mapping ψ_n in Chapter 8, §7, formula **LS 7**, namely

$$\begin{aligned}\langle x, x_n\rangle &= [T_n(\lambda(x)\psi_n(x_n)]_A(x_n),\\ &= [T_n(x\psi_n(x_n)]_A(x_n).\end{aligned}$$

Indeed, $\lambda(x) \equiv x \bmod \pi^{n\tau(n)}$ and $T_n(\pi^{2\tau(n)}\psi_n(x_n)) \equiv 0 \bmod \pi^{n+1}$, because the image of ψ_n is contained in $\lambda A(p_n)^\perp$, and

$$\lambda A(\mathfrak{p}_n) \supset \pi\mathfrak{o}_n \quad \text{so} \quad \lambda A(\mathfrak{p}_n)^\perp \subset (\pi\mathfrak{o}_n)^\perp = \pi^{-1}\mathfrak{D}_n^{-1}.$$

Therefore we need

$$T_n\left(\frac{x^2}{\pi\mathfrak{D}_n}\right) \equiv 0 \bmod \pi^{n+1}.$$

For this it suffices that

$$(2) \qquad 2\tau(n) \geq n + 2.$$

Again $\tau(n) \geq (n/2) + 2$ suffices.

Since $\lambda'(x_n)$ is a unit, we may replace x by $\lambda'(x_n)x$ to obtain

$$\langle \lambda'(x_n)x, x_n \rangle = [T_n(\lambda'(x_n)x\psi_n(x_n)]_A(x_n).$$

By Taylor's formula, using the fact that $\lambda'(X)$ has integral coefficients, and $\lambda(x_n) = 0$ we get

$$\lambda(x + x_n) \equiv \lambda'(x_n)x \bmod \frac{\pi^{2\tau(n)}}{2}.$$

provided that also $\tau(n) \geq e$. So let us make

$$\tau(n) \geq \frac{n}{2} + 1 + 2e.$$

As we have already seen that

$$T_n(\tfrac{1}{2}\pi^{2\tau(n)}\psi_n(x_n)) \equiv 0 \bmod \pi^{n+1},$$

we may replace $\lambda'(x_n)x$ by $\lambda(x + x_n)$ to conclude the proof of Step 1.

Step 2. Formally, this is just the alternating property **LS 6** of the symbol $\langle \alpha, \alpha \rangle = 0$, but the proof has to be adjusted because the groups involved on the right and left do not play a symmetric role. We have:

$$\begin{aligned} 0 = \langle x_n + x, x_n + x \rangle &= \left\langle x_n + x, 1 + \frac{x}{x_n} \right\rangle [+] \langle x_n + x, x_n \rangle \\ &= \left\langle (x_n + x) [-] x_n, 1 + \frac{x}{x_n} \right\rangle \\ &\quad [+] \left\langle x_n, 1 + \frac{x}{x_n} \right\rangle [+] \langle x_n + x, x_n \rangle \\ &= \left\langle (x_n + x) [-] x_n, 1 + \frac{x}{x_n} \right\rangle \\ &\quad [+] \langle x_n, x_n + x \rangle [+] \langle x_n + x, x_n \rangle. \end{aligned}$$

There remains to prove that the first term on the right is 0, and for this it suffices to show that

$$(x_n + x) [-] x_n \in [\pi^r]A(\mathfrak{p}_n) \quad \text{and} \quad 1 + \frac{x}{x_n} \in K_n^{*p^d}$$

with positive integers r, d such that

$$r + ed \geq n + 1. \tag{3}$$

Let $F(X, Y)$ be the group law on A. Since $F(0, Y) = Y$ and $F(X, 0) = X$ we see that

$$F(X, Y) \equiv X + Y \bmod XY.$$

It follows at once that

$$(x + x_n) [-] x_n \equiv 0 \bmod x \equiv 0 \bmod \pi^{\tau(n)}.$$

But then we may take

$$r = \tau(n) - 1. \tag{4}$$

To solve $1 + y = u^{p^d}$ with some unit u, for $y = x/x_n$, we use the fact that the ordinary log and exp preserve the order on the disc of elements z such that $\operatorname{ord}_p z > 1/(p - 1)$. It follows that we can take any integer d satisfying

$$0 \leq d < \operatorname{ord}_p y - \frac{1}{p - 1},$$

or in terms of π,

$$0 \leq d < \frac{\tau(n)}{e} - \frac{1}{eq^n(q - 1)} - \frac{1}{p - 1}. \tag{5}$$

For instance it suffices that

$$0 \leq d < \frac{1}{e}\tau(n) - 1.$$

Picking $\tau(n) = [n/2] + 1 + 2e$ suffices. This concludes Step 2.

Step 3. Let $y = x/x_n$, and $\alpha = 1 + y$. We have:

$$\begin{aligned} \langle x_n, x_n + x \rangle &= \langle x_n, 1 + y \rangle \\ &= \left[\frac{1}{\pi^{n+1}} (N_n(1 + y)^{-1} - 1)\right]_A (x_n). \end{aligned}$$

We contend that

$$N_n(1 + y)^{-1} \equiv -T_n(y) \bmod \pi^{2(n+1)},$$

whence it follows that

$$\langle x_n, x_n + x \rangle = \left[-\frac{1}{\pi^{n+1}} T_n(y)\right]_A (x_n),$$

thereby concluding the proof of the main lemma.

To prove the contention, write $N_n(1 + y)^{-1} = 1 + z$, with

$$z \equiv 0 \bmod \pi^{n+1}.$$

(Cf. Theorem 1.1.) Then

$$\begin{aligned} \log N_n(1 + y)^{-1} &= T_n(\log(1 + y)^{-1}) \\ &\equiv -T_n(y) \bmod T_n(\tfrac{1}{2}y^2\mathfrak{o}_n). \end{aligned}$$

This amounts to the same type of congruence as before, and is obviously satisfied. This concludes the proof of the main lemma.

§5. The Main Theorem for the Symbol $\langle x, x_n \rangle_n$

Theorem 5.1. *We have the equality*

$$\langle x, x_n \rangle_n = [x, x_m]_m^A(x_n)$$

under either of the following conditions:

(i) $x \in A(\mathfrak{p}_n)$ *and* $m \geq 2n + 1$

(ii) $x \in A(\mathfrak{p}_n^{2q^n}\mathfrak{D}_0)$ *and* $m \geq n$.

Proof. By **LS 5** of §5 in the preceding chapter, we have for $m \geq n$

$$\langle x, x_n \rangle_n = \langle [\pi^{m-n}]x, x_m \rangle_m.$$

This shifts the burden of the proof to level m, and $[\pi^{m-n}]x$ is divisible approximately of order m (asymptotically for $m \to \infty$). Specifically, to apply the main lemma of the preceding section, we want to take m so large that

$$\operatorname{ord}_\pi [\pi^{m-n}]x \geq [m/2] + 1 + 2e.$$

Since

$$[\pi^n]A(\mathfrak{p}_n) \subset \mathfrak{p}_n^{q^n} \quad \text{and} \quad [\pi^{m-n}]x \in \pi^{m-2n}\mathfrak{p}_n^{q^n},$$

it suffices to take $m \geq 4(n + 1) + 4e$.

Let $x' = [\pi^{m-n}]x$. We apply the main lemma of the preceding section to x' instead of x and m instead of n, to conclude that

$$\begin{aligned} \langle x, \alpha_n \rangle_n &= [[\pi^{m-n}]x, \alpha_m]_A(x_m) \\ &= \left[\pi^{m-n}\frac{1}{\pi}T_m(\lambda(x)\delta_m(x_m)\right]_A(x_m). \end{aligned}$$

We write this as

$$[\pi^{m-n}a]_A(x_m), \quad \text{where} \quad a = \frac{1}{\pi} T_m(\lambda(x)\delta_m(x_m)).$$

This means in particular that $\pi^{m-n}a$ is integral. On the other hand, there exists $b \in \mathfrak{o}$ such that

$$\langle x, \alpha_n \rangle_n = [b]_A(x_n) = [b]_A \circ [\pi^{m-n}]_A(x_m) = [\pi^{m-n}b]_A(x_m).$$

Hence we have the congruence

$$\pi^{m-n}a \equiv \pi^{m-n}b \bmod \pi^{m+1},$$

whence it follows that a is integral and also that

$$a \equiv b \bmod \pi^{n+1},$$

which concludes the proof of the equality between the symbols under either condition (i) or (ii) according to Lemma 3.2.

The next theorem shows how one can refine the conditions of Theorem 4.1, with a more precise definition of the symbol $\delta_n(x_n)$ for the special case $\alpha = x_n$.

Theorem 5.2. *Assume that the Frobenius power series associated with the Lubin–Tate group A has the form*

$$f(X) = X^q + \cdots + \pi X,$$

i.e., is a polynomial of degree q with leading coefficient 1. *Define more precisely*

$$[x, -x_n] = \frac{1}{\pi^{n+1}} T_n\left(\lambda(x) \frac{1}{\lambda'(x_n)x_n}\right).$$

Then for $x \in A(\mathfrak{p}_n)$ *we have*

$$\langle x, -x_n \rangle_n = [x, -x_n]_A(x_n).$$

Proof. First observe that the elements $-x_m$ form a vector

$$(-x_0, -x_1, \ldots) \in T(K_\infty^*),$$

i.e., each is the norm of the successive one. Instead of using Lemma 3.2, however, which relied on **DL 5**, we may now use directly the more precise relation **DL 6** which gives an equality implying the stronger statement:

$$\begin{aligned}
[x, x_m]_m &= \frac{1}{\pi} T_m(\lambda(x)\delta_m(-x_m)) \\
&= \frac{1}{\pi} T_n(\lambda(x)T_{m,n}\delta_m(-x_m)) \\
&= \frac{1}{\pi} T_n(\lambda(x)\delta_n(-x_n)), \\
&= [x, -x_n].
\end{aligned}$$

The value $\delta_m(-x_m)$ is here taken to be specifically $(1/\pi^m)(1/\lambda'(x_m)x_m)$, rather than up to a congruence.

Example 1. Take $A = \mathbf{G}_m$ to be the formal multiplicative group. Then it satisfies the hypothesis of Theorem 5.2, and we obtain another reciprocity law of Artin–Hasse:

$$(x, -x_n) = \zeta^{[x, -x_n]}$$

where

$$[x, -x_n] = \frac{1}{p^{n+1}} T_n\left(\frac{\zeta}{x_n} \log(1 + x)\right)$$

$$x_n = \zeta - 1,$$

and ζ is a primitive p^{n+1}th root of unity.

Example 2. Take again $A = \mathbf{G}_m$, let $p \neq 2$ and let B be the special Lubin–Tate group, corresponding to the Frobenius polynomial

$$X^{p-1} + p = 0.$$

We contend that:

$$\langle x_0^i, x_0 \rangle = 0 \text{ if } i \text{ is an integer prime to } p, \text{ or } i > p.$$

$$\langle x_0^p, x_0 \rangle = x_0.$$

Proof. For the first statement, we have by multiplicativity:

$$0 = \langle x_0^i, x_0^i \rangle = [i]\langle x_0^i, x_0 \rangle.$$

If i is prime to p, this proves our assertion because A_p is a p-group. If $i > p$, then one sees from the formula with the trace that the symbol gives 0.

The more interesting case is $\langle x_0^p, x_0 \rangle$. We could work directly on the multiplicative group as was done classically, but the functorial formula **LS 2** of §5 in Chapter 8 shows that it suffices to prove the result on the special group, where

$$x_0^{p-1} = -p.$$

By Lemma 2 of §6 in Chapter 7 (the logarithm for the special group) we know that

$$\lambda_B(x) = x + \text{terms of degree} \geq p.$$

Hence

$$\begin{aligned}[x_0^p, x_0] &= \frac{1}{p} T_0\left(\frac{1}{1 + O(x_0^{p-1})} \frac{1}{x_0} \log(1 + x_0^p)\right) \pmod p \\ &= \frac{1}{p} T_0\left(\frac{1}{x_0} x_0^p\right) \pmod p \\ &= T_0(-1) \pmod p \\ &= 1 \pmod p.\end{aligned}$$

This proves the assertion.

§6. The Main Theorem for Divisible x and α = unit

Theorem 6.1. *Let $x \in A(\mathfrak{p}_n)$ and suppose*

$$\operatorname{ord}_\pi x \geq [n/2] + 2, \qquad \text{if } p \text{ is odd}$$
$$\operatorname{ord}_\pi x \geq \max\{[n/2] + 2, e + 1\} \quad \text{if } p = 2.$$

Then for any unit α we have

$$\langle x, \alpha\rangle_n = [x, \alpha]_n^A(x_n).$$

Proof. The units are generated by μ_{q-1} and the units $\equiv 1 \bmod \mathfrak{p}_n$. If $\alpha \in \mu_{q-1}$ then both sides of the equation are trivially equal to 0. So we deal with the units $\equiv 1 \bmod \mathfrak{p}_n$. A (topological) set of generators for these units consists of the elements

$$1 - \varepsilon x_n^j, \quad \text{with } j = 1, 2, \ldots,$$

where ε is a $(q - 1)$th root of unity. The elements x_n^i form additive generators for $A(\mathfrak{p}_n)$ over $\mathfrak{o}$, and in our case, we need consider just those powers with i satisfying

$$\operatorname{ord}_\pi x_n^i \geq [n/2] + 2.$$

It suffices to prove the lemma for the symbol $\langle x_n^i, 1 - \varepsilon x_n^j \rangle_n$ with such values of i. We shall reduce the proof to the case $\langle x, x_n \rangle$.

We start with the symbol $[x_n^i, 1 - \varepsilon x_n^j]$. Since

$$\delta_n(1 - \varepsilon x_n^j) = \frac{-j\varepsilon x_n^{j-1}}{\lambda'(x_n)(1 - x_n^j)} = \frac{-j}{\lambda'(x_n)x_n} \sum_{r=1}^{\infty} \varepsilon^r x_n^{rj},$$

we find

$$\begin{aligned}[x_n^i, 1 - \varepsilon x_n^j] &= \frac{-j}{\pi^{n+1}} \sum_{r=1}^{\infty} T_n\left(\lambda(x_n{}^i) \frac{1}{\lambda'(x_n)x_n} \varepsilon^r x_n^{rj}\right) \\ &= \frac{-j}{\pi^{n+1}} \sum_{r=1}^{\infty} T_n\left(x_n^i \frac{1}{\lambda'(x_n)x_n} \varepsilon^r x_n^{rj}\right) \\ &= -j \sum_{r=1}^{\infty} [\varepsilon^r x_n^{i+rj}, x_n].\end{aligned}$$

The above formal steps are obviously justified. First the sums taken mod π^{n+1} are actually finite, and second we have replaced $\lambda(y)$ by y and vice versa twice in the range where this applies. The equality takes place in $K/\pi^{n+1}\mathfrak{o}$ where the symbol $[x, \alpha]$ takes its values.

By Theorem 5.1 we know that

$$[\varepsilon^r x_n^{i+rj}, x_n]_A(x_n) = \langle \varepsilon^r x_n^{i+rj}, x_n \rangle.$$

Therefore

$$[x_n^i, 1 - \varepsilon x_n^j]_A(x_n) = [-j] \sum_{r=1}^{\infty}{}_A \langle \varepsilon^r x_n^{i+rj}, x_n \rangle,$$

and this latter sum is taken on A. Since $\langle x, -1 \rangle = 0$ by **LS 6** if $\operatorname{ord}_\pi x$ is big enough, it will therefore suffice to prove the next and final result.

Theorem 6.2. *Suppose*

$$\operatorname{ord}_\pi x_n^i \geq [n/2] + 2 \qquad \textit{if } p \textit{ is odd}$$

$$\operatorname{ord}_\pi x_n^i \geq \max\{[n/2] + 2, e + 1\} \qquad \textit{if } p = 2.$$

Let $j \geq 1$. Then

$$\langle x_n^i, \varepsilon x_n^j - 1 \rangle_n = [-j] \sum_{r=1}^{\infty}{}_A \langle \varepsilon^r x_n^{i+rj}, x_n \rangle_n.$$

Proof. Let F be the group law on A. Since

$$F(X, Y) \equiv X + Y \bmod XY,$$

we obtain for $x, y \in \mathfrak{p}_n$,

$$x\,[+]\,y \equiv x + y \bmod xy \quad \text{and} \quad x\,[-]\,y \equiv x - y \bmod xy.$$

This will be applied when $\operatorname{ord}_\pi x$ and $\operatorname{ord}_\pi y \geq [n/2] + 2$, so that

$$\operatorname{ord}_\pi xy \geq n + 3.$$

In that case, $A(\pi^{n+3}\mathfrak{o}_n) \subset [\pi^{n+1}]A(\mathfrak{p}_n)$, so addition on A and addition on $\mathbf{G}_a$ are interchangeable on the left of the symbol

$$\langle x, \alpha \rangle_n$$

under the condition $\operatorname{ord}_\pi x \geq [n/2] + 2$.

This being said, we find:

$$\begin{aligned} 0 &= \langle x_n^i(\varepsilon x_n^j - 1), x_n^i(\varepsilon x_n^j - 1)\rangle \\ &= \langle \varepsilon x_n^{i+j} - x_n^i, x_n^i\rangle\,[+]\,\langle \varepsilon x_n^{i+j} - x_n^i, \varepsilon x_n^j - 1\rangle \\ &= \langle \varepsilon x_n^{i+j}, x_n^i\rangle\,[+]\,\langle \varepsilon x_n^{i+j}, \varepsilon x_n^j - 1\rangle\,[-]\,\langle x_n^i, \varepsilon x_n^j - 1\rangle \end{aligned}$$

whence

$$\langle x_n^i, \varepsilon x_n^j - 1\rangle = \langle \varepsilon x_n^{i+j}, \varepsilon x_n^j - 1\rangle\,[+]\,\langle \varepsilon x_n^{i+j}, x_n^i\rangle.$$

Note that $\langle x, \varepsilon\rangle = 0$ for all $\varepsilon \in \mu_{q-1}$.

Recursively we obtain

$$\begin{aligned} \langle x_n^i, \varepsilon x_n^j - 1\rangle &= \sum_{r=1}^{\infty}{}_A \langle \varepsilon^r x_n^{i+rj}, x_n^{i+(r-1)j}\rangle \\ &= \sum_{r=1}^{\infty}{}_A \langle \varepsilon^r x_n^{i+rj}, x_n^{-j}\rangle \\ &= [-j] \sum_{r=1}^{\infty}{}_A \langle \varepsilon^r x_n^{i=rj}, x_n\rangle. \end{aligned}$$

Observe that the sums are in fact finite since for large r the left-hand side of each symbol in the sum is highly divisible, so the symbol is 0. Hence the above formal steps are valid, and the theorem is proved.

The special case when $n = 0$ is often interesting for its own sake. The next lemma may be omitted in the proof of Theorem 7.1 or Theorem 1.2, but is useful for the proof of Theorem 7.2.

Lemma 6.3. *Assume that B is the special Lubin–Tate group with Frobenius power series $X^q + \pi X$. Let $w_0 \in B_\pi$. Then for $i \geq 2$ and $j \geq 1$ we have the same formula with $n = 0$ as in the previous lemma, namely*

$$\langle w_0^i, w_0^j - 1\rangle_0 = [-j] \sum_{r=1}^{\infty}{}_B \langle w_0^{i+rj}, w_0\rangle_0.$$

Proof. By Theorem 2.3(ii) of Chapter 8, we know that the group law on B satisfies

$$F(X, Y) = X + Y + \text{terms of degree} \geq q.$$

If $x, y \in \mathfrak{p}_0^2$ it follows that

$$x + y \equiv x \,[+]\, y \bmod \mathfrak{p}_0^{2q}.$$

But $\mathfrak{p}_0^{2q} \subset [\pi]B(\mathfrak{p}_0)$. Hence again addition on A and addition on $\mathbf{G}_a$ are interchangeable on the left of the symbol $\langle x, \alpha\rangle_0$ under the stated conditions, and the rest of the proof is then identical with that of Theorem 6.2.

§7. End of the Proof of the Main Theorems

Theorem 7.1. *Let $x \in A(\mathfrak{p}_n)$ and let*

$$\alpha = (\alpha_n, \alpha_{n+1}, \ldots, \alpha_k) \quad \textit{with } \alpha_m = N_{k,m}\alpha_k.$$

Assume that

$$k \geq [n/2] + 2(n + 1) \qquad \textit{if } p \textit{ is odd}$$

$$k \geq \max\{[n/2] + 2(n + 1), e + 1\} \quad \textit{if } p = 2.$$

Then $[x, \alpha_m]$ lies in $\mathfrak{o}/\pi^{n+1}\mathfrak{o}$ for $2n + 1 \leq m \leq k$, and we have

$$\langle x, \alpha_n\rangle_n = [x, \alpha_m]_m^A(x_n)$$

for such m.

Proof. The theorem has already been proved when α is a power of x_n or when $\alpha \in \mu_{q-1}$. We may therefore assume that α is a unit $\equiv 1 \bmod \mathfrak{p}_n$. We reduce the theorem to the result of the preceding section in exactly the same manner that Theorem 5.1 was reduced to the main Lemma 4.1. This time we need

$$\operatorname{ord}_\pi[\pi^{k-n}]x \geq [n/2] + 2 \qquad \text{if } p \text{ is odd}$$

$$\operatorname{ord}_\pi[\pi^{k-n}]x \geq \max\{[n/2] + 2, e + 1\} \quad \text{if } p = 2.$$

Clearly then, the lower bound on k given above is sufficient. The proof is then identical with that already given for Theorem 5.1, as desired.

For the case $n = 0$, special features arise, and we next give a generalization to Lubin–Tate groups of another explicit formula of Artin–Hasse.

Theorem 7.2. *Assume $p \neq 2$. Let $x \in A(\mathfrak{p}_0^2)$ and $\alpha \in K_0^*$. Then*

$$[x, \alpha]_0 = \frac{1}{\pi} T_0(\lambda(x)\delta_0(\alpha))$$

is well defined mod π, *lies in $\mathfrak{o}/\pi\mathfrak{o}$, and we have*

$$\langle x, \alpha \rangle_0 = [x, \alpha]_0^A(x_0).$$

Proof. Since $\lambda(x_0) = 0$ it follows that $\lambda(\mathfrak{p}_0) = \lambda(\mathfrak{p}_0^2) = \mathfrak{p}_0^2$. This shows that $[x, \alpha]_0$ is well defined mod π because $\delta_0(\alpha)$ is well defined in $\mathfrak{p}_0^{-1}$ mod$\mathfrak{D}_0$.

By formula **LS 2** of Chapter 8, §5 it suffices to prove the theorem when $A = B$ is the special Lubin–Tate group. In that case we already know the result when $\alpha = x_0$ or when $\alpha \in \mu_{q-1}$. We may therefore assume that α is a unit $\equiv 1$ mod $\mathfrak{p}_0$. In that case, we follow the same arguments as in the proof of Theorem 6.1 using Lemma 6.3 instead of Theorem 6.2. The reader can check that each step is valid, to conclude the proof.

10 Measures and Iwasawa Power Series

This chapter gives a number of complements to Chapter 4. In §1 we extend the formalism of the associated power series to the change of variables

$$x \leftrightarrow \gamma^x$$

for $x \in \mathbf{Z}_p$ and γ equal to a topological generator of $1+p\mathbf{Z}_p$. A measure on $1+p\mathbf{Z}_p$ then corresponds to a measure on $\mathbf{Z}_p$, and we give relations between their associated power series. This is then applied to express Bernoulli numbers $B_{k,\chi}$ as values of power series. We write

$$\chi = \theta\omega^{-k}\psi = \theta_k\psi,$$

where first θ is an even character on $\mathbf{Z}(dp)^*$ (d prime to p), ω is the Teichmuller character, and ψ is a character on $1 + p\mathbf{Z}_p$. Let $\zeta = \psi(\gamma)$. Then

$$\frac{1}{k} B_{k,\chi} = f_{\theta,k}(\zeta - 1),$$

where $f_{\theta,k}$ depends only on θ and k. This allows a partial asymptotic determination of $\operatorname{ord}_p B_{k,\chi}$ when θ is fixed, and the conductor of ψ tends to infinity, due to Iwasawa [Iw 14], §7. This gives rise to the corresponding asymptotic estimate for the minus part of class numbers of cyclotomic extensions.

The Iwasawa expressions for the Bernoulli numbers gives an asymptotic value for their orders:

$$\operatorname{ord}_p B_{k,\theta_k\psi} = mp^n + \lambda n + c$$

for n sufficiently large, cond $\psi = p^{n+1}$. In order that $m \neq 0$, Iwasawa showed that a system of congruences had to be satisfied (essentially that the coefficients of the appropriate power series are $\equiv 0 \pmod{p}$). We derive these congruences here in each case successively. The next chapter is devoted to the proofs by Ferrero–Washington that these congruences cannot all be satisfied, whence the Iwasawa invariant m is equal to 0.

At the end of their paper, Ferrero–Washington conjecture that the invariant λ_p for the cyclotomic $\mathbf{Z}_p$-extension of $\mathbf{Q}(\boldsymbol{\mu}_p)$ satisfies a bound

$$\lambda_p \ll \frac{\log p}{\log \log p}.$$

I am much indebted to Washington for communicating to me the exposition of the steps which lead to this conjecture, and which were omitted from their paper.

§1. Iwasawa Invariants for Measures

We let p be an odd prime for simplicity. The multiplicative group $1 + p\mathbf{Z}_p$ is then topologically cyclic, and we let γ denote a fixed topological generator. Then $\gamma \bmod p^n$ generates the finite cyclic group $1 + p\mathbf{Z}_p \bmod p^n$ for each positive integer n. For instance, we may take

$$\gamma = 1 + p.$$

[*Note*: If $p = 2$, then one has to consider $1 + 4\mathbf{Z}_2$ instead of $1 + 2\mathbf{Z}_2$.]

There is an isomorphism

$$\mathbf{Z}_p \to 1 + p\mathbf{Z}_p$$

given by

$$x \mapsto \gamma^x.$$

Its inverse is denoted by α, so that by definition

$$\alpha(\gamma^x) = x.$$

Let $d \geq 1$ be a positive integer prime to p. We shall consider measures on the projective system of groups

$$Z_n = \mathbf{Z}(dp^n) = \mathbf{Z}/dp^n\mathbf{Z} = \mathbf{Z}(d) \times \mathbf{Z}(p^n).$$

The projective limit is simply denoted by

$$Z = \mathbf{Z}(d) \times \mathbf{Z}_p.$$

A measure is then determined by a family of functions μ_n on Z_n, as in Chapter 2, §2. We let

$$Z^* = \mathbf{Z}(d) \times \mathbf{Z}_p^* \quad \text{and} \quad Z^{**} = \mathbf{Z}(d)^* \times \mathbf{Z}_p^*.$$

An element $z \in Z^*$ can be written uniquely in the form

$$z = (z_0, \eta\gamma^x) = (z_0, z_p) \qquad \text{with } z_0 \in \mathbf{Z}(d), \eta \in \boldsymbol{\mu}_{p-1}, x \in \mathbf{Z}_p.$$

We define the homomorphism

$$\alpha: Z^* \to \mathbf{Z}_p \quad \text{by} \quad \alpha(z_0, \eta\gamma^x) = x.$$

We define as usual

$$\langle z \rangle_p = \langle z \rangle = \langle z_p \rangle = \gamma^x,$$

so that $\alpha(z) = \alpha(\langle z \rangle)$. As above, we usually omit the index p on $\langle z \rangle_p$.

A continuous function on $\mathbf{Z}_p$ gives rise to a continuous function on $1+p\mathbf{Z}_p$ by composition with α, and conversely.

As in Chapter 2, §1 we let $\mathfrak{o}$ be the ring of p-integers in $\mathbf{C}_p$, and we let μ be an $\mathfrak{o}$-valued distribution, i.e. a measure.

By the basic correspondence between functionals and measures, we obtain the following theorem.

Theorem 1.1. *Let μ be a measure on Z with support in Z^*. Then there exists a unique measure $\alpha_*\mu$ on $\mathbf{Z}_p$ such that for any continuous function φ on $1+p\mathbf{Z}_p$ we have*

$$\int_{Z^*} \varphi(\langle a \rangle)\, d\mu(a) = \int_{\mathbf{Z}_p} \varphi(\gamma^x)\, d(\alpha_*\mu)(x).$$

We now describe the power series associated with $\alpha_*\mu$ modulo the polynomial

$$h_n(X) = (1 + X)^{p^n} - 1.$$

Thus we fix a value of $n \geq 0$, and for each $a \in Z^*$ we let $r(a)$ be the unique integer such that

$$0 \leq r(a) < p^n \quad \text{and} \quad r(a) \equiv \alpha(a) \bmod p^n.$$

Theorem 1.2. *Let f be the power series associated with $\alpha_* \mu$. Let*

$$Z_{n+1}^* = \mathbf{Z}(d) \times \mathbf{Z}(p^{n+1})^*$$

Then

$$f(X) \equiv \sum_{a \in Z_{n+1}^*} \mu_{n+1}(a)(1+X)^{r(a)} \bmod h_n(X).$$

Proof. By the definition of the associated power series, we have

$$f(X) \equiv \sum_{r=0}^{p^n-1} (\alpha_* \mu)(r)(1+X)^r.$$

But letting char denote the characteristic function, we have:

$$(\alpha_* \mu)(r \bmod p^n) = \int_{\mathbf{Z}_p} (\text{char of } r \bmod p^n)\, d(\alpha_* \mu)$$

$$= \int_{Z^*} (\text{char of } Z(d) \times \boldsymbol{\mu}_{p-1} \times \gamma^{r+p^n\mathbf{Z}_p})\, d\mu$$

(by Theorem 1.1)

$$= \sum_{\eta} \mu_{n+1}(\eta\gamma^r \bmod p^{n+1})$$

where this last sum is taken over $\eta \in \mathbf{Z}(d) \times \boldsymbol{\mu}_{p-1}$. This proves the theorem.

Corollary 1. *Let ψ be a nontrivial character of $1+p\mathbf{Z}_p$, with conductor p^{n+1}. Define $\psi(a) = \psi(\langle a \rangle)$. Let*

$$\psi(\gamma) = \zeta = \textit{primitive } p^n\text{-th } \textit{root of unity}.$$

Let f be the power series associated with $\alpha_* \mu$. Then

$$\int_{\mathbf{Z}_p^*} \psi\, d\mu = f(\zeta - 1).$$

Proof. We have

$$\int_{Z^*} \psi \, d\mu = \int_{\mathbf{Z}_p} \psi(\gamma^x) \, d(\alpha_* \mu)(x) \qquad \text{(by Theorem 1.1)}$$

$$= \int_{\mathbf{Z}_p} \zeta^x \, d(\alpha_* \mu)(x)$$

$$= f(\zeta - 1). \qquad \text{(by Theorem 1.2 of Chapter 4).}$$

This proves the corollary.

We continue with the same notation as in the theorem. We shall use the notation

$$B(\psi, \mu) = \int_{Z^*} \psi \, d\mu = f(\zeta_\psi - 1).$$

Suppose that there exists a rational number m such that the power series f can be written in the form

$$f(X) = p^m(c_0 + c_1 X + \cdots + c_{\lambda-1} X^{\lambda-1} + c_\lambda X^\lambda + \cdots)$$

where c_λ is a unit in $\mathfrak{o}$, and $c_0, \ldots, c_{\lambda-1} \in \mathfrak{m}$, the maximal ideal of $\mathfrak{o}$. We call m, λ the **Iwasawa invariants of** μ, or f. If the measure μ has values in the maximal ideal of the integers in a field where the valuation is discrete (which is the case in applications), then f has coefficients in that ring, and such m, λ exist if $f \neq 0$. If $m = 0$, then λ is the Weierstrass degree of f. In any case, λ is the Weierstrass degree of $p^{-m} f$.

As usual, we shall write

$$x \sim y$$

to mean that x, y have the same order at p.

Corollary 2. *There exists a positive integer n_0 (depending only on f) such that if $n \geq n_0$ and* cond $\psi = p^n$, *then*

$$B(\psi, \mu) \sim p^m (\zeta - 1)^\lambda$$

where ζ is a primitive p^n-th root of unity.

Proof. As $n \to \infty$, the values $|\zeta - 1|$ approach 1, and so the term $c_\lambda(\zeta - 1)^\lambda$ dominates in the power series $f(\zeta - 1)$ above.

Corollary 3. *For some constant $c = c(f)$, we have*

$$\operatorname{ord}_p \prod_{\substack{\text{cond}\,\psi = p^t \\ n_0 \le t \le n}} B(\psi, \mu) = mp^n + \lambda n + c(f)$$

Proof. Since

$$\prod_{\substack{\zeta^{p^n}=1 \\ \zeta \ne 1}} (\zeta - 1) = p^n,$$

the formula is immediate, since the product taken for $n_0 \le t \le n$ differs by only a finite number of factors (depending on n_0) from the product taken over all t, and we can apply Corollary 2 to get the desired order.

In the light of Corollary 3, we shall call m the **exponential invariant**, and λ the **linear invariant**.

Let f be as above, the power series associated with $\alpha_* \mu$, and put

$$\boxed{c_r^{(n)} = \sum_{\eta} \mu_{n+1}(\eta\gamma^r \bmod p^{n+1}).}$$

Then

$$\begin{aligned} f(X) &\equiv \sum_{r=0}^{p^n - 1} c_r^{(n)}(1+X)^r \bmod h_n \\ &\equiv \sum_{r=0}^{p^n - 1} a_r^{(n)} X^r \qquad \bmod h_n, \end{aligned}$$

where the coefficients $a_r^{(n)}$ are obtained from the change of basis from

$$1, X, \ldots, X^{p^n - 1}$$

to

$$1, 1+X, \ldots, (1+X)^{p^n - 1}.$$

We can rewrite $c_r^{(n)}$ in terms of the variable $u = \gamma^r$, namely

$$\boxed{c^{(n)}(u) = \sum_{\eta} \mu_{n+1}(\eta u \bmod p^{n+1}).}$$

These coefficients $c^{(n)}(u)$ will be called the **Iwasawa coefficients**.

Theorem 1.3. *Let n be an integer ≥ 0 such that $c_r^{(n)}$ is a p-unit for some integer r with*

$$0 \leq r \leq p^n - 1.$$

Then the exponential Iwasawa invariant m of μ is equal to 0, and we have $\lambda \leq p^n$.

Proof. Some coefficient $a_r^{(n)}$ must also be a p-unit with r in the same range, and we can write

$$f(X) = \sum_{r=0}^{p^n - 1} a_r^{(n)} X^r + g_1(X) X^{p^n} + p g_2(X),$$

where $g_1(X), g_2(X) \in \mathfrak{o}[[X]]$. Hence the coefficient a_r of $f(X)$ is itself a p-unit, whence the theorem follows.

We shall sometimes deal with certain measures derived by the following operation from μ. Let $s \in \mathbf{Z}_p$. We define the s-th **twist** of μ to be the measure defined on Z^* by

$$\mu^{(s)}(a) = \langle a \rangle^s \mu(a),$$

and equal to 0 outside Z^*. In that case, the coefficients $c_r^{(n)}$ should be indexed by s, i.e.

$$c_{r,s}^{(n)} = c_r^{(n)} \gamma^{rs}.$$

Since γ^{rs} is a p-adic unit, it follows that the same power of p divides all $c_{r,s}^{(n)}$ as divides $c_r^{(n)}$. Thus Theorem 1.3 also applies to the twisted measure and the power series f_s associated with $\alpha_*(\mu^{(s)})$ instead of f in the theorem, and we find:

Theorem 1.4. *Let m_s, λ_s be the Iwasawa invariants of $\mu^{(s)}$. If $m_s = 0$ for some s, then $m_s = 0$ for all s. Suppose this is the case, and let n be the positive integer such that*

$$p^{n-1} \leq \lambda_0 < p^n.$$

Then we also have

$$p^{n-1} \leq \lambda_s < p^n$$

for all s.

§2. Application to the Bernoulli Distributions

Let $\mathbf{B}_k$ be the k-th Bernoulli polynomial (cf. Chapter 2). We had defined the distribution E_k at level N by

$$E_k^{(N)}(x) = N^{k-1}\frac{1}{k}\mathbf{B}_k\left(\left\langle\frac{x}{N}\right\rangle\right).$$

We shall now use

$$N = dp^n,$$

where d is a positive integer prime to the prime number p.

We continue using the notation of the preceding section. An element of $Z = \mathbf{Z}(d) \times \mathbf{Z}_p$ is described by its two components

$$x = (x_0, x_p).$$

Let $c \in \mathbf{Z}(d)^* \times \mathbf{Z}_p^* = \lim \mathbf{Z}(dp^n)^*$. We **define**

$$\boxed{E_{k,c}^{(N)}(x) = E_k^{(N)}(x) - c_p^k E_k^{(N)}(c^{-1}x)}$$

for $x \in \mathbf{Z}(N)$. The multiplication $c^{-1}x$ is defined in $\mathbf{Z}(N)^*$.

Note. In Chapter 2, we took c to be a rational number. This is not necessary, and restricts possible applications too much. When c occurs as a coefficient in Chapter 2, we must use c_p instead of c, i.e. we must use its projection on $\mathbf{Z}_p^*$. When c occurs inside a diamond bracket, then no change is to be made for the present case. For instance, we have

E 1. $$E_{1,c}^{(N)}(x) = \left\langle\frac{x}{N}\right\rangle - c_p\left\langle\frac{c^{-1}x}{N}\right\rangle + \frac{1}{2}(c_p - 1).$$

Similarly, formula **E 2** and Theorem 2.2 of Chapter 2 yield the relation

E 2. $$E_{k,c}(x) = x_p^{k-1}E_{1,c}(x)$$

symbolically for $x \in Z$. We then obtain the integral representations of the Bernoulli numbers as follows.

$$\frac{1}{k}B_k = \frac{1}{1 - c_p^k}\int_Z x_p^{k-1}\, dE_{1,c}(x),$$

provided only that $c_p^k \neq 1$. Furthermore, if χ is a character of conductor $m = m_\chi$ dividing dp^n for some n, then χ defines in the usual way a function on $\mathbf{Z}(N)$ for $m|N$ by composition

$$\mathbf{Z}(N) \to \mathbf{Z}(m) \xrightarrow{\chi} \mathfrak{o}^*,$$

and χ is defined to be 0 on elements of $\mathbf{Z}(m)$ not prime to m. Then we define

$$\frac{1}{k} B_{k,\chi} = \int_Z \chi \, dE_k.$$

Note. This definition made by taking into account the conductor of χ is more appropriate than that of Chapter 2, §2. There we dealt only with characters of $\mathbf{Z}_p^*$, so it made little difference, only for the trivial character.

More generally, if φ is a locally constant function (step function) on Z, then we can **define**

$$\frac{1}{k} B_{k,\varphi} = \int_Z \varphi \, dE_k.$$

Then

$$\int_Z \varphi(x_0, x_p) x_p^{k-1} \, dE_{1,c}(x) = \frac{1}{k} B_{k,\varphi} - c_p^k \frac{1}{k} B_{k,\varphi \circ c}. \tag{1}$$

In particular, if φ is a character χ, then

$$\int_Z \chi(x) x_p^{k-1} \, dE_{1,c}(x) = (1 - \chi(c) c_p^k) \frac{1}{k} B_{k,\chi}.$$

We define the **p-adic L-function** by the integral

$$L_p(1 - s, \chi) = \frac{-1}{1 - \chi(c)\langle c \rangle_p^s} \int_{Z^*} \chi(a) \langle a \rangle_p^s a_p^{-1} \, dE_{1,c}(a).$$

If the conductor of χ is dp^n for some $n \geq 0$, then the support of the integral is really on the set

$$Z^{**} = \mathbf{Z}(d)^* \times \mathbf{Z}_p^*.$$

Let $\omega = \omega_p$ be the Teichmuller character, and put

$$\chi_k = \chi \omega^{-k}.$$

Theorem 2.1. *For every integer $k \geq 1$ and character χ of conductor dp^n with $n \geq 0$, we have*

$$L_p(1-k, \chi) = -(1-\chi_k(p)p^{k-1})\frac{1}{k}B_{k,\chi_k}.$$

Proof. We have:

$$-(1-\chi_k(c)c_p^k)L_p(1-k, \chi) = \int_{Z^*} \chi_k(a)a_p^{k-1}\, dE_{1,c}(a).$$

Write

$$\int_{Z^*} = \int_Z - \int_{pZ}.$$

Let $N = dp^{n+1}$. Then

$$\begin{aligned}\int_{pZ} &= \lim_{n\to\infty} \sum_{y=0}^{(N/p)-1} \chi_k(p)p^{k-1}\chi_k(y)y^{k-1}E_{1,c}\left(\left\langle\frac{py}{N}\right\rangle\right)\\ &= \chi_k(p)p^{k-1}\lim_{n\to\infty} \sum_{y=0}^{(N/p)-1} \chi_k(y)y^{k-1}E_{1,c}\left(\left\langle\frac{y}{N/p}\right\rangle\right)\\ &= \chi_k(p)p^{k-1}(1-\chi_k(c)c_p^k)\frac{1}{k}B_{k,\chi_k}.\end{aligned}$$

The theorem follows at once.

We now let

θ = even character on $\mathbf{Z}(dp)^*$, $\theta \neq 1$, cond $\theta = d$ or dp.
$\chi = \theta\psi$ where ψ is a character on $1+p\mathbf{Z}_p$.

Then

$$\begin{aligned}(1-\chi_k(p)p^{k-1})\frac{1}{k}B_{k,\chi_k} &= \frac{1}{1-\chi(c)\langle c\rangle_p^k}\int_{Z^{**}} \psi(a)\theta\omega^{-k}(a)a_p^{k-1}\, dE_{1,c}(a)\\ &= \frac{1}{1-\chi(c)\langle c\rangle_p^k}\int_{Z^{**}} \psi(\langle a_p\rangle)\, d\mu(a)\end{aligned}$$

where μ is the measure given by

$$\mu(a) = \theta(a)\omega^{-k}(a)a_p^{k-1}E_{1,c}(a).$$

Therefore by Corollary 1 of Theorem 1.2 we find that

$$(1-\chi_k(p)p^{k-1})\frac{1}{k}B_{k,\chi_k}=\frac{1}{1-\chi(c)\langle c\rangle_p^k}f_{\theta,k}(\zeta_\psi-1)$$

where $f_{\theta,k}(X)$ is a power series given mod h_n by Theorem 1.2.

We may use formula **E 1** of Chapter 2, §2 to give the value of μ at intermediate levels, namely

$$\mu_{n+1}(a)=\theta(a)\omega^{-k}(a)a_p^{k-1}\left[\left\langle\frac{a}{dp^{n+1}}\right\rangle-c_p\left\langle\frac{c^{-1}a}{dp^{n+1}}\right\rangle+\tfrac{1}{2}(c_p-1)\right].$$

Starting with the general formula of Theorem 1.2, we shall derive a slightly simpler expression for the coefficients of $f_{\theta,k}$, which can be written in the form

$$f_{\theta,k}(X)\equiv\sum_u c_k^{(n)}(u)(1+X)^{r(u)} \bmod h_n, \tag{2}$$

where

$$c_k^{(n)}(u)=\sum_\eta \theta\omega^{-k}(\eta)(\eta_p u)^{k-1}\left[\left\langle\frac{\eta u}{dp^{n+1}}\right\rangle-c_p\left\langle\frac{c^{-1}\eta u}{dp^{n+1}}\right\rangle+\tfrac{1}{2}(c_p-1)\right].$$

The sums are taken for $u\in 1+p\mathbf{Z}_p \bmod p^{n+1}$ and $\eta\in\mathbf{Z}(d)^*\times\boldsymbol{\mu}_{p-1}$. The component η_p is just $\omega(\eta)$. The character $\theta\omega^{-1}$ is odd, and in particular is not trivial. Hence the sum over η times the factor $(c_p-1)/2$ is equal to 0, and that term can be omitted.

We now select $c\in\mathbf{Z}(d)^*\times\boldsymbol{\mu}_{p-1}$, so that $\langle c\rangle_p=1$. Furthermore $\chi(c)=\theta(c)$. We can select c such that $\chi(c)\neq 1$. We change variables in the sum over η, with respect to the second term involving $c^{-1}\eta$, letting $\eta\mapsto c\eta$. Then we may combine the sums over both terms, with a factor

$$1-\chi(c)$$

which cancels $1-\chi(c)\langle c\rangle_p^k=1-\chi(c)$ in front. In other words, we find:

$$c_k^{(n)}(u)=\sum_\eta \theta\omega^{-1}(\eta)u^{k-1}\left\langle\frac{\eta u}{dp^{n+1}}\right\rangle. \tag{3}$$

We are interested in applying Theorem 1.3. In other words, we are interested in proving the Iwasawa conjecture that some coefficient of $f_{\theta,k}$ is a

p-unit. Clearly the power u^{k-1} can be disregarded for this purpose. Thus the expressions (3) for the coefficients of the Iwasawa power series give rise to the following criterion.

Theorem 2.2 (Iwasawa congruences). *Let d be an integer ≥ 1 and prime to p. Let θ be an even character $\neq 1$ of conductor d or dp. If no coefficient of $f_{\theta,k}$ is a p-unit, then we have the congruences (independent of k):*

$$\sum_{\eta} \theta\omega^{-1}(\eta)\left\langle \frac{\eta\alpha}{dp^{n+1}} \right\rangle \equiv 0 \bmod \mathfrak{p}$$

for all $\alpha \in \mathbf{Z}_p^*$, and all integers $n \geq 0$.

Proof. We have proved the assertion when α lies in $1+p\mathbf{Z}_p$. However, for any fixed $\eta_0 \in \boldsymbol{\mu}_{p-1}$ we can make the change of variables

$$\eta \mapsto \eta\eta_0,$$

leading to the congruences as stated above.

Theorem 2.3 (Ferrero–Washington). *For $\theta \neq 1$, not all these congruences are satisfied, and therefore some coefficient of $f_{\theta,k}$ is a p-unit.*

The proof that not all these congruences are satisfied will be given in the next chapter. Here, we first give formulations for these congruences which are more easily dealt with. Then in the next section, we indicate how this result applies to the divisibility of class numbers in the cyclotomic $\mathbf{Z}_p$-extension.

The case d = 1. We shall give an alternative version of Iwasawa's congruences adapted for the Ferrero–Washington proof. Write any element $z \in \mathbf{Z}_p^*$ as a series

$$z = z_0 + z_1 p + z_2 p^2 + \cdots$$

with integers z_i satisfying $0 \leq z_i \leq p-1$. Let

$$s_n(z) = z_0 + z_1 p + \cdots + z_n p^n$$

be the n-th partial sum. In the above congruences, we may replace $\eta\alpha$ by $s_n(\eta\alpha)$, and then omit the brackets giving the representative as a rational number. Furthermore, let us write

$$\theta\omega^{-1} = \omega^{\nu},$$

where v is a positive integer, necessarily odd since we assumed that θ is an even power of the Teichmuller character. Furthermore, $v \not\equiv -1 \bmod p-1$ because $\theta \neq 1$. Multiplying the congruence by p^{n+1} yields

$$\sum_{\eta \in \mu_{p-1}} s_n(\eta\alpha)\eta^v \equiv 0 \bmod p^{n+2}$$

where v is a positive odd integer, $v \not\equiv -1 \bmod p-1$.

Now in the p-adic expansion of z we let

$$z_n = t_n(z).$$

We shall express the above congruence in terms of t_n.

Theorem 2.4. *Let $\theta \neq 1$ be an even character of conductor p. Then the Iwasawa congruences imply that there exists an odd integer*

$$v \not\equiv -1 \bmod p-1$$

such that, for all $\alpha \in \mathbf{Z}_p^*$ and all integers $n \geq 1$ we have

$$2\sum_{\eta \in \mathscr{R}} t_n(\alpha\eta)\eta^v \equiv (p-1)\sum_{\eta \in \mathscr{R}} \eta^v \bmod p,$$

where $\mathscr{R}$ is a system of representatives for $\boldsymbol{\mu}_{p-1}$ mod ± 1. In particular the congruence class on the left-hand side is independent of α and n.

Proof. We have

$$s_n(\alpha\eta) = s_{n+1}(\alpha\eta) - t_{n+1}(\alpha\eta)p^{n+1}.$$

Furthermore

$$s_{n+1}(\alpha\eta) \equiv \alpha\eta \bmod p^{n+2}$$

and

$$\sum_{\eta \in \mu_{p-1}} \eta^{v+1} = 0$$

because $v \not\equiv -1 \bmod p-1$. Hence the congruence of the theorem is equivalent to

$$\sum_{\eta} t_{n+1}(\alpha\eta)\eta^v \equiv 0 \bmod p.$$

Since $t_0(\alpha\eta) \equiv \alpha\eta \bmod p$, we always have

$$\sum_{\eta} t_0(\alpha\eta)\eta^v \equiv 0 \bmod p.$$

Finally, since $t_{n+1}(px) = t_n(x)$, we are led to the congruence

$$\sum_{\eta} t_n(\alpha\eta)\eta^{\nu} \equiv 0 \bmod p$$

for all n and all α. But since $0 = p + (p-1)p + (p-1)p^2 + \dots$,

$$t_n(-\alpha\eta) = p-1 - t_n(\alpha\eta) \quad \text{for } n \geq 1.$$

Therefore

$$\sum_{\eta} t_n(\alpha\eta)\eta^{\nu} = 2\sum_{\eta\in\mathfrak{R}} t_n(\alpha\eta)\eta^{\nu} - (p-1)\sum_{\eta\in\mathfrak{R}} \eta^{\nu},$$

thus proving the theorem.

The case d > 1.

Theorem 2.5. *Let $\theta \neq 1$ be an even character of conductor d or dp with $d > 1$ prime to p. Let $\theta_1 = \theta\omega^{-1}$. Then the Iwasawa congruences imply that for all $\alpha \in \mathbf{Z}_p^*$ and all $n \geq 0$ we have*

$$\sum_{\eta\in\mathfrak{R}} \sum_{i=0}^{d-1} i\theta_1(s_n(\alpha\eta) + ip^{n+1}) \equiv 0 \bmod \mathfrak{p}.$$

Proof. In Theorem 2.2 we may rewrite the congruence in the form

$$\frac{1}{dp^{n+1}} \sum a\theta_1(a) \equiv 0 \bmod \mathfrak{p}$$

where the sum is taken over a prime to dp, such that

$$0 < a < dp^{n+1} \quad \text{and} \quad \langle a\rangle_p \equiv \langle\alpha\rangle_p \bmod p^{n+1}.$$

We can also replace these elements a by elements of the form

$$\eta\alpha + ip^{n+1} \quad \text{with } i = 0, \dots, d-1,$$

and η is some $(p-1)$th root of unity. The sum is then taken over η and i. The sum over i with the factor $\eta\alpha$ is then equal to 0, and we are left only with a sum having ip^{n+1} as a factor. Combining terms with η and $-\eta$, and using the fact that θ_1 is odd yields the desired formula.

§3. Class Numbers as Products of Bernoulli Numbers

We continue to let p be an odd prime. We write $x \sim y$ to mean that $x = yu$ where u is a p-unit. We let:

θ = even character on $\mathbf{Z}(dp)^*$.
ψ = character on $1 + p\mathbf{Z}_p$ of conductor dividing p^{n+1}.

The characters on $\mathbf{Z}(d)^* \times \mathbf{Z}_p^*$ of the same parity as k of conductor dividing dp^{n+1} can be written uniquely in the form

$$\psi\theta\omega^{-k} = \psi\theta_k.$$

For any integer k with $1 \leq k \leq p - 1$, we **define**

$$h_n^{(k)} = p^{n+1} \prod_{\theta \text{ even}} \prod_{\psi} \frac{1}{k} B_{k,\psi\theta_k}.$$

In particular,

$$h_0^{(k)} = p \prod_{\theta \text{ even}} \frac{1}{k} B_{k,\theta_k}.$$

We can simplify these expressions in so far as p-divisibility is concerned. We need a lemma of von Staudt type.

Lemma 1. *Let k be an integer with $1 \leq k \leq p-1$. Then*

$$\frac{1}{k} B_{k,\omega^{-k}} \equiv -\frac{1}{kp} \bmod \mathbf{Z}_p.$$

Proof. The proof is entirely similar to that of the Von Staudt congruence, Corollary 2 of Theorem 2.3, Chapter 2, combined with the expression for the Bernoulli number as an integral in Theorem 2.4 of Chapter 2. We leave it to the reader.

Lemma 2. *Let $1 \leq k \leq p-1$. Then*

$$h_0^{(k)} \sim \prod_{\theta \neq 1} \frac{1}{k} B_{k,\theta\omega^{-k}}.$$

Proof. The case when $\theta = 1$ combined with Lemma 1 shows that the factor p in the definition of $h_0^{(k)}$ cancels the pole of order 1 at p of the single term with $\theta = 1$ in the product. What remains is the desired expression.

Lemma 3. *Let* $1 \le k \le p-1$. *Then*

$$h_n^{(k)} \sim h_0^{(k)} \prod_{\theta \neq 1} \prod_{\psi \neq 1} \frac{1}{k} B_{k,\psi\theta\omega^{-k}}.$$

Proof. Write $\chi = \theta\psi$. Then

$$(1-\chi_k(p)p^{k-1}) \frac{1}{k} B_{k,\chi\omega^{-k}} = \frac{1}{1-\chi(c)\langle c\rangle_p^k} \int_{Z^*} \chi(a)\langle a\rangle_p^k a_p^{-1}\, dE_{1,c}(a).$$

We distinguish three cases, for the terms in the product defining $h_n^{(k)}$.

If $\theta = 1$ and $\psi = 1$, then we apply Lemma 1. We use one factor of p from p^{n+1} multiplied with

$$\frac{1}{k} B_{k,\omega^{-k}}$$

to find a p-unit.

If $\theta \neq 1$ and $\psi = 1$, then we use Lemma 2 to get the $h_0^{(k)}$ on the right-hand side of the formula to be proved.

The proof of Lemma 3 is concluded by the next lemma.

Lemma 4. *If* $\theta = 1$ *and* $\psi \neq 1$ *then*

$$\frac{1}{k} B_{k,\psi\omega^{-k}} \sim \frac{1}{\zeta - 1},$$

where $\gamma = 1+p$ *and* $\zeta = \psi(\gamma)$. *Furthermore*

$$p^n \prod_{\psi \neq 1} \frac{1}{k} B_{k,\psi\omega^{-k}} \sim 1.$$

Proof. We also take $c = 1+p$. Then

$$1 - \psi(c)\langle c\rangle^k \sim 1-\zeta \quad \text{and} \quad \prod_{\substack{\zeta^{p^n}=1 \\ \zeta \neq 1}} (1-\zeta) = p^n.$$

We note that $\chi(c) = \psi(c)$, and we obtain

$$p^n \prod_{\psi \neq 1} \frac{1}{1-\psi(c)\langle c\rangle_p^k} \sim 1.$$

Finally we wish to show that

$$\int_{\mathbf{Z}_p^*} \psi(a)\langle a\rangle_p^k a_p^{-1}\, dE_{1,c}(a) \sim 1,$$

i.e. the above integral is a p-unit. Since

$$\psi(a) \equiv 1 \bmod 1-\zeta \quad \text{and} \quad \langle a\rangle \equiv 1 \bmod p,$$

it suffices to prove that

$$\int_{\mathbf{Z}_p^*} a_p^{-1}\, dE_{1,c}(a)$$

is a p-unit. This is immediate by writing down the first approximation at level p, and concludes the proof.

For each $\theta \neq 1$ we let $\lambda(\theta, k)$ be the linear Iwasawa invariant of the power series $f_{\theta,k}$ in §2, and we let

$$\lambda(k) = \sum_{\theta \neq 1} \lambda(\theta, k).$$

From the Ferrero–Washington theorem and Lemma 3, we then obtain:

Theorem 3.1. *There is a constant c_k such that for all n sufficiently large, we have*

$$\operatorname{ord}_p h_n^{(k)} = \lambda(k)n + c_k.$$

This is merely a special case of Corollary 3 of Theorem 1.2, applied to the Bernoulli distributions, as discussed in §2.

We can then apply the theorem to the class number.

Theorem 3.2. *Let h_n be the class number of $\mathbf{Q}(\boldsymbol{\mu}_{p^{n+1}})$. Then there is a constant c such that for all n sufficiently large, we have*

$$\operatorname{ord}_p h_n^- = \lambda(1)n + c.$$

Proof. The classical class number formula asserts that

$$h_n^- = 2p^{n+1} \prod_{\chi \text{ odd}} -\tfrac{1}{2}B_{1,\chi},$$

so that we can apply Theorem 3.1 with $k = 1$ to conclude the proof.

Theorem 3.3. *Let K be a cyclotomic extension of the rationals (i.e. a subfield of a cyclotomic field). Let K_∞ be the cyclotomic $\mathbf{Z}_p$-extension of K, and let h_n be the class number of K_n. Then there exists a constant c' such that for all n sufficiently large, we have*

$$\operatorname{ord}_p h_n^- = \lambda(1)n + c'.$$

Proof. It is an easy exercise from the class number formula of Chapter 3 to show that the minus part of the class number differs from the product giving $h_n^{(1)}$ only by a finite number of factors. Hence the same estimate holds as in Theorem 3.2.

In Theorem 2.3 of Chapter 12 we shall prove Iwasawa's inequality bounding the order of h_n in terms of the order of h_n^-. We then obtain:

Theorem 3.4. *Notation being as in Theorem 3.3, there exist constants c_1, c_2 (depending on K) such that for all n sufficiently large, we have*

$$\operatorname{ord}_p h_n = c_1 n + c_2.$$

Remark. Iwasawa developed his theory with the point of view that $\mathbf{Z}_p$-extensions are analogous to constant field extensions for curves over finite fields. The formula

$$h_n^- = h_0^- \prod_\zeta f(\zeta - 1)$$

is analogous for the function field case of the class number formula. The fact that $\operatorname{ord}_p h_n^-$ is linear in n follows at once from the existence of the Jacobian in the function field case. Kubert–Lang theory suggests the possibility of using the analogous theory in the modular case to analyze the Bernoulli numbers $B_{2,\chi}$ and obtain a bound for the invariant λ in terms of the dimensions of abelian subvarieties of the modular curves.

Appendix by L. Washington: Probabilities

We shall give a heuristic argument which estimates the size of $\lambda_p = \lambda_p(\mathbf{Q}(\boldsymbol{\mu}_p))$. The contribution from λ_p^+ will be ignored, since Vandiver's conjecture says it should be zero. In any case, $\lambda_p^+ \leq \lambda_p^-$, so we could alternatively double our final estimate.

Let $\mathrm{i}(p)$ = index of irregularity = number of Bernoulli numbers $B_2, B_4, \ldots, B_{p-3}$ which are divisible by p. The idea will be to show that usually

$$\lambda_p = \mathrm{i}(p),$$

and that one should expect

$$\lambda_p \leq \mathrm{i}(p) + 1$$

for all but a finite number of p.

There are $(p-3)/2$ relevant power series. We assume that each coefficient is random mod p, and that these coefficients behave independently of each other. The first coefficients of these power series correspond to the Bernoulli numbers in such a way that a first coefficient is divisible by p exactly when the corresponding Bernoulli number is divisible by p. The numerical evidence bears out the assumption that the Bernoulli numbers are random mod p. However, we are also assuming that the higher coefficients are random and independent of each other. This is a more dangerous assumption, and I know of no supporting numerical evidence.

Suppose $\lambda_p \geq \mathrm{i}(p) + 2$. Then we have two cases.

Case 1. *Some power series has its first three coefficients divisible by p.* The probability that at least one of the first three coefficients for a given power series is not divisible by p is $1-1/p^3$. The probability that for all $(p-3)/2$ power series we have one of the first three coefficients not divisible by p is

$$\left(1 - \frac{1}{p^3}\right)^{(p-3)/2}.$$

Therefore the probability that at least one power series has its first three coefficients divisible by p is

$$1 - \left(1 - \frac{1}{p^3}\right)^{(p-3)/2} = O\left(\frac{1}{p^2}\right).$$

The expected number of times this should happen is therefore finite, since $\sum 1/p^2 < \infty$.

Case 2. *At least two different series have their first two coefficients divisible by p.* Reasoning as in Case 1, we see that the probability that none of the power series has both of the first two coefficients divisible by p is

$$\left(1 - \frac{1}{p^2}\right)^{(p-3)/2}.$$

The probability that exactly one has its first two coefficients divisible by p is

$$\binom{(p-3)/2}{1}\left(1 - \frac{1}{p^2}\right)^{((p-3)/2)-1}\left(\frac{1}{p^2}\right).$$

Therefore, the probability that at least two power series have their first two coefficients divisible by p is

$$1-\left[\left(1-\frac{1}{p^2}\right)^{((p-3)/2)}+\binom{(p-3)/2}{1}\left(1-\frac{1}{p^2}\right)^{((p-3)/2)-1}\left(\frac{1}{p^2}\right)\right]=O\left(\frac{1}{p^2}\right).$$

So again one expects only finitely many occurrences.

We therefore expect

$$\mathrm{i}(p) \le \lambda_p \le \mathrm{i}(p)+1$$

for all but finitely many p. Therefore estimating λ_p is equivalent to estimating $\mathrm{i}(p)$, which we shall do.

However, first we shall show that usually one should expect $\lambda_p = \mathrm{i}(p)$, as was the case in Wagstaff's calculations for $p \le 125{,}000$.

If $\lambda_p \ge \mathrm{i}(p) + 1$, then at least one power series has its first two coefficients divisible by p. The probability is

$$1-\left(1-\frac{1}{p^2}\right)^{((p-3)/2)}=\frac{1}{2p}+O\left(\frac{1}{p^2}\right).$$

Therefore the number of expected occurrences of $\lambda_p \ge \mathrm{i}(p)+1$ for $p \le x$ should be

$$\sum_{p\le x}\frac{1}{2p}\sim\frac{1}{2}\log\log x.$$

Since $\frac{1}{2}\log\log(125{,}000) \sim 1.2$, it is not very surprising that $\lambda_p = \mathrm{i}(p)$ for $p < 125{,}000$. In fact, one might expect to search rather far before finding a counterexample. A reasonable bound might be 10^{24} since $\frac{1}{2}\log\log 10^{24} \sim 2$. Also note that the fact that

$$1 < \tfrac{1}{2}\log\log 125{,}000$$

is really caused by the first few primes. If one considers, for example,

$$\sum_{30<p<x}\frac{1}{2p},$$

then the expected number is much less than 1. Starting the sum at $p = 31$ is perhaps justified by the fact that the early Bernoulli numbers, etc., are too small to be random mod p. In fact, even though 39% of primes are irregular, 37 is the first one.

We now estimate $\mathrm{i}(p)$. The probability that $\mathrm{i}(p) = i$ is

$$\binom{(p-3)/2}{i}\left(1-\frac{1}{p}\right)^{((p-3)/2)-i}\left(\frac{1}{p}\right)^{i} \xrightarrow[\text{as } p \to \infty]{} e^{-1/2}\frac{(\frac{1}{2})^i}{i!}.$$

The right-hand side is the Poisson distribution. The probability is as stated because i of the Bernoulli numbers are divisible by p, each with probability $1/p$. There are $(p-3)/2 - i$ of them not divisible by p, each with probability $1 - 1/p$. Finally, there are

$$\binom{(p-3)/2}{i}$$

ways of choosing the i Bernoulli numbers which are divisible by p.

For $i = 0$, we obtain the "result" that the fraction of regular primes is $e^{-1/2} \sim 61\%$.

The number of occurrences of $\mathrm{i}(p) = i$ for $p \leq x$ should be approximately

$$\frac{x}{\log x} e^{-1/2} \frac{(\frac{1}{2})^i}{i!}.$$

We should therefore expect the first occurrence to be when

$$\frac{x}{\log x} e^{-1/2} \frac{(\frac{1}{2})^i}{i!} \sim 1.$$

Taking logarithms and ignoring lower order terms, we find, with the help of Stirling's formula:

$$\log x \sim i \log i \quad \text{so} \quad \log\log x \sim \log i \quad \text{and} \quad \frac{\log x}{\log\log x} \sim i.$$

Since x was the first occurrence of i, we obtain approximately

$$\mathrm{i}(p) \leq \frac{\log p}{\log\log p} \quad \text{therefore} \quad \lambda_p \leq \frac{\log p}{\log\log p}.$$

For $p \sim 125{,}000$, this yields $\lambda_p \leq 4.8$ which is close to the truth, namely $\lambda_p \leq 5$. Of course, most of the time λ_p will be much less than this bound: 61% of the time we should have $\lambda_p = 0$. The "average value" of λ_p is

$$\begin{aligned}
\lim_{x\to\infty} \frac{\sum_{p\leq x} \lambda_p}{\sum_{p\leq x} 1} &= \lim\left(\frac{\sum \mathrm{i}(p)}{\sum 1} + \frac{O(\log\log x)}{x/\log x}\right) \\
&= \lim \frac{\sum \mathrm{i}(p)}{\sum 1} \\
&= \sum_{i=0}^{\infty} (\mathrm{i})\ (\text{probability that } \mathrm{i}(p) = i) \\
&= \sum_{i=0}^{\infty} ie^{-1/2}\frac{(\frac{1}{2})^i}{i!} \\
&= \tfrac{1}{2}.
\end{aligned}$$

Bibliography. D. H. Lehmer seems to have been the first to use probability arguments such as the above, since he mentioned that $1 - e^{-1/2} = 39\%$ of primes are irregular in [Leh]. Later, Siegel published a probability argument giving this result in [Si 2]. Numerical evidence appears in Johnson [Jo] and Wagstaff [Wag]. Kummer (last page of vol. I of his collected works) claimed that a simple probability argument yields the ratio of irregular to regular primes is $\frac{1}{2}$, but it appears he was mistaken.

§4. Divisibility by l Prime to p: Washington's Theorem

Theorem 4.1. *Let l, p be distinct primes. If $p \neq 2$, let $q = p$ and if $p = 2$, let $q = 4$. Let χ be an odd Dirichlet character of conductor dividing dq with $(d, p) = 1$. If ψ is a character on $1 + q\mathbf{Z}_p$ with conductor p^{n+1} sufficiently large (depending on l and χ), then the Bernoulli number $\frac{1}{2}B_{1,\chi\psi}$ is an l-unit.*

Before proving the theorem, we give its application to the class numbers of cyclotomic fields.

Theorem 4.2. *Let K be a cyclotomic extension of $\mathbf{Q}$. Let K_∞ be the cyclotomic $\mathbf{Z}_p$-extension of K. Let l be a prime $\neq p$. Then*

$$\mathrm{ord}_l |C(K_n)|$$

is bounded.

Proof. By lemma 2, §1 of Chapter 13 it suffices to prove the theorem when $K = \mathbf{Q}(\boldsymbol{\mu}_{dq})$ for some positive integer d not divisible by p. Furthermore, we may also adjoin an l-th root of unity to the ground field, and thus assume without loss of generality that d is divisible by l. Theorem 3.2 of Chapter 13 then shows that it suffices to prove that

$$\mathrm{ord}_l\, h_n^-$$

is bounded. But we have the formula

$$h_n^- = Q_n w_n \prod_{\chi} \prod_{\psi} -\tfrac{1}{2} B_{1,\chi\psi},$$

where χ ranges over all odd characters of $\mathbf{Z}(dq)^*$ and ψ ranges over all characters of $1+q\mathbf{Z}_p$ of conductor dividing p^{n+1}. The factor Q_n is Hasse's index, equal to 1 or 2, and w_n is l-bounded. Hence we may apply Theorem 4.1 to conclude the proof.

In the rest of this section, we reduce Theorem 4.1 to congruences similar to those which we have already met. To avoid using notation involving 4 in case $p = 2$, we assume that p is odd. Actually, the case $p = 2$ is easier and was solved by Washington before the general case.

Let $\psi = \psi_n$ have conductor p^{n+1}. Let

$$F_n = \mathbf{Q}(\chi, \psi) = \mathbf{Q}(\chi, \boldsymbol{\mu}_{p^n})$$

be the field obtained by adjoining the values of χ and ψ_n to $\mathbf{Q}$. Then $B_{1,\chi\psi}$ belongs to F_n. Let $T_{n,m}$ be the trace from F_n to F_m. Let $m_0 \geq 1$ be a positive integer such that if

$$n > m \geq m_0$$

then $F_n \neq F_m$. Finally, let $\mathscr{R}$ be a set of representatives of the group of $(p-1)$-th roots of unity in $\mathbf{Z}_p$, modulo ± 1.

Given the prime $l \neq p$, there exist only a finite number of prime ideals in $\mathbf{Q}(\boldsymbol{\mu}^{(p)})$ lying above l, as is seen immediately from the structure of the decomposition group. It follows at once that if

$$F_\infty = \bigcup F_n,$$

then there is only a finite number of prime ideals in F_∞ lying above l. Let $\mathfrak{L}$ be one of these primes. We shall now indicate how to prove that given χ, for all ψ with sufficiently large conductor p^{n+1}, we have

$$\tfrac{1}{2} B_{1,\chi\psi} \not\equiv 0 \bmod \mathfrak{L}.$$

We can choose an integer m such that all primes of F_m above l remain prime in F_∞, and we take $m \geq m_0$. Let $n \geq m$.

Suppose that $\frac{1}{2}B_{1,\chi\psi} \equiv 0 \bmod \mathfrak{L}$. Since $\mathfrak{L}$ is the only prime of F_∞ lying above its restriction to F_{m0}, we conclude that

$$T_{n,m}(\tfrac{1}{2}\psi(\alpha)^{-1}B_{1,\chi\psi}) \equiv 0 \bmod \mathfrak{L}$$

for all $\alpha \in \mathbf{Z}_p^*$.

We shall now transform this congruence into more explicit terms.

Lemma 4.3. *If $\frac{1}{2}B_{1,\chi\psi} \equiv 0 \bmod \mathfrak{L}$ for infinitely many ψ (so of arbitrarily large conductor p^{n+1}), then there exist infinitely many n such that for such ψ and all $\alpha \in \mathbf{Z}_p^*$ we have*

$$\frac{\psi(\alpha)^{-1}}{d} \sum_{\eta \in \mathfrak{R}} \sum_{r=0}^{dp^m-1} r\chi\psi(s_{n-m}(\alpha\eta) + rp^{n-m+1}) \equiv 0 \bmod \mathfrak{L}.$$

Proof. Abbreviate $T = T_{n,m}$. From the irreducible equation of a p-power root of unity, we see at once that $T(\varepsilon) = 0$ for any p-power root of unity ε which does not lie in F_m. Thus if $\beta \in \mathbf{Z}_p^*$ and we write

$$\beta = \omega(\beta)\langle\beta\rangle_p$$

where ω is the Teichmuller character and $\langle\beta\rangle_p \equiv 1 \bmod p$, we get:

$$\begin{aligned} T(\psi(\beta)) \neq 0 &\Leftrightarrow \psi(\beta)^{p^m} = 1 \\ &\Leftrightarrow \langle\beta\rangle_p{}^{p^m} \equiv 1 \bmod p^{n+1} \\ &\Leftrightarrow \langle\beta\rangle_p \equiv 1 \bmod p^{n-m+1}. \end{aligned}$$

Consequently we find

$$T(\tfrac{1}{2}\psi(\alpha)^{-1}B_{1,\chi\psi}) = \tfrac{1}{2}p^{n-m} \sum \chi(a)\psi(a\alpha^{-1}) \frac{a}{dp^{n+1}}$$

where the sum is taken for

$$0 < a < dp^{n+1} \quad \text{and} \quad \langle a\rangle_p \equiv \langle\alpha\rangle_p \bmod p^{n-m+1}.$$

This can be rewritten

$$T(\tfrac{1}{2}\psi(\alpha)^{-1}B_{1,\chi\psi}) = \frac{\psi(\alpha)^{-1}}{2\,dp^{m+1}} \sum_{\eta \in \mathfrak{R}} S(\eta, \alpha)$$

where

$$S(\eta, \alpha) = \sum_{a \equiv \pm\eta\alpha(p^{n-m+1})} \chi(a)\psi(a)a.$$

Note that given η, elements a satisfying

$$0 < a < dp^{n+1} \quad \text{and} \quad a \equiv \eta\alpha \bmod p^{n-m+1},$$

can be paired with elements a' such that

$$0 < a' < dp^{n+1} \quad \text{and} \quad a' \equiv -\eta\alpha \bmod p^{n-m+1}$$

of the form

$$a' = dp^{n+1} - a.$$

Since χ is odd and ψ is even, we therefore find

$$S(\eta, \alpha) = \sum_{a \equiv \eta\alpha(p^{n-m+1})} [2\chi(a)\psi(a)a - \chi(a)\psi(a)\, dp^{n+1}].$$

But the integers a satisfying the above conditions are precisely those of the form

$$s_{n-m}(\alpha\eta) + rp^{n-m+1} \quad \text{with } 0 \leq r \leq dp^m - 1.$$

This concludes the proof of the lemma.

To prove Theorem 4.1, we still have to deal with the possibility that $\frac{1}{2}B_{1,\chi\psi}$ has a pole at $\mathfrak{L}$. However, we note that $\mathfrak{L}$ has finite ramification index over l. Consequently for each χ there is only a finite number of ψ such that

$$\tfrac{1}{2}B_{1,\chi\psi}$$

has a pole at $\mathfrak{L}$, because such terms contribute negative l-order to the class number h_n^-, which can be cancelled only by a finite number of other factors in light of the congruence

$$\tfrac{1}{2}B_{1,\chi\psi} \not\equiv 0 \bmod \mathfrak{L}$$

proved for all but a finite number of ψ. This concludes the proof of the assertion in Theorem 4.1 that in fact, all but a finite number of $\frac{1}{2}B_{1,\chi\psi}$ are l-units.

The Ferrero–Washington Theorems 11

In this chapter we prove that the Iwasawa congruences cannot be satisfied, thus giving a bound for the divisibility of Bernoulli numbers with characters, and hence a bound for the divisibility of the corresponding class numbers with respect to certain primes.

The proofs closely follow Ferrero–Washington [Fe–W] and Washington [Wa 2], except that Gillard [Gi 2] gave a simplification which we take into account.

§1. Basic Lemma and Applications

The impossibility of the congruences derived in the preceding chapter will follow from the next lemma, valid for some choice of representatives $\mathcal{R}$ for $\boldsymbol{\mu}_{p-1}$ (mod ± 1).

Lemma 1.1. *Let d, m be positive integers with d prime to p. For all n sufficiently large, there exists $\alpha_1, \alpha_2 \in \mathbf{Z}_p$, with $\alpha_1, \alpha_2 \equiv 1 \bmod p^m$, and an element $\eta_0 \in \mathcal{R}$, having the following properties.*

(i) $s_n(\alpha_1 \eta) = s_{n-m}(\alpha_1 \eta) \equiv 0 \bmod d$ *for all* $\eta \in \mathcal{R}$;
(ii) $s_n(\alpha_2 \eta) = s_{n-m}(\alpha_2 \eta) \equiv 0 \bmod d$ *for all* $\eta \neq \eta_0$;
(iii) $s_n(\alpha_2 \eta_0) = s_{n-m}(\alpha_2 \eta_0) + p^{n-m+1} \equiv 0 \bmod d.$

The proof of this lemma will be given in the next sections.

Although Ferrero–Washington used similar lemmas for their theorems, Gillard [Gi 2] observed that the above single statement suffices in all cases. We now show its applications in each of the three cases under consideration.

The congruences of Theorem 2.4 of Chapter 10. We recall that these congruences are:

(1) $$\sum_{\eta \in \mathfrak{R}} t_n(\alpha\eta)\eta^{\nu} \bmod p \text{ is independent of } \alpha \text{ and } n.$$

In this case, we can take $m = d = 1$, and the congruences mod d become irrelevant, so that

$$t_n(\alpha_1\eta) = t_n(\alpha_2\eta) = 0 \quad \text{for all } \eta \neq \eta_0;$$

$$t_n(\alpha_1\eta_0) = 0 \quad \text{and} \quad t_n(\alpha_2\eta_0) = 1.$$

Subtracting the corresponding expressions in the congruences yields the contradiction.

The congruences of Theorem 2.5 of Chapter 10. We recall that these congruences are:

(2) $$\sum_{\eta \in \mathfrak{R}} \sum_{i=0}^{d-1} i\chi(s_n(\alpha\eta) + ip^{n+1}) \equiv 0 \bmod \mathfrak{p}.$$

The character χ is odd, of conductor d or dp, and $(d, p) = 1$, while $d \neq 1$. We take $m = 2$ in Lemma 1.1. Then we obtain for $\eta \neq \eta_0$:

$$s_n(\alpha_1\eta) \equiv \eta \equiv s_n(\alpha_2\eta) \bmod p$$

$$s_n(\alpha_1\eta) \equiv 0 \equiv s_n(\alpha_2\eta) \bmod d,$$

and hence $s_n(\alpha_1\eta) \equiv s_n(\alpha_2\eta) \bmod dp$. Since χ has conductor d or dp, we get for $i = 0, \ldots, d-1$:

$$\chi(s_n(\alpha_1\eta) + ip^{n+1}) = \chi(s_n(\alpha_2\eta) + ip^{n+1}).$$

Similarly,

$$s_n(\alpha_2\eta_0) \equiv s_n(\alpha_1\eta_0) - p^{n+1} \bmod dp,$$

Let $a = s_n(\alpha_1\eta_0)$. Then the congruence (2) yields

$$\sum_{i=0}^{d-1} i\chi(a+ip^{n+1}) \equiv \sum_{i=0}^{d-1} i\chi(a+(i-1)p^{n+1}) \bmod \mathfrak{p}.$$

But

$$\sum_{i=0}^{d-1} \chi(a+ip^{n+1}) = 0$$

because the conductor of χ does not divide p^{n+1}. Hence

$$\sum_{i=0}^{d-1}(i+1)\chi(a+ip^{n+1}) \equiv \sum_{i=0}^{d-1} i\chi(a+(i-1)p^{n+1}) \bmod \mathfrak{p}.$$

Subtracting yields

$$d\chi(a+(d-1)p^{n+1}) \equiv 0 \bmod \mathfrak{p}.$$

This is a contradiction because $a+(d-1)p^{n+1}$ is prime to dp. Indeed, it is obviously prime to p because it is $\equiv \eta_0 \bmod p$, and on the other hand, since $a \equiv 0 \bmod d$, we get

$$a+(d-1)p^{n+1} \equiv -p^{n+1} \bmod d.$$

This concludes the proof.

The congruences of Lemma 4.3 of Chapter 10. We recall that these congruences are:

$$\frac{1}{d}\sum_{\eta\in\mathscr{R}}\sum_{r=0}^{dp^m-1} r\chi\psi(s_{n-m}(\alpha\eta)+rp^{n-m+1}) \equiv 0 \bmod \mathfrak{L}. \tag{3}$$

For $\alpha = \alpha_1$ or $\alpha = \alpha_2$, we get for all $\eta \in \mathscr{R}$:

$$\alpha^{-1}(s_n(\alpha\eta)+rp^{n-m+1}) \equiv \eta + rp^{n-m+1} \bmod p^{n+1}.$$

Let $a = s_n(\alpha_1\eta_0)$. Arguing as in the preceding case, we get

$$\begin{aligned}\frac{1}{d}\sum_{i=0}^{dp^m-1} & i\chi(a+ip^{n-m+1})\psi(\eta_0+ip^{n-m+1})\\ &\equiv \frac{1}{d}\sum_{i=0}^{dp^m-1} i\chi(a+(i-1)p^{n-m+1})\psi(\eta_0+(i-1)p^{n-m+1}) \bmod \mathfrak{L}.\end{aligned}$$

Again changing i to $i+1$ as in the preceding case, we get

$$\frac{1}{d}dp^m\chi(a+(dp^m-1)p^{n-m+1})\psi(\eta_0+(dp^m-1)p^{n-m+1}) \equiv 0 \bmod \mathfrak{L}.$$

But $a+(dp^m-1)p^{n-m+1}$ is prime to dp and $\eta_0+(dp^m-1)p^{n-m+1}$ is prime to p, so we get the desired contradiction.

§2. Equidistribution and Normal Families

We recall some facts about equidistribution of sequences on $\mathbf{R}/\mathbf{Z}$. Let $\mathscr{F}$ be a family of Riemann integrable functions of $\mathbf{R}/\mathbf{Z}$. Let $\mathbf{C}\langle\mathscr{F}\rangle$ be the vector space generated by $\mathscr{F}$. We say that $\mathbf{C}\langle\mathscr{F}\rangle$ is **Riemann dense** (in the space of all Riemann integrable functions) if given ε and given a real Riemann integrable function g, there exist real functions $f_1, f_2 \in \mathbf{C}\langle\mathscr{F}\rangle$ such that

$$f_1 \leq g \leq f_2$$

and

$$\int_{\mathbf{R}/\mathbf{Z}} (f_2 - f_1) < \varepsilon.$$

(In other words, we can approximate g above and below by functions from $\mathbf{C}\langle\mathscr{F}\rangle$.)

Let $\{y_n\}$ be a sequence of elements in $\mathbf{R}/\mathbf{Z}$. Consider the condition:

EQU. *Let $\mathscr{F}$ be a family of (complex valued) Riemann-integrable functions on $\mathbf{R}/\mathbf{Z}$ such that the vector space generated by $\mathscr{F}$ is Riemann dense. Then for every function f in $\mathscr{F}$, we have*

$$\int_{\mathbf{R}/\mathbf{Z}} f(x)\,dx = \lim_{N\to\infty} \frac{1}{N}(f(y_1) + \cdots + f(y_N)).$$

By a three epsilon argument, one sees that if the sequence satisfies **EQU** for one family $\mathscr{F}$, then it satisfies **EQU** for every such family. If that is the case, we say that the sequence is **equidistributed**, or **uniformly distributed**. Examples of such families which we shall use are as follows.

The most classical family is the family of characteristic functions of intervals $[a, b)$ contained in $[0, 1)$. Then equidistribution means that the density of n such that y_n lies in $[a, b)$ exists and is equal to the measure of $[a, b)$. In other words,

$$\lim_{N\to\infty} \frac{1}{N}(\text{number of } n \leq N \text{ such that } y_n \in [a, b)) = b - a.$$

In the application we deal with an even more restricted family, when the end points a, b of the intervals satisfy additional restrictions (rational numbers whose denominators are powers of a prime p), but the sequence still satisfies **EQU** with respect to this family.

We shall also deal with the family of characters on $\mathbf{R}/\mathbf{Z}$, i.e., functions of type $e^{2\pi imy}$, which satisfies **EQU**. Then **EQU** is known as **Weyl's criterion**.

The above criteria apply mutatis mutandis to r-space $\mathbf{R}^r/\mathbf{Z}^r$, using cubic boxes instead of intervals.

Let $\{\beta_1, \ldots, \beta_r\}$ be a family of p-adic integers. The following two conditions are equivalent, and define what it means for this family to be **normal**.

NOR 1. *For every positive integer k and every $r \times k$ matrix (c_{ij}) of integers with $0 \leq c_{ij} \leq p-1$, there exists $n \geq -1$ such that*

$$t_{n+j}(\beta_i) = c_{ij}$$

for $i = 1, \ldots, r$ and $j = 1, \ldots, k$, and in fact the asymptotic density of such n is p^{-rk}.

The condition means that every possible block of coefficients appears in the p-adic expansions of $\beta_1, \ldots, \beta_r$ with the expected frequency.

NOR 2. *The sequence*

$$\left\{\left(\frac{1}{p^{n+1}} s_n(\beta_1), \ldots, \frac{1}{p^{n+1}} s_n(\beta_r)\right)\right\}, \qquad n = 1, 2, \ldots$$

is uniformly distributed mod $\mathbf{Z}^r$.

We shall now prove that these two conditions are equivalent.

Let $\beta \in \mathbf{Z}_p$ and let $C = (c_1, \ldots, c_k)$ be a k-tuple of integers with

$$0 \leq c_i \leq p-1.$$

Denote the principal part

$$\mathrm{Pr}_k(C) = \frac{c_1}{p^k} + \cdots + \frac{c_k}{p}.$$

Let $I_k(C)$ be the interval of real numbers $[a, a + 1/p^k)$ where $a = \mathrm{Pr}_k(C)$. Write

$$s_n(\beta) - s_{n-k}(\beta) = z_{n-k+1}p^{n-k+1} + \cdots + z_n p^n.$$

Define the principal part

$$\mathrm{Pr}_{n,k}(\beta) = \frac{1}{p^{n+1}}(s_n(\beta) - s_{n-k}(\beta)) = \frac{z_{n-k+1}}{p^k} + \cdots + \frac{z_n}{p}.$$

Then one verifies at once that

$$\boxed{\mathrm{Pr}_{n,k}(\beta) = \mathrm{Pr}_k(C) \quad \text{if and only if} \quad \frac{1}{p^{n+1}} s_n(\beta) \in I_k(C).}$$

Applying this criterion to an r-tuple $\beta_1, \ldots, \beta_r$, we see that if the sequence in **NOR 2** is equidistributed, then $\{\beta_1, \ldots, \beta_r\}$ satisfies **NOR 1**. Conversely, we also see from the above criterion that if $\{\beta_1, \ldots, \beta_r\}$ satisfies **NOR 1**, then that sequence is equidistributed over intervals of type $[a, a + 1/p^k)$, with $a = \mathrm{Pr}_k(C)$ as above. We can then apply **EQU**, with the family of characteristic functions of such intervals.

Remark. As a special case, we also see that $t_n(\beta)$ depends only on the interval $[a/p, (a+1)/p)$ in which $p^{-(n+1)}s_n(\beta)$ lies. This remark will be used in the applications. For instance if $p^{-(n+1)}s_n(\beta)$ lies in the interval $[0, 1/p)$, then $t_n(\beta) = 0$. If it lies in the interval $[1/p, 2/p)$ then $t_n(\beta) = 1$, and so forth.

We use Haar measure on $\mathbf{Z}_p$ (normalized to have total measure 1), and the expression "almost all" refers to all elements except on a set of measure 0 for that measure.

Lemma 2.1. *Let $\{\beta_1, \ldots, \beta_r\}$ be elements of $\mathbf{Z}_p$ which are linearly independent over the rationals. Then for almost all $\alpha \in \mathbf{Z}_p$ the family $\{\alpha\beta_1, \ldots, \alpha\beta_r\}$ is normal.*

Proof. By Weyl's criterion, we must show that for every r-tuple of integers $(a_1, \ldots, a_r)$ not all 0, and almost all α, we have

$$\lim_{N\to\infty} \frac{1}{N} \sum_{n=1}^{N} e\left(\sum_{i=1}^{r} \frac{1}{p^{n+1}} s_n(\alpha\beta_i)a_i \right) = 0,$$

where $e(x) = e^{2\pi i x}$. (For each r-tuple we exclude a set of measure 0, but there are only countably many r-tuples.) Let

$$\beta = \sum_{i=1}^{r} a_i \beta_i.$$

Since

$$s_n(\alpha\beta) \equiv \alpha\beta \equiv \sum_{i=1}^{r} s_n(\alpha\beta_i)a_i \bmod p^{n+1}$$

it suffices to show that

$$\lim_{N\to\infty} \frac{1}{N} \sum_{n=1}^{N} e\left(\frac{1}{p^{n+1}} s_n(\alpha\beta)\right) = 0$$

for almost all α. Let

$$S(N, \alpha) = \frac{1}{N} \sum_{n=1}^{N} e\left(\frac{1}{p^{n+1}} s_n(\alpha\beta)\right).$$

Then writing $|S(N, \alpha)|^2 = S(N, \alpha)\overline{S(N, \alpha)}$ we find:

$$\begin{aligned} \int_{\mathbf{Z}_p} |S(N, \alpha)|^2 \, d\alpha &= \frac{1}{N} + \frac{1}{N^2} \sum_{m \neq n} \int_{\mathbf{Z}_p} (\text{cross terms}) \, d\alpha \\ &= \frac{1}{N}, \end{aligned}$$

because the integral of the cross terms is equal to 0, since the integral of a non-trivial character over the group $\mathbf{Z}_p$ is equal to 0. (In this case, the integral is a sum of roots of unity.) Thus we obtain

$$\sum_{m=1}^{\infty} \int |S(m^2, \alpha)|^2 \, d\alpha = \sum_{m=1}^{\infty} \frac{1}{m^2} < \infty.$$

By Fubini's theorem, we can put the summation sign on the left inside the integral, and thus conclude that

$$\lim_{m\to\infty} S(m^2, \alpha) = 0$$

for almost all α.

For arbitrary N, choose m such that $m^2 \leq N < (m+1)^2$. Trivial estimates show that

$$|S(N, \alpha)| \leq |S(m^2, \alpha)| + \frac{2m}{N} \to 0 \quad \text{as } N \to \infty.$$

This concludes the proof.

§3. An Approximation Lemma

Lemma 3.1. *Let $\{\beta_1, \ldots, \beta_r\}$ be p-adic integers, linearly independent over the rationals. Suppose we are given $\varepsilon > 0$; an integer $m > 0$; an integer d with $(p, d) = 1$; real numbers $x_1, \ldots, x_r \in (0, 1)$. Then for all n sufficiently large, there exists $\alpha \in \mathbf{Z}_p$ satisfying:*

(i) $\alpha \equiv 1 \bmod p^m$;
(ii) $|p^{-(n+1)}s_n(\alpha\beta_i) - x_i| < \varepsilon$ *for* $i = 1, \ldots, r$;
(iii) $s_n(\alpha\beta_i) \equiv 0 \bmod d$ *for* $i = 1, \ldots, r$.

Proof. We use vector notation and put $x = (x_1, \ldots, x_r)$, $\beta = (\beta_1, \ldots, \beta_r)$. We let $\| \;\|$ be the sup norm on the torus $\mathbf{R}^r/\mathbf{Z}^r$. For each n we define the **residue**

$$\operatorname{res}_n(\beta) = p^{-(n+1)}(s_n(\beta_1), \ldots, s_n(\beta_r)).$$

We may assume that ε is so small that the intervals $[x_i - \varepsilon, x_i + \varepsilon]$ are contained in the open interval $(0, 1)$ for all i. Select N sufficiently large so that $1/N < \varepsilon/2d$. By Lemma 2.1, for each r-tuple

$$k = (k_1, \ldots, k_r)$$

of integers k_i with $0 \leq k_i \leq N-1$ we can find a p-adic integer α_k and an integer n_k such that

$$\left\| \operatorname{res}_{n_k}(\alpha_k \beta) - \frac{k}{N} \right\| < \varepsilon/2d.$$

Let $n_0 = m + \max_k n_k$ and let $n \geq n_0$.

There exists some k such that

$$\left\| \frac{x}{d} - \operatorname{res}_n\left(\frac{\beta}{d}\right) - \frac{k}{N} \right\| < \varepsilon/2d.$$

Let $\alpha' = (1/d) + p^{n-n_k}\alpha_k$. Then $\alpha' \equiv 1/d \bmod p^m$, and

$$\left\| \operatorname{res}_n\left(\frac{\beta}{d}\right) + \operatorname{res}_{n_k}(\alpha_k \beta) - \operatorname{res}_n(\alpha'\beta) \right\| = 0.$$

Hence the above inequalities yield

$$\left\| \frac{x}{d} - \operatorname{res}_n(\alpha'\beta) \right\| < 2\varepsilon/2d = \varepsilon/d.$$

It follows that

$$p^{-(n+1)}s_n(\alpha'\beta_i) \in \left[\frac{x_i-\varepsilon}{d}, \frac{x_i+\varepsilon}{d}\right]$$

for all $i = 1, \ldots, r$.

Finally we let $\alpha = d\alpha'$. Then $s_n(\alpha\beta_i) = ds_n(\alpha'\beta_i)$, and α satisfies the required conditions.

The above proof, considerably simpler than the original proof, is taken from the Bourbaki report by Oesterle (Bourbaki Seminar, February 1979).

§4. Proof of the Basic Lemma

This section contains the proof of Lemma 1.1.

Let η^1 be a primitive $(p-1)$th root of unity, and let its powers be

$$\eta^j \quad \text{for } j = 1, \ldots, R \quad \text{where} \quad R = \frac{p-1}{2}.$$

Then these powers η^j represent the elements of $\boldsymbol{\mu}_{p-1}/\pm 1$.

We shall write $\eta_1, \ldots, \eta_r$ for $\pm\eta^1, \ldots, \pm\eta^r$, with any choice of sign, and $r = \phi(p-1)$. We can express

$$\eta^j = \eta_j = \sum_{i=1}^{r} a_{ji}\eta_i \quad \text{for } j = r+1, \ldots, R,$$

with integral coefficients a_{ji}. Thus we obtain an $(R-r) \times r$ matrix

$$A = (a_{ji}), \qquad (r+1 \leq j \leq R; \quad 1 \leq i \leq r).$$

We let $x_1, \ldots, x_r$ be real numbers, and we then let

$$x_j = \sum_{i=1}^{r} a_{ji}x_i \quad \text{for } j = r+1, \ldots, R.$$

Observe that changing the signs of $\eta_1, \ldots, \eta_r$ amounts to changing the signs of the columns in the matrix A. Changing the signs of $\eta_{r+1}, \ldots, \eta_R$ amounts to changing the signs of the rows. Such changes of signs will be called admissible.

Since η_i/η_j is not rational for $i \neq j$, it follows that in each row, there are at least two non-zero elements. Furthermore, it is clear that no two rows of the matrix A are equal. We now prove:

Lemma 4.1. *Let $A = (a_{ji})$ be a real matrix such that no row equals + or − another, and in every row there are at least two non-zero elements. After an admissible change of sign, we can find a vector $(x_1, \ldots, x_r) = {}^tX^{(r)} \in \mathbf{R}^r$ such that, if we put*

$$X^{(R-r)} = AX^{(r)} \quad \text{and} \quad X^{(R-r)} = {}^t(x_{r+1}, \ldots, x_R),$$

then:

(i) *We have $x_j > 0$ for $j = 1, \ldots, R$;*
(ii) *We have $x_{j'} \neq x_j$, for all $j \neq j'$.*

Proof. In R-space, we consider the conditions:

$x_j = 0$ for some $j = 1, \ldots, R$;
$x_j - x_{j'} = 0$ or $x_j + x_{j'} = 0$ for some pair (j, j') with $j \neq j'$.

Each such condition defines a hyperplane. We want some $X^{(r)} \in \mathbf{R}^r$ such that

$$(X^{(r)}, AX^{(r)})$$

does not lie in the union of these hyperplanes. Let V be the vector space of all vectors $(X^{(r)}, AX^{(r)})$, i.e. the graph of the linear map represented by A. Then V is not contained in any one of the above hyperplanes because of the two assumptions on the matrix A. Hence V is not contained in the finite union of these hyperplanes. Let V' be the complement of these hyperplanes in V, and let V'_0 be the projection of V' on the first r coordinates. Take $(x_1, \ldots, x_r)$ in the positive 2^r-quadrant intersected with V'_0, and let $(x_1, \ldots, x_R)$ be a point in V' above it. Since V is symmetric with respect to sign changes on the last $R - r$ coordinates, we can then make such sign changes on $x_{r+1}, \ldots, x_R$ to achieve the desired positivity condition. (I am indebted to Roger Howe for the above proof.)

We let j_0 be the index such that $x_j < x_{j_0}$ for all $j \neq j_0$. After replacing $x_1, \ldots, x_R$ by $cx_1, \ldots, cx_R$ for some real number $c > 0$, we may assume that

$$0 < x_j < p^{-m} \quad \text{for all } j = 1, \ldots, R.$$

We then apply Lemma 3.1, (i) and (ii) with $\beta_i = \eta_i$ for $i = 1, \ldots, r$. Let n and α be as in that lemma. If ε is small enough, we obtain

$$0 < \sum_{i=1}^{r} a_{ji} p^{-(n+1)} s_n(\alpha\eta_i) < p^{-m} \quad \text{for } j = r+1, \ldots, R.$$

Hence

$$0 < \sum a_{ji} s_n(\alpha\eta_i) < p^{n+1-m} < p^{n+1}.$$

Therefore

$$s_{n-m}(\alpha\eta_j) = s_n(\alpha\eta_j) = \sum a_{ji} s_n(\alpha\eta_i).$$

We take $\alpha_1 = \alpha$ to satisfy the first part of Lemma 1.1.

For Lemma 1.1 (ii) and (iii), we select the scaling factor $c > 0$ such that

$$0 < x_j < p^{-m} \quad \text{for } j \neq j_0 \quad \text{but} \quad p^{-m} < x_{j_0} < 2p^{-m}.$$

We select $\alpha = \alpha_2$ in Lemma 3.1. Then the desired conditions are satisfied for $j \neq j_0$ as before, but for j_0, Lemma 3.1 now shows

$$s_{n-m}(\alpha\eta_{j_0}) = s_n(\alpha\eta_{j_0}) - p^{n-m+1}.$$

This concludes the proof.

12 Measures in the Composite Case

In Chapter 4 we developed the formalism of associated power series for measures on $\mathbf{Z}_p$. It is necessary to develop it in general. We do this in the present chapter, which could (and should) have been done immediately following Chapter 4. In all this work, a prime p is given a special role. Values of functions lie in $\mathbf{C}_p$. In dealing with this composite case, it is also useful to follow Katz, and associate to a measure not only a power series, but an analytic function on the "formal multiplicative group." This is explained in §2. The introduction of additional notation to handle this composite case, however, made it worthwhile to separate the two cases. Measures on $\mathbf{Z}_p$ itself, without the extra d, occur both in their own right, and as auxiliaries to the composite case, so it is useful to have their properties tabulated separately.

The present chapter is independent of everything else in the book, and can be omitted by those who wish to read at once the results of Chapter 13, needed to bound the plus part of the class number in terms of the minus part, for the Ferrero–Washington theorems.

§1. Measures and Power Series in the Composite Case

In Chapter 4, we dealt only with the formalism of measures and power series on $\mathbf{Z}_p$. To handle characters with conductor dp^n where d is a positive integer prime to p, one has to deal with the composite case. Thus we now give an exposition of the formalism in this more general context.

Let Z be a profinite group, equal to the projective limit of its quotients

$$Z = \lim_{H} Z/H,$$

where H ranges over the open subgroups of finite index. Let $\mathfrak{o}$ be a complete valuation subring of the p-integers in $\mathbf{C}_p$. We let the **Iwasawa algebra** be

$$\Lambda_{\mathfrak{o}}(Z) = \lim_{H} \mathfrak{o}\,[Z/H],$$

where $\mathfrak{o}[Z/H]$ is the group ring of Z/H over $\mathfrak{o}$. We recall that an $\mathfrak{o}$-valued measure on the projective system $\{Z/H\}$ is a family of $\mathfrak{o}$-valued functions $\{\mu_H\}$, which is a distribution. This means: given $H' \subset H$, we have

$$\mu_H(x) = \sum_{y} \mu_{H'}(y),$$

where $x \in Z/H$ and the sum is taken over $y \in Z/H'$ lying above x under the canonical map $Z/H' \to Z/H$. The association

$$\mu_H \mapsto \sum_{x \in Z/H} \mu_H(x)x \in \mathfrak{o}[Z/H]$$

lifts to an isomorphism between the additive group of $\mathfrak{o}$-valued measures and the Iwasawa algebra.

Observe that the product in $\Lambda_{\mathfrak{o}}(Z)$ corresponds to the convolution of measures.

If φ is an $\mathfrak{o}$-valued function on Z, factoring through some factor group Z/H, in other words, if φ is a step function, then we define its **integral**

$$\int \varphi \, d\mu = \sum_{x \in Z/H} \varphi(x)\mu_H(x).$$

This value is independent of the choice of H, by the distribution relation. The integral extends to the space of $\mathfrak{o}$-valued continuous functions on Z by uniform approximation with step functions.

Let $C(Z, \mathfrak{o})$ be the $\mathfrak{o}$-space of continuous functions of Z into $\mathfrak{o}$, with sup norm. There is a bijection between $\mathfrak{o}$-valued measures on Z and $\mathfrak{o}$-valued $\mathfrak{o}$-linear functionals

$$\lambda : C(Z, \mathfrak{o}) \to \mathfrak{o}.$$

Indeed, the measure μ gives rise to the functional

$$\varphi \mapsto \int \varphi \, d\mu.$$

Conversely, given λ, if $x \in Z/H$ and φ_x is the characteristic function of the set of all $y \in Z$ such that $y \equiv x \bmod H$, then we define

$$\mu_H(x) = \lambda(\varphi_x).$$

It is easily verified that these associations are inverse to each other, and establish a norm-preserving bijection, where

$$\|\mu\| = \sup_{x, H} |\mu_H(x)|$$

is the sup norm.

Now suppose that there is a finite subgroup Z_0 such that

$$Z = Z_0 \times \mathbf{Z}_p.$$

Let Γ be a multiplicative group isomorphic to $\mathbf{Z}_p$, with topological generator γ, so the isomorphism is given by

$$z \mapsto \gamma^z, \quad \text{with } z \in \mathbf{Z}_p.$$

Let $H_n = \{1\} \times p^n\mathbf{Z}_p$, and let $\gamma_n = \gamma \bmod \Gamma^{p^n}$. Then

$$Z_n = Z/H_n \cong Z_0 \times \Gamma_n,$$

where $\Gamma_n = \Gamma/\Gamma^{p^n}$ is cyclic, with generator γ_n.

Let X be a variable, $T = 1+X$, and

$$h_n(X) = (1+X)^{p^n} - 1.$$

Then the element of $\mathfrak{o}[Z_n]$ corresponding to μ_{H_n} is a linear combination

$$(1) \qquad \begin{aligned} P_n(X) &= \sum_{\sigma \in Z_0} \sum_{r=0}^{p^n-1} \mu_{n,\sigma}(r)\sigma\gamma_n^r \\ &\equiv \sum_{\sigma \in Z_0} \sum_{r=0}^{p^n-1} \sum_{k=0}^{r} \mu_{n,\sigma}(r)\binom{r}{k}\sigma X^k \bmod h_n(X). \end{aligned}$$

where $\gamma_n = T \bmod h_n$. For each σ, the map

$$r \mapsto \mu_{n,\sigma}(r)$$

determines a measure μ_σ on $\mathbf{Z}_p$ as discussed in Chapter 4. This makes it possible to extend results concerning measures on $\mathbf{Z}_p$ to measures on the more general groups now being considered. If we let

$$P_n(X) = \sum_{k=0}^{p^n-1} c_{n,k}X^k,$$

then the coefficients

$$c_{n,k} = \sum_{\sigma} \sum_{r} \mu_{n,\sigma}(r) \binom{r}{k} \sigma = \sum_{\sigma} c_{n,k}(\sigma)\sigma \tag{2}$$

lie in the group ring $\mathfrak{o}[Z_0]$.

We let $P(X)$ be the projective limit of the polynomials $P_n(X)$. Then $P(X)$ is an element of $\mathfrak{o}[Z_0][[X]]$, and we shall write the correspondence between μ and the power series P by

$$f = P\mu, \qquad \mu = \mu_f.$$

Let us write

$$P(X) = \sum_{k=0}^{\infty} c_k X^k = \sum_{k=0}^{\infty} \sum_{\sigma \in Z_0} c_{k,\sigma} \sigma X^k.$$

Then the coefficients $c_{k,\sigma}$ are given by the integrals

$$c_{k,\sigma} = \int_{\mathbf{Z}_p} \binom{x}{k} d\mu_\sigma(x), \tag{3}$$

because we can apply Theorem 1.1 of Chapter 4.

Let φ be a continuous function on Z. Then for each σ, we get a function

$$\varphi_\sigma(r) = \varphi(\sigma, r) \quad \text{with } r \in \mathbf{Z}_p.$$

If φ factors through Z_n, then

$$\begin{aligned} \int_Z \varphi \, d\mu &= \sum_{\sigma} \sum_{r \in \mathbf{Z}(p^n)} \varphi(\sigma, r)\mu_n(\sigma, r) \\ &= \sum_{\sigma \in Z_0} \int_{Z_p} \varphi_\sigma \, d\mu_\sigma . \end{aligned} \tag{4}$$

Any continuous function φ on Z can be viewed as a family of continuous functions $\{\varphi_\sigma\}$ on $\mathbf{Z}_p$. Hence

$$\int_Z \varphi \, d\mu = \sum_{\sigma \in Z_0} \int_{\mathbf{Z}_p} \varphi_\sigma \, d\mu_\sigma. \tag{5}$$

Using Mahler's Theorem 1.3 of Chapter 4, if we write

$$\varphi_\sigma(x) = \sum a_{n,\sigma} \binom{x}{n}$$

and the power series associated with μ_σ is

(6) $$f_\sigma(X) = \sum c_{n,\sigma} X^n,$$

then

(7) $$\int_Z \varphi \, d\mu = \sum_\sigma \sum_n c_{n,\sigma} a_{n,\sigma}.$$

The sum formula (5) in principle allows us to reduce the study of any measure μ_f to the individual measures associated with the power series f_σ on the fiber $\{\sigma\} \times \mathbf{Z}_p$. The formulas of Chapter 4 apply to each f_σ. In addition, we have trivial ones relating to the extra factor Z_0. For instance:

(8) *If φ is a function on Z_0, then*

$$\varphi\mu_f = \mu_g \quad \textit{where } g_\sigma = \varphi(\sigma) f_\sigma.$$

This applies in particular to the characteristic function φ_α of a single element $\alpha \in Z_0$. Then

$$\varphi_\alpha \mu_f = \mu_{f_\alpha}$$

where f_α is identified with the element $f_\alpha \cdot \alpha$ in $\mathbf{Z}_p[Z_0][[X]]$. In this last example, we have of course the Fourier expansion of φ_α given by

$$\varphi_\alpha = \frac{1}{|Z_0|} \sum_\psi \bar{\psi}(\alpha)\psi,$$

where the sum is taken over all characters ψ of Z_0.

At any finite level, that is on any one of the groups

$$Z_0 \times \mathbf{Z}(p^n),$$

the space of functions is generated by the product functions

$$\varphi_0 \otimes \varphi_p,$$

where φ_0 is a function on Z_0 and φ_p is a function on $\mathbf{Z}(p^n)$. By definition,

$$(\varphi_0 \otimes \varphi_p)(\sigma, x_p) = \varphi_0(\sigma)\varphi_p(x_p).$$

Thus to test whether two measures are equal, it suffices to test whether their integrals on such functions are equal. These integrals decompose as simple sums according to (4) and (5), namely

$$\int_{Z} \varphi_0(\sigma)\varphi_p(x_p)\, d\mu(\sigma, x_p) = \sum_{\sigma} \varphi_0(\sigma) \int_{\mathbf{Z}_p} \varphi_p(x_p)\, d\mu_\sigma(x_p). \tag{9}$$

This applies for instance if φ is a character of Z which factors through a finite level.

Let f be the power series associated with μ. Given the function φ_0 on Z_0, the expression of (9) defines a measure μ_{φ_0} on $\mathbf{Z}_p$ whose associated power series is

$$f_{\varphi_0} = \sum_{\sigma} \varphi_0(\sigma) f_\sigma, \quad \text{while} \quad \mu_{\varphi_0} = \sum_{\sigma} \varphi_0(\sigma)\mu_\sigma.$$

In other words, we have the formula

$$\int_{Z} \varphi_0 \otimes \varphi_p\, d\mu_f = \int_{\mathbf{Z}_p} \varphi_p\, d\mu_g \quad \text{where } g = f_{\varphi_0}. \tag{10}$$

If $Z_0 = \mathbf{Z}(d) = \mathbf{Z}/d\mathbf{Z}$, then we may take for φ_0 an additive character, which is of the form

$$\psi_0 : x_0 \mapsto \zeta_0^{x_0}, \qquad x_0 \in \mathbf{Z}(d),$$

for some d-th root of unity ζ_0. In that case, the measure μ_g above will also be denoted by

$$\mu_{\zeta_0} \quad \text{or} \quad \mu_{\psi_0}$$

if ψ_0 is the above character.

Let $N = dp^n$. An N-th root of unity ζ has a unique product decomposition

$$\zeta = \zeta_0 \zeta_p$$

where ζ_0 is a d-th root of unity, and ζ_p is a p^n-th root of unity. Then we have, by definition and (10):

$$\int_{Z} \zeta_0^{x_0}\zeta_p^{x_p}\varphi_p(x_p)\, d\mu(x) = \int_{\mathbf{Z}_p} \zeta_p^{x_p}\varphi_p(x_p)\, d\mu_{\zeta_0}(x_p). \tag{11}$$

In particular, for $\varphi_p = 1$,

$$\int_Z \zeta_0^{x_0}\zeta_p^{x_p}\, d\mu(x) = \int_{\mathbf{Z}_p} \zeta_p^{x_p}\, d\mu_{\zeta_0}(x_p) \tag{12}$$

where μ_{ζ_0} is the measure whose associated power series is

$$f_{\zeta_0} = \sum_{x_0 \in \mathbf{Z}(d)} \zeta_0^{x_0} f_{x_0}. \tag{13}$$

For this formalism it is therefore useful to define $x \in Z$ by its two components, $x = (x_0, x_p)$, where $x_0 \equiv x \bmod d$, and $x_p \equiv x \bmod p^n$. Then

$$\zeta^x = \zeta_0^{x_0}\zeta_p^{x_p},$$

so that the above integral can be written more simply with ζ^x.

By the orthogonality of characters, it is then immediate that we can recover f_{x_0} from f_{ζ_0} by means of the formula

$$f_{x_0} = \frac{1}{d}\sum_{\zeta_0} \zeta_0^{-x_0} f_{\zeta_0}. \tag{14}$$

§2. The Associated Analytic Function on the Formal Multiplicative Group

Katz [Ka 3] has shown that in addition to the power series associated with a measure, it is also useful to associate an analytic function on the "formal multiplicative group." The formalism is similar to the one of power series developed in Chapter 4, but it is more convenient in some situations, especially when dealing with the extra factor $\mathbf{Z}(d)$ in the composite case. Again, Katz's formalism makes certain constructions due to Iwasawa and Leopoldt appear completely natural, and we reproduce this formalism below.

We fix a positive integer d prime to p. In the sequel we let N denote any positive integer of the form dp^n, and again we let $Z = \mathbf{Z}(d) \times Z_p$, so that $Z_0 = \mathbf{Z}(d)$. Let

$$T = 1 + X$$

as usual. Let $\mathfrak{m}$ be the maximal ideal in $\mathfrak{o}_{\mathbf{C}_p}$. A function R on $1 + \mathfrak{m}$ is called **analytic** if there exists a power series f with coefficients in $\mathbf{C}_p$, converging on $\mathfrak{m}$, such that

$$R(1 + z) = f(z) \quad \text{for all } z \in \mathfrak{m}.$$

Let η be a d-th root of unity. We say that R is **analytic** on $\eta(1 + \mathfrak{m})$ if there exists a power series f as above such that

$$R(\eta(1 + z)) = f(z) \quad \text{for all } z \in \mathfrak{m}.$$

Then f is uniquely determined by R, and is said to be **associated** with R.

Example. The function $\log T = \log(1 + X)$ is analytic in the above sense. In the applications, we shall deal principally with this log, or with rational functions.

We let $\mathbf{G}_p(d)$ be the (disjoint) union

$$\mathbf{G}_p(d) = \bigcup_{\eta^d = 1} \eta(1 + \mathfrak{m}).$$

A function R is called **analytic** on $\mathbf{G}_p(d)$ if it is analytic on each "component" $\eta(1+\mathfrak{m})$. Then R consists of a family $\{R_\eta\}$ where each R_η is analytic on the corresponding component. If $\zeta \in \eta(1+\mathfrak{m})$, we use the notation

$$\zeta = \eta u, \qquad \eta = \zeta_0 = \omega(\zeta), \qquad u = \zeta_p = \langle \zeta \rangle_p.$$

Then

$$R(\zeta) = R_{\omega(\zeta)}(\zeta) = f_{\omega(\zeta)}(\zeta_p - 1).$$

We shall call $\mathbf{G}_p(d)$ the **formal multiplicative group** (at p, of level d).

Remark. The group $\mathbf{G}_p(d)$ may be viewed as the group of (continuous) $\mathbf{C}_p^*$-valued characters on $Z = \mathbf{Z}(d) \times \mathbf{Z}_p$, by the mapping

$$\psi_\zeta : x \mapsto \zeta^x = \zeta_0^{x_0} \zeta_p^{x_p},$$

for each $\zeta \in \mathbf{G}_p(d)$.

Let μ be a measure on Z, with values in $\mathfrak{o}$. An analytic function R on $\mathbf{G}_p(d)$ will be said to be **associated with** μ if the power series associated with R have coefficients in $\mathfrak{o}$, and if we have the formula

$$\int_Z \zeta^x \, d\mu(x) = R(\zeta)$$

for all N-th roots of unity ζ, $N = dp^n$ (all n). The Weierstrass preparation theorem shows that an associated analytic function R, if it exists, is uniquely determined by μ. Of course, we want the above formula to be true also for all $\zeta \in \mathbf{G}_p(d)$, but we can prove it later, when ζ is not a root of unity.

We shall say that μ is **rational** if R is a rational function of T (and hence of X). Thus the right-hand side $R(\zeta)$ is just the value of this rational function at $T = \zeta$. We then call R the **associated rational function**. We also observe that the rational function is the same (if it exists) as that associated with the restriction of μ to

$$\{0\} \times \mathbf{Z}_p.$$

Indeed, we have from the definition of a distribution:

$$R(\zeta_p) = \int_Z \zeta_p^{x_p}\, d\mu(x) = \int_{\mathbf{Z}_p} \zeta_p^{x_p}\, d\mu(0, x_p).$$

R 1. *A measure always has an associated analytic function R. If $d = 1$ and μ is a measure on $\mathbf{Z}_p$, then*

$$R(T) = f(X)$$

is the associated power series of Chapter 4.

Proof. For a measure on $\mathbf{Z}_p$ with $d = 1$, this follows from **Meas 0** and **Meas 2** of Chapter 4, §2. The general case then follows at once. Note that in **Meas 0**, we have

$$f(0) = R(1).$$

Furthermore, if μ is a measure on $Z = \mathbf{Z}(d) \times \mathbf{Z}_p$, then the measure μ_{ζ_0} at the end of the last section is a measure on $\mathbf{Z}_p$ to which we can apply **R 1**, with $d = 1$.

R 2. *If a measure μ on Z has an associated rational function $R(T)$, then the measure μ_{ζ_0} has an associated rational function, given by*

$$R_{\zeta_0}(T) = R(\zeta_0 T) = f_{\zeta_0}(X).$$

Proof. This is immediate from the definition of the associated rational function, and the integral formula (12) of §1.

Lemma 2.1. *Assume that μ is a measure on Z such that each μ_{ζ_0} is rational, for every d-th root of unity ζ_0. Assume also that the functions*

$$R_{\zeta_0}(\zeta_0^{-1} T) = R(T)$$

are independent of ζ_0. *Then* μ *is rational, and its associated rational function is* $R(T)$.

Proof. This is again immediate from the integral formula (12) of §1.

R 3. *Let* μ *be a measure having associated analytic function* R. *Then the formula*

$$\int_{\mathbf{Z}} \zeta^x \, d\mu(x) = R(\zeta)$$

holds for any $\zeta \in \mathbf{G}_p(d)$.

Proof. This is a direct consequence of **R 1**, formula (12) of §1, and the analogous result for measures on $\mathbf{Z}_p$ stated in Theorem 1.2 of Chapter 4.

R 4. *Let* $\zeta \in \mathbf{G}_p(d)$. *Let* R *be the associated analytic function of* μ. *Then*

$$\zeta^x \mu(x) \textit{ has associated analytic function } R(\zeta T).$$

Proof. Immediately from **R 3**, since $\zeta_1^x \zeta_2^x = (\zeta_1 \zeta_2)^x$.

Just as with associated power series, **R 4** allows us to use the Fourier expansion of any step function to obtain its associated analytic function. Thus let φ be a step function of level M which is a divisor of dp^n (possibly a pure power of p). Then

$$\varphi(x) = \sum_{\zeta^M = 1} \phi(\zeta)\zeta^x$$

and

$$\phi(\zeta) = \frac{1}{M} \sum_{x \in \mathbf{Z}(M)} \varphi(x)\zeta^{-x}.$$

We shall now give a first example of the use of such expansions in connection with the **unitization operator U** defined by the formula

$$\mathbf{U}R(T) = R(T) - \frac{1}{p} \sum_{\zeta^p = 1} R(\zeta T).$$

Let

$$Z^* = \mathbf{Z}(d) \times \mathbf{Z}_p^*.$$

R 5. *Let φ be the characteristic function of Z^*. If R is associated with μ, then*

$$UR \text{ is associated with } \varphi\mu.$$

Proof. The Fourier expansion of the characteristic function of $\mathbf{Z}_p^*$ was already computed in Chapter 4, §2 and is trivially determined to be given by

$$\hat{\varphi}(\zeta) = \begin{cases} -1/p & \text{if } \zeta \neq 1. \\ \dfrac{p-1}{p} & \text{if } \zeta = 1, \end{cases}$$

and ζ ranges over p-th roots of unity. Property **R 5** then follows from **R 4** and the Fourier expansion.

R 6. *Let $N = dp^n$ where $n \geq 0$. Let χ be a Dirichlet character of conductor N, and let ζ be a primitive N-th root of unity. If R is the analytic function associated with μ, then the analytic function associated with $\chi\mu$ is*

$$\frac{S(\chi, \zeta)}{N} \sum_{a \in \mathbf{Z}(N)^*} \bar{\chi}(a) R(\zeta^{-a} T).$$

where

$$S(\chi, \zeta) = \sum \chi(a) \zeta^a.$$

Proof. The computation of the Fourier transform of χ is routine and is left to the reader. (One has of course to use Theorem 1.1 of Chapter 3, §1.)

R 7. *Let R be the analytic function associated with μ. Let $D = TD_T$ and let k be an integer ≥ 0. Then*

$$x_p^k \mu(x) \text{ has associated analytic function } D^k R(T).$$

The same statement holds if μ is rational, and "analytic" is replaced by "rational" in the above statement.

Proof. We may use $\varphi_p(x_p) = x_p^k$ in (11) of §1, and then apply **Meas 6** of Chapter 4, §2 to each one of the measures μ_{ζ_0}, after using **R 1**. Thus the general case is reduced to the special case of measures on $\mathbf{Z}_p$.

R 8. *Let $R = \mathbf{U}R$ be the associated analytic function of a measure μ with support in Z^*. Then the measure*

$$x_p^{-1}\mu(x) \text{ has associated analytic function } \mathbf{U}H,$$

where H is any analytic function on $\mathbf{G}_p(d)$ such that

$$DH = R.$$

Proof. Since $x_p^{-1}\mu(x)$ is a measure on Z^*, there exists an analytic function F associated with it. Then **R 7** yields

$$DF = R.$$

We can let $H = F$. Since the kernel of D consists of the functions which are constant on each component of $\mathbf{G}_p(d)$, if we select any H such that $DH = R$, then $\mathbf{U}H$ will have the same value, independent of the choice of H, thus proving the desired property.

The "integration" of the analytic function $R(T)$ can be performed formally in terms of the variable X. Indeed, say on the coset $1 + \mathfrak{m}$, if R is defined by a power series $f(X)$ then an analytic function H such that $DH = R$ is given in terms of X by

$$H(T) = \int F(X) \frac{dX}{1+X} = \int \frac{F(X)}{1+X} dX = \int R(T) \frac{dT}{T}.$$

Remark. The formalism of this section is set up to extend at once to its adelization over d prime to p.

§3. Computation of $L_p(1, \chi)$ in the Composite Case

Let $E_{1,c}$ be the measure defined in Chapter 10, §2, giving rise to the p-adic L-function and the Bernoulli distribution, regularized with c so as to be integral valued. We shall apply the previous considerations to this measure.

By definition, we have for $s = 0$ the value

$$L_p(1, \chi) = \frac{-1}{1 - \chi(c)} \int_Z \chi(a) a_p^{-1} \, dE_{1,c}(a).$$

The conductor of χ being $dp^n = N$, the integral might as well be taken over

$$Z^{**} = \mathbf{Z}(d)^* \times \mathbf{Z}_p^*.$$

In any case, if we find (as we shall do below) that $E_{1,c}$ has an associated rational function, then the general formalism also yields successively the corresponding analytic function for the measure

$$\chi(a)a_p^{-1}E_{1,c}(a).$$

We may then evaluate this analytic function at $T = 1$ to get the value $L_p(1, \chi)$.

We shall now carry out on $\mathbf{Z}(d) \times \mathbf{Z}_p$ the same analysis that we did for $\mathbf{Z}_p$ in Chapter 4, concerning the associated power series, and the Leopoldt formula for the value of the L-function at $s = 1$.

We recall that if c is a positive integer prime to dp, then the measure $E_{1,c}$ on $\mathbf{Z}_p$ has an associated rational function (equal to its associated power series by **R 1**), which is

$$R_{1,c}(T) = \frac{1}{T-1} - \frac{c}{T^c - 1}$$

according to Proposition 3.4 of Chapter 4.

Proposition 3.1. *Let c be a positive integer prime to dp. Then $E_{1,c}$ on $\mathbf{Z}(d) \times \mathbf{Z}_p$ has an associated rational function, equal to $R_{1,c}(T)$ above.*

Proof. To extend the above result from $\mathbf{Z}_p$ to $\mathbf{Z}(d) \times \mathbf{Z}_p$, it suffices to prove that for every root of unity $\zeta \in \boldsymbol{\mu}_N$, and $N = dp^n$, we have

$$\int_Z \zeta^x \, dE_{1,c}(x) = R_{1,c}(\zeta).$$

By definition, essentially (cf. formula **B 6** of Chapter 2, §2) we know that for any function φ on $\mathbf{Z}(N)$ we have

$$\sum_{k=0}^{\infty} B_{k,\varphi} \frac{Z^k}{k!} = \sum_{a=0}^{N-1} \varphi(a) \frac{Ze^{aZ}}{e^{NZ} - 1}.$$

We apply this to $\varphi(x) = \zeta^x$ and $\varphi(x) = \zeta^{cx}$. Summing a geometric series from 0 to $N-1$ then yields

$$\sum_{k=1}^{\infty} \left(\frac{1}{k} B_{k,\varphi} - c^k \frac{1}{k} B_{k,\varphi \circ c} \right) \frac{Z^{k-1}}{(k-1)!} = \frac{1}{\zeta T - 1} - \frac{c}{\zeta^c T^c - 1},$$

where $T = e^Z$. The right-hand side at $T = 1$ is the same as $R_{1,c}(\zeta)$, and is also the same as the left-hand side at $Z = 0$. But that is precisely the value of the desired integral for $k = 1$. This proves Proposition 3.1.

Proposition 3.2. *Let χ have conductor $N = dp^n$ with $n \geq 0$. Then $\chi E_{1,c}$ has an associated rational function $R_{\chi,c}$ given by the formula*

$$R_{\chi,c}(T) = G_\chi(T) - c\chi(c)G_\chi(T^c),$$

where for any primitive N-th root of unity ζ,

$$G_\chi(T) = \frac{S(\chi, \zeta)}{N} \sum_{a \in \mathbf{Z}(N)^*} \bar{\chi}(a) \frac{1}{\zeta^{-a}T - 1}.$$

Proof. Special case of **R 6**.

We can write $R_{\chi,c}(T)$ in full in the form:

$$R_{\chi,c}(T) = \frac{S(\chi, \zeta)}{N} \sum_a \bar{\chi}(a)\left[\frac{\zeta^a}{T - \zeta^a} - \frac{c\chi(c)\zeta^a}{T^c - \zeta^a}\right].$$

This puts us in the position of applying **R 8**, and of finding an analytic function $H_{\chi,c}$ such that $H_{\chi,c}(1) = 0$, and

$$DH_{\chi,c} = R_{\chi,c}.$$

We let λ range over c-th roots of unity, and we let

$$H_{\chi,c}(T) = \frac{-S(\chi, \zeta)}{N} \sum_{\lambda \neq 1} \sum_{a \in \mathbf{Z}(N)^*} \bar{\chi}(a) \log \frac{T - \lambda\zeta^a}{1 - \lambda\zeta^a}.$$

Exactly the same verification as in Chapter 4, §3 (or a direct integration using partial fractions) shows that this value for $H_{\chi,c}(T)$ satisfies our requirements.

By **R 6**, the analytic function associated with the measure $\chi(a)a_p^{-1}E_{1,c}(a)$ is $\mathrm{U}H_{\chi,c}(T)$.

Theorem 3.3. *Let χ have conductor $N = dp^n$, $n \geq 0$. Let ζ be a primitive N-th root of unity. Then*

$$L_p(1, \chi) = -\left(1 - \frac{\chi(p)}{p}\right) \frac{S(\chi, \zeta)}{N} \sum_{a \in \mathbf{Z}(N)^*} \bar{\chi}(a) \log_p(1 - \zeta^a).$$

Proof. Let ξ range over the p-th roots of unity. By **R 6**,

$$-(1 - \chi(c))L_p(1, \chi) = \mathrm{U}H_{\chi,c}(1).$$

Hence, following exactly the proof of Chapter 4, Theorem 3.6:

$$-(1-\chi(c))L_p(1,\chi) = \frac{1}{p}\frac{S(\chi,\zeta)}{N}\sum_{\xi}\sum_{a}\sum_{\lambda\neq 1}\bar{\chi}(a)\log_p\left(\frac{\xi-\lambda\zeta^a}{1-\lambda\zeta^a}\right)$$

$$= \frac{1}{p}\frac{S(\chi,\zeta)}{N}\sum_{a}\bar{\chi}(a)\log_p\left(\prod_{\lambda\neq 1}\prod_{\xi}\frac{\xi-\lambda\zeta^a}{1-\lambda\zeta^a}\right).$$

But

$$\prod_{\lambda\neq 1}\prod_{\xi}\frac{\xi-\lambda\zeta^a}{1-\lambda\zeta^a} = \prod_{\lambda\neq 1}\frac{1-\lambda\xi^{ap}}{(1-\lambda\zeta^a)^p}.$$

If $p \mid N$, then as in Theorem 3.6 of Chapter 4 we find that

$$\sum_{a}\bar{\chi}(a)\log_p(1-\lambda\zeta^{ap}) = 0,$$

and the rest of the proof is identical with the previous one. If $p \nmid N$, then we change variables, letting $a \mapsto p^{-1}a \bmod N$. We then find

$$\sum_{a}\bar{\chi}(a)\log_p(1-\lambda\zeta^{ap}) = \chi(p)\sum_{a}\bar{\chi}(a)\log_p(1-\lambda\zeta^{a}).$$

Thus we get an extra term besides that of Chapter 4, which gives rise to the factor $(1-\chi(p)/p)$ as stated in the theorem. Except for that, the proof is again identical with the previous one.

Divisibility of Ideal Class Numbers 13

Classical results of Kummer give bounds for certain class numbers and estimate, for instance, the plus part in terms of the minus part for cyclotomic fields.

Such estimates have been carried out more systematically and generally by Iwasawa, to include estimates for his invariants, and will be given in the first two sections of this chapter.

Next we deal with the l-part of the class number in a cyclic extension of degree p^n when p is a prime $\neq l$. We also give examples of Iwasawa of non-cyclotomic $\mathbf{Z}_p$-extensions when the order of the class number grows exponentially. These examples are based on a classical formula (Takagi–Chevalley) expressing the fact that in a highly ramified extension, the ramified primes have a strong tendency to generate independent ideal classes.

We conclude with a lemma of Kummer which is still somewhat isolated. In this connection, cf. [Wa 5].

§1. Iwasawa Invariants in $\mathbf{Z}_p$-extensions

Let K be a number field and K_∞ a $\mathbf{Z}_p$-extension. We let K_n denote the subfield of degree p^n over K, so $K_0 = K$.

Let C_n be the p-primary part of the ideal class group of K_n. We also write $C(K)$ for the *p-primary part* of the ideal class group of K. Then by Iwasawa theory of Chapter 5, we have

$$|C_n| = p^{e_n} \quad \text{where } e_n = mp^n + \lambda n + O(1).$$

We call m, λ the **Iwasawa invariants of the $\mathbf{Z}_p$-extension**. We call m the **exponential invariant**, and λ the **linear invariant**. We indicate the dependence of m and λ on the $\mathbf{Z}_p$-extension by the notation.

$$m = m(K_\infty/K) \quad \text{and} \quad \lambda = \lambda(K_\infty/K).$$

If K_∞ is the cyclotomic $\mathbf{Z}_p$-extension, then these invariants depend only on K. We may then write more briefly

$$m = m(K) \quad \text{and} \quad \lambda = \lambda(K).$$

Put

$$V_K = C(K_\infty/K) = \varprojlim C(K_n).$$

From the structure theorem of Chapter 5, we have a quasi-isomorphism

$$\boxed{V_K \sim \prod \Lambda/p^{m_i} \oplus \prod \Lambda/f_j.}$$

For any abelian group V we let $V^{(p)}$ be its p-primary part, i.e. the subgroup of elements annihilated by a power of p. Then we have quasi-isomorphisms

$$V_K^{(p)} \sim \prod \Lambda/p^{m_i} \quad \text{and} \quad V_K/V_K^{(p)} \sim \prod \Lambda/f_j.$$

From the structure Theorems 1.2 and 1.3 of Chapter 5, we know that

$$\boxed{\lambda(K_\infty/K) = \sum \deg f_j.}$$

Furthermore, we have the characterization:

λ is the rank of $V_K/V_K^{(p)}$ as a finitely generated $\mathbf{Z}_p$-module.

For the exponential invariant, we have

$$\boxed{m = \sum m_i.}$$

We shall now compare these invariants in a $\mathbf{Z}_p$-extension and its lifting over a finite extension.

First note that if $F \subset K$ and F_∞ is a $\mathbf{Z}_p$-extension of F, then $K_\infty = KF_\infty$ is a $\mathbf{Z}_p$-extension of K, and there is some integer r such that

$$K_n = KF_{n+r}.$$

Lemma 1. *Let $F \subset K$ be number fields. Let F_∞ be a $\mathbf{Z}_p$-extension of F. Then*

$$m(F_\infty/F) \leq m(KF_\infty/K) \quad \text{and} \quad \lambda(F_\infty/F) \leq \lambda(KF_\infty/K).$$

Proof. The degree $[K_n : KF_n]$ is bounded by a fixed power p^r. By class field theory, we have

$$(C(F_n) : N_{K_n/F_n} C(K_n)) \leq [K : F]p^r.$$

Hence

$$|C(K_n)| \gg |C(F_n)|,$$

and therefore

$$\operatorname{ord}_p C(K_n) \geq \operatorname{ord}_p C(F_n) - O(1).$$

This proves the first inequality in light of the formula for the orders.

For the second, let us use the functorial notation

$$C(F_\infty/F) = \varprojlim C(F_n).$$

The norm maps

$$N_{K_n/F_n} : C(K_n) \to C(F_n)$$

are compatible with the norm maps in the projective limits in the tower over F_0 and K_0 respectively, and thus induce a homomorphism

$$N : C(K_\infty/K) \to C(F_\infty/F).$$

Since the index of the image of N_{K_n/F_n} is bounded for all n (as in the first part of the proof), it follows that the image of N in $C(F_\infty/F)$ is of finite index, in other words that N is quasi-surjective.

Now put $V_F = C(F_\infty/F)$ and similarly for V_K. Let $V^{(p)}$ as usual denote the subgroup of elements annihilated by a power of p. Then we have a quasi-surjective homomorphism

$$V_K/V_K^{(p)} \to V_F/V_F^{(p)}.$$

The inequality for the λ-invariants then follows at once since λ is the rank of the above modules as finitely generated $\mathbf{Z}_p$-modules.

We have been concerned with the p-divisibility of the order of the ideal class group in $\mathbf{Z}_p$-extensions. An analogue of Lemma 1 can be given for any prime number.

Lemma 2. *Let F_∞ be a $\mathbf{Z}_p$-extension of a number field F. Let K be a finite extension of F and let $K_\infty = KF_\infty$. Let l be any prime number. If* $\operatorname{ord}_l |C(K_n)|$ *is bounded, then* $\operatorname{ord}_l |C(F_n)|$ *is bounded.*

Proof. The argument using the norm index in the first part of the proof of Lemma 1.1 applies equally well to prove the result stated in Lemma 1.2.

Next we study another Iwasawa invariant, the rank.

If A is an abelian group, we define the p-**rank**,

$$\mathbf{rank}_p(A) = \text{dimension of } A/pA \text{ over the prime field } \mathbf{F}_p.$$

We have a criterion for the vanishing of the exponential Iwasawa invariant in terms of this p-rank of the groups C_n. We phrase the criterion to apply in general to torsion modules over the Iwasawa algebra.

We use the notation of Chapter 5, Theorem 1.3, where we dealt with a module of Iwasawa type, and say

$$V \sim \prod_{i=1}^{r} \Lambda/p^{m_i} \oplus \prod_{j=1}^{t} \Lambda/f_j,$$

where the f_j are distinguished polynomials. We let

$$r_1(V) = r$$

be the number of factors of type Λ/p^m for some m. As in Chapter 5, we let $v_1, \ldots, v_s$ be elements of V such that if we put

$$U_0 = \mathbf{Z}_p\text{-submodule of } V \text{ generated by } (\gamma - 1)V \text{ and } v_1, \ldots, v_s,$$
$$U_n = g_n U_0 \text{ with } g_n = 1 + \gamma + \cdots + \gamma^{p^n - 1},$$

then

$$V_n = V/U_n$$

is finite for all n. We put $e_n = e(V_n) = \operatorname{ord}_p V_n$. The proofs of Theorem 1.2 and Theorem 1.3 of Chapter 5 distinguish the two cases of modules of type

$$\Lambda/p^m \quad \text{and} \quad \Lambda/f$$

where f is distinguished. They immediately show the following result.

Lemma 3. *We have* $\operatorname{rank}_p V_n = r_1(V)p^n + O(1)$, *where* $r_1(V)$ *is the number of factors of type* Λ/p^m *above. Furthermore, the following conditions are equivalent.*

(i) *All the factors* Λ/p^{m_i} *are equal to* 0, *so* $r_1(V) = 0$ *and*

$$V \sim \prod \Lambda/(f_j).$$

(ii) *The p-rank of* V_n *is bounded independently of n.*
(iii) *In the order formula* $e_n = mp^n + \lambda n + O(1)$, *the exponential invariant m is equal to* 0.

§2. CM Fields, Real Subfields, and Rank Inequalities

We now wish to make a comparison of the behavior of the ideal class group in a field and a real subfield. We prove Kummer's theorem (Theorem 2.2) and Kummer type theorems in the following context.

A number field K is said to be a CM **field** (complex multiplication field) if it is a totally imaginary quadratic extension of a totally real field. We leave it as an exercise to prove:

K is a CM *field if and only if the following condition is satisfied. Let* ρ *be complex conjugation. Then* $\rho\sigma = \sigma\rho$ *for all embeddings* σ *of* K *into the complex numbers, and* K *is not real.*

The totally real subfield of K is then uniquely determined, and denoted by K^+. It also follows that a composite of CM fields is a CM field.

Although we are primarily interested in the cases when the CM field is abelian over the rationals, it is just as easy to deal with the more general case. However, we have to restate some results of Chapter 3 in this context.

Let K be a CM field. Let W_K be the group of roots of unity in K. The **Hasse index** Q_K can be defined as for abelian fields over the rationals, namely

$$Q_K = (E_K : W_K E_K^+),$$

where E_K is the group of units in K, while $E_K^+ = E(K^+)$ is the group of units in the real subfield.

We have:

Lemma 1. $Q_K = 1$ *or* 2.

The proof given for Theorem 4.1 of Chapter 3 applies here. In fact one verifies immediately that the map $u \mapsto \bar{u}/u$ gives an injection

$$E_K/W_K E_K^+ \to W_K/W_K^2.$$

Lemma 2. *Let K be a CM field and p a prime. Let $C^{(p)}(K)$ be the p-primary part of the ideal class group of K.*

(i) *The natural map $C^{(p)}(K^+) \to C^{(p)}(K)$ is injective if p is odd, and its kernel has order 1 or 2 if $p = 2$.*
(ii) *The norm map*

$$N_{K/K^+} : C(K) \to C(K^+)$$

is surjective.

Proof. For p odd this is a special case of the similar (trivial) lemma concerning ideal classes in extensions whose degree is prime to p. The proof will be given in full in the next section when we deal with this case for its own sake.

Suppose $p = 2$. Let $\mathfrak{a}$ be an ideal of K^+ and suppose $\mathfrak{a} = (\alpha)$ with α in K. Then $\bar{\alpha}/\alpha$ is a unit, and in fact a root of unity (the absolute value of all its conjugates is 1). We had defined above the map $\varphi : E_K \to W_K$ by $u \mapsto \bar{u}/u$. The association $\mathfrak{a} \mapsto \bar{\alpha}/\alpha$ then gives a well defined map

$$\operatorname{Ker}[C(K^+) \to C(K)] \to W_K/\varphi(E_K),$$

which is immediately verified to be injective. Hence the kernel has order 1 or 2. This proves (i).

The proof of (ii) is identical with that given for Theorem 4.3 of Chapter 3. It is a routine lemma of class field theory.

The eigenspace $C(K)^-$ is the kernel of the norm map N_{K/K^+} in $C(K)$, and for an odd prime p, we can identify the p-primary parts

$$C^{(p)}(K^+) = C^{(p)}(K)^+.$$

Unless otherwise specified, we continue to assume that p is an odd prime.

Let $K = K_0$ be a CM field, and let K_∞ be a $\mathbf{Z}_p$-extension such that each K_n is a CM field.

Remark. As Washington observes [Wa 2], if Leopoldt's conjecture is true, then K_∞ is necessarily the cyclotomic $\mathbf{Z}_p$-extension, as follows from Theorem 6.2 of Chapter 5.

The real subfield K_∞^+ is a $\mathbf{Z}_p$-extension of K_0^+. It is then clear that

$$K_\infty = KK_\infty^+,$$

so K_∞ is the lifting over K of the $\mathbf{Z}_p$-extension K_∞^+ of K_0^+.

Let $C_n = C_n^{(p)}$ be the p-primary part of the ideal class group of K_n. We have the orders

$$\operatorname{ord}_p C_n = e_n, \qquad \operatorname{ord}_p C_n^- = e_n^-, \qquad \operatorname{ord}_p C_n^+ = e_n^+,$$

and $e_n = e_n^+ + e_n^-$. Similarly, we have the Iwasawa invariants associated with the Λ-modules

$$C = \varprojlim C_n, \qquad C^- = \varprojlim C_n^-, \qquad C^+ = \varprojlim C_n^+,$$

and we denote these invariants by m, λ with a plus or minus sign as superscript corresponding to the two cases. Then

$$m = m^- + m^+ \quad \text{and} \quad \lambda = \lambda^- + \lambda^+.$$

Indeed, we have for an odd prime p,

$$C = C^- \oplus C^+.$$

Remark. For $p = 2$ one has to be more careful, and for instance, one has to distinguish $C(K^+)$ from its natural image in $C(K)$, which we might denote by C^+. Cf. Lemma 2, and also Lemma 4.1 below.

Theorem 2.1. *Let p be an odd prime. Let K be a* CM *field, and assume that the p-th roots of unity are in K. Let C be the p-primary part of the ideal class group of K. Then*

(i) $$\operatorname{rank}_p C^+ \leq \operatorname{rank}_p C^- + 1.$$

Let W_K be the group of roots of unity in K. If $K(W_K^{1/p})$ is ramified over K, then

(ii) $$\operatorname{rank}_p C(p)^+ \leq \operatorname{rank}_p C(p)^-.$$

Proof. Let L be the maximal abelian extension of K of exponent p. Let $G = \operatorname{Gal}(L/K)$. By class field theory,

$$G \approx C(p) \quad \text{and} \quad G^+ \approx C(p)^+.$$

Since the p-th roots of unity are assumed to be in K, the extension L is a Kummer extension. Let B be its Kummer group containing K^{*p}, so that

$$K^* \supset B \supset K^{*p} \quad \text{and} \quad L = K(B^{1/p}).$$

We have the exact Kummer duality

$$G \times B/K^{*p} \to \boldsymbol{\mu}_p.$$

Since p is assumed to be odd, we have a direct product decomposition

$$G = G^+ \times G^-.$$

For simplicity of notation, let

$$V = B/K^{*p} = V^+ \times V^-.$$

Since $\boldsymbol{\mu}_p$ is a (-1)-eigenspace for complex conjugation, it follows that we have an exact pairing

$$G^+ \times V^- \to \boldsymbol{\mu}_p,$$

and therefore

$$\operatorname{rank}_p G^+ = \operatorname{rank}_p C(p)^+ = \operatorname{rank}_p V^-.$$

On the other hand, if $b \in B$ we know that $K(b^{1/p})$ is unramified, and hence there exists an ideal $\mathfrak{b}$ of K such that $(b) = \mathfrak{b}^p$. The map $b \mapsto \mathrm{Cl}(\mathfrak{b})$ gives rise to a homomorphism

$$\varphi : V = B/K^{*p} \to C_p,$$

which induces a homomorphism

$$\varphi^- : V^- \to C_p^-$$

of V^- into C_p^- because φ commutes with complex conjugation. One then verifies immediately that we obtain an injective map

$$\operatorname{Ker} \varphi^- \to E_K(p)^-,$$

by writing $b \in B$ as $b = a^p u$ with $u \in E_K$ and mapping $b \mapsto u$. Since $(E_K : W_K E_K^+) = 1$ or 2, we obtain

$$E_K(p)^- = W_K(p)^- = W_K(p)$$

because $W_K = W_K^-$. Therefore

$$\operatorname{rank}_p W_K(p) = 1 = \operatorname{rank}_p E_K(p)^-,$$

and

$$\operatorname{rank}_p C(p)^+ = \operatorname{rank}_p V^- \leq 1 + \operatorname{rank}_p C_p^-.$$

This proves the first assertion. Furthermore, if $K(W_K^{1/p})$ is ramified over K, then B does not contain W_K, and consequently

$$\operatorname{Ker} \varphi^- = 1.$$

Hence the last inequality on the right can be replaced by the stronger inequality

$$\operatorname{rank}_p V^- \leq \operatorname{rank}_p C^-,$$

thus proving the theorem.

The next two results are really corollaries of the theorem, but we label them theorems in their own right in view of their importance. The first is a classical result of Kummer.

Theorem 2.2. *Let $K = \mathbf{Q}(\boldsymbol{\mu}_p)$, and let h_p be the class number of K. If h_p^- is prime to p, then h_p^+ is prime to p, and so h_p is prime to p.*

Proof. Obvious, since the p-rank is 0, and the stronger of the two inequalities applies.

Let K_∞ be a $\mathbf{Z}_p$-extension of K_0 such that each K_n is a CM field. Then C, C^+, C^- are modules over the Iwasawa algebra, and thus we have the invariants

$$r_1(C) = r_1(C^+) + r_1(C^-) = r_1^+ + r_1^-.$$

as defined in §1.

Theorem 2.3. *Let K_∞ be a $\mathbf{Z}_p$-extension of K_0 (p odd), such that each K_n is a* CM *field. Then*

$$r_1^+ \leq r_1^-.$$

In particular, if $m^- = 0$ then $m^+ = 0$.

Proof. Immediate from Theorem 2.1 and Lemma 3 of §1.

For the prime $p = 2$ the estimates are not as good, but one has a result of Greenberg (cf. Washington [Wa 2]).

Proposition 2.4. *Let K be a* CM *field, and $C = C(K)$. Then*

$$\operatorname{rank}_2 C_{K^+} \leq \operatorname{ord}_2 |C_K^-| + 1.$$

If K_∞ is a $\mathbf{Z}_2$-extension of K_0 such that each K_n is a CM field, and if $m^- = 0$ then $m^+ = 0$ also.

Proof. Let $r = \operatorname{rank}_2 C_{K^+}$. Then by Lemma 2,

$$2^r = (C_{K^+} : C_{K^+}^2) = (NC_K : NC^+),$$

where C^+ denotes the image of $C(K^+)$ in $C(K)$. This last index divides

$$(C_K : C^+) = \frac{|C_K|}{|C^+|},$$

which by Lemma 2 divides

$$\frac{2|C_K|}{|C(K^+)|} = 2h_K^-.$$

This proves the inequality. The statement about $m^- = 0$ then follows from the structure theorem, cf. Lemma 3 of §1, as for p odd.

§3. The l-primary Part in an Extension of Degree Prime to l

In this section we prove a lemma of Iwasawa used by Washington [Wa 2]. We begin by a trivial remark.

Lemma 1. *Let F be a number field and K a Galois extension of degree d. Let l be a prime number not dividing d. Let C_F denote the l-primary part of the ideal class group. Then the natural homomorphism $C_F \to C_K$ is injective, the norm $N_{K/F} : C_K \to C_F$ is surjective, and the following conditions are equivalent:*

(i) $C_F = C_K$.
(ii) $\operatorname{rank}_l C_F = \operatorname{rank}_l C_K$.
(iii) *The norm $N_{K/F} : C_K(l) \to C_F(l)$ is an isomorphism.*

Proof. For the first assertion, suppose an ideal $\mathfrak{a}$ of F becomes principal in K, say $\mathfrak{a} = (\alpha)$. Taking the norm yields

$$\mathfrak{a}^d = (N_{K/F}\alpha),$$

and since d is prime to l, it follows that $\mathfrak{a}$ is also principal. Abbreviate the norm by N. Since

$$NC_F = C_F^d = C_F,$$

it follows that the norm is surjective.

It is clear that (i) implies (ii). Assume (ii). Then the norm in (iii) is an isomorphism because it is surjective. It is also a G-isomorphism, where $G = \text{Gal}(K/F)$. Hence G acts trivially on $C_K(l)$, and therefore

$$N = d \cdot \text{identity on } C_K(l).$$

It follows that $C_F(l) = C_K(l)$. Nakayama's lemma or the structure theorem for abelian groups concludes the proof that $C_F = C_K$.

Lemma 2. *Let p be a prime number $\neq l$. Let K_n be a cyclic extension of a number field K_0, of degree p^n. Let f be the order of l mod p^n. Let C_ν be the l-primary part of the ideal class group in the subfield of degree p^ν. Let D_n be the kernel in the exact sequence*

$$0 \to D_n \to C_n(l) \xrightarrow{\text{Norm}} C_{n-1}(l) \to 0.$$

Then $D_n \neq 0$ if and only if $C_n \neq C_{n-1}$, and in that case,

$$\dim D_n \geq f.$$

Proof. Let $G = \text{Gal}(K_n/K_0)$. We have a representation of G on the $\mathbf{Z}(l)$-vector space $D = D_n$, and we first show that if $D \neq 0$, then the representation is faithful. If not, it is not injective on the unique cyclic subgroup of order p, namely $\text{Gal}(K_n/K_{n-1})$. Hence it is trivial on that subgroup, and therefore

$$N_{n,n-1} D = pD = 0,$$

whence $D = 0$.

Now assume $D \neq 0$. Then the tensor product of D with the algebraic closure of $\mathbf{F}_l$ splits as a G-direct sum

$$D \otimes \mathbf{F}_l^a = \bigoplus D_i,$$

where the D_i are irreducible components, of dimension 1. If the operation of G on every D_i is not faithful, then it is not faithful on the unique subgroup of order p, so not faithful on D itself. By what we have shown, it follows that G operates faithfully on some D_{i_0}. Hence G operates on this D_{i_0} by a representation into the p^n-th roots of unity, and a generator of G goes to a primitive p^n-th root of unity ζ. But D is defined over $\mathbf{F}_l$, so the conjugate representations by means of the conjugates $\zeta^l, \zeta^{l^2}, \ldots$ occur. So exactly f distinct conjugates occur among the D_i, so $\dim D \geq f$. This proves the lemma.

This last part of the argument proves the general statement:

Let D be a finite dimensional representation of a cyclic group G of order p^n, over the prime field $\mathbf{F}_l$. If D is faithful, then

$$\dim D \geq f,$$

where f is the order of l mod p^n.

Washington's application of the lemma then runs as follows.

Theorem 3.1. *Let K_∞ be a $\mathbf{Z}_p$-extension of K_0, and let C_n be the l-primary part of the ideal class group in K_n, where l is a prime $\neq p$. If the l-ranks of C_n are bounded, then the orders of C_n are also bounded for all n.*

Proof. Otherwise, we must have $D_n \neq 0$ for arbitrarily large n, and the order f_n of l mod p^n satisfies

$$f_n \gg p^n.$$

The preceding lemma would then imply that the ranks tend to infinity, a contradiction, which proves the theorem.

Remark. The lemma need not only be applied to a $\mathbf{Z}_p$-extension. For instance, given a number field F, and a prime number l, if K is cyclic of degree p over F, then either the l-rank of C_K tends to infinity, or $C_F = C_K$ as $p \to \infty$. It would be interesting to investigate for what sequence of cyclic extensions of degree p does the l-rank remain bounded, or tends to infinity.

Theorem 3.2. *Let K_∞ be a $\mathbf{Z}_p$-extension of K_0. Assume that each K_n is a* CM *field. Let l be a prime number $\neq p$, and assume that the l-th roots of unity are in K_0. If $\operatorname{ord}_l|C_n^-|$ is bounded, then $\operatorname{ord}_l|C_n^+|$ is also bounded.*

Proof. By Theorem 2.1 the l-rank of C_n^+ is bounded, so the l-rank of C_n is bounded. Then Theorem 3.1 concludes the proof.

§4. A Relation between Certain Invariants in a Cyclic Extension

The result of this section will be used later to give examples due to Iwasawa, of $\mathbf{Z}_p$-extensions in which orders of ideal class groups tend rapidly to infinity. Results like the first lemma are classical. The prime power case was known in the last century, and the general cyclic case is in Takagi and Chevalley's thesis on class field theory.

Let K be a cyclic extension of a number field F. Let $G = \mathrm{Gal}(K/F)$. For each (normalized) absolute value v of F we let $e(v)$ be the ramification index of v in K. If $v = v_{\mathfrak{p}}$ for some prime ideal $\mathfrak{p}$ of F, then $e(v) = e(\mathfrak{p})$ is the usual ramification index of $\mathfrak{p}$ in K. If v is Archimedean, then $e(v) = 1$ or 2 according as the local extension is trivial or of degree 2 (complex numbers over the real numbers). We put

$$e(K/F) = \prod_{v} e(v).$$

Then $e(K/F) = e_0(K/F)e_{\infty}(K/F)$, where

$$e_0(K/F) = \prod_{\mathfrak{p}} e(\mathfrak{p}) \quad \text{and} \quad e_{\infty}(K/F) = \prod e(v_{\infty}).$$

We let E denote the group of units and C the group of ideal classes. If G acts on a module A, we let A^G be the submodule of elements fixed under G. The next lemma implies that highly ramified primes have a tendency to generate independent ideal classes, and that the obstruction to this is contained in some cohomology.

Lemma 4.1. *Let K/F be a cyclic extension with Galois group G. Then*

$$|C_K^G| = \frac{h(F)e(K/F)}{[K:F](E_F : N_{K/F}K^* \cap E_F)}.$$

Proof. We assume that the reader is acquainted with a minimum of Galois cohomology for cyclic groups. What is needed is covered for instance in Chapter IX, §1 of my *Algebraic Number Theory*, referred to as *ANT*.

Let I denote the ideal group and P the principal ideal group. The group C_K^G occurs naturally at the beginning of the cohomology sequence associated with the exact sequence

$$0 \to P_K \to I_K \to C_K \to 0.$$

Note that I_K is the direct sum of its semilocal components over primes of F, and by semilocal theory, we have

$$H^1(G, I_K) = \bigoplus_{\mathfrak{p}} H^1(G_{\mathfrak{p}}, \mathbf{Z}),$$

where $G_{\mathfrak{p}}$ is the decomposition group. But $H^1(G_{\mathfrak{p}}, \mathbf{Z}) = 0$, so

$$H^1(G, I_K) = 0.$$

The exact cohomology sequence then yields

$$0 \to P_K^G \to I_K^G \to C_K^G \to H^1(G, P_K) \to 0,$$

whence an exact sequence

$$0 \to I_K^G/P_K^G \to C_K^G \to H^1(G, P_K) \to 0$$

which yields the index relation

$$|C_K^G| = (I_K^G : P_K^G)|H^1(P_K)|. \tag{1}$$

We analyze the two indices on the right-hand side.

From the inclusions

$$I_K^G \supset P_K^G \supset P_F$$

we obtain the index

$$\begin{aligned}(I_K^G : P_K^G) &= \frac{(I_K^G : P_F)}{(P_K^G : P_F)} \\ &= \frac{(I_K^G : I_F)(I_F : P_F)}{(P_K^G : P_F)} \\ &= e_0(K/F)\frac{h_F}{(P_K^G : P_F)}. \qquad (2)\end{aligned}$$

We now use the second exact sequence

$$0 \to E_K \to K^* \to P_K \to 0.$$

We get the beginning of the cohomology sequence

$$0 \to E_F \to F^* \to P_K^G \to H^1(E_K) \to 0$$

because $H^1(K^*) = 0$ by Hilbert's Theorem 90. Hence

$$\begin{aligned}(P_K^G : P_F) &= |H^1(E_K)| \\ &= |H^0(E_K)| \frac{[K:F]}{e_\infty(K/F)}\end{aligned}$$

(by Corollary 2 of Theorem 1, *ANT* Chapter IX), so by definition,

$$(P_K^G : P_F) = (E_F : N_{K/F}E_K)[K : F]/e_\infty(K/F). \tag{3}$$

Once more, from the second exact sequence, we have another portion of the cohomology sequence

$$0 = H^1(K^*) \to H^1(P_K) \to H^0(E_K) \to H^0(K^*).$$

The map from $H^0(E_K)$ to $H^0(K^*)$ is the natural homomorphism

$$E_F/N_{K/F}E_K \to F^*/N_{K/F}K^*.$$

By exactness, we find

$$\begin{aligned} |H^1(P_K)| &= |\mathrm{Ker}(E_F/N_{K/F}E_K \to F^*/N_{K/F}K^*)| \\ &= (N_{K/F}K^* \cap E_F : N_{K/F}E_K). \end{aligned} \tag{4}$$

Using the inclusions

$$E_F \supset (N_{K/F}K^* \cap E_F) \supset N_{K/F}E_K,$$

and putting (1), (2), (3), (4) together proves the lemma.

We shall apply the lemma as does Iwasawa [Iw 15] to get an example of a field with a highly divisible class number.

Lemma 4.2. *Let l be an integer ≥ 2. Let K_d be an extension of a number field K of degree d. Let $\mathfrak{q}_1, \ldots, \mathfrak{q}_t$ be prime ideals of K which split completely in K_d. Let K' be a cyclic extension of K, of degree l, in which $\mathfrak{q}_1, \ldots, \mathfrak{q}_t$ are totally ramified. Let $K'_d = K'K_d$. Then*

$$\frac{|C(K'_d)|}{|C(K_d)|} \quad \textit{is divisible by} \quad l^{(t-[K:\mathbf{Q}])d-1}.$$

Proof. We have the diagram

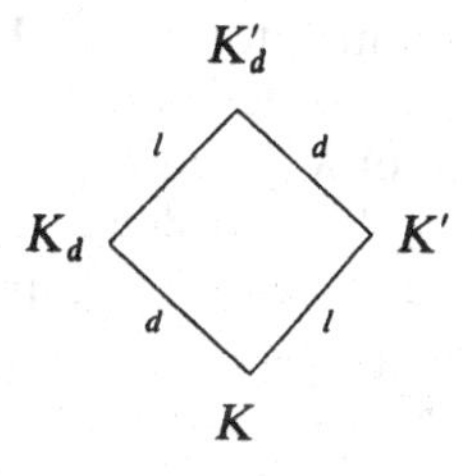

The extensions K_d and K' are linearly disjoint because of the way any one of the primes $\mathfrak{q}_i$ splits in them. Thus K'_d has degree l over K_d. We apply Lemma 4.1 to the cyclic extension K'_d over K_d. This yields:

$$\frac{|C(K'_d)|}{|C(K_d)|} \quad \text{is divisible by} \quad \frac{e(K'_d/K_d)}{[K'_d : K_d](E_d : E_d^l)},$$

where $E_d = E(K_d)$ is the group of units of K_d. Indeed, E_d^l is contained in the norm group from K'_d. All we have to do is now estimate the divisibility of the expression on the right.

Each ideal $\mathfrak{q}_i$ splits into d primes in K_d, and each factor in K_d is then totally ramified in K'_d. Consequently

$$e(K'_d/K_d) \quad \text{is divisible by} \quad l^{td}.$$

On the other hand,

$$[K'_d : K_d] = l.$$

Thirdly, E_d mod roots of unity has rank bounded by

$$[K_d : \mathbf{Q}] - 1 = d[K : \mathbf{Q}] - 1,$$

and consequently

$$(E_d : E_d^l) \text{ divides } l^{d[K:\mathbf{Q}]}.$$

Putting these three estimates together proves the lemma.

§5. Examples of Iwasawa

Let K be a number field and let K_∞ be a $\mathbf{Z}_p$-extension. Let $\mathfrak{q}$ be a prime ideal of K. The decomposition group of $\mathfrak{q}$ in $\Gamma = \text{Gal}(K_\infty/K)$ is either trivial or closed of finite index. In the first case, we say that $\mathfrak{q}$ **splits completely**, and in the second case, we say that $\mathfrak{q}$ is **finitely decomposed** in K_∞.

Let l be a prime number, which may be equal to p. We are interested in giving examples of Iwasawa [Iw 15] for which the l-primary part of the ideal class group C_n grows exponentially. We shall use the formula of the last section, applied to extensions where the ramification indices grow faster than the unit index in the denominator, times the degree. This implies that the class number on the left-hand side of the relation grows equally rapidly.

Let K' be a finite extension of K. Then $K_\infty K' = K'_\infty$ is a $\mathbf{Z}_p$-extension of K'. If $\mathfrak{q}$ splits completely in K_∞, then any divisor $\mathfrak{q}'$ of $\mathfrak{q}$ in K' splits completely in K'_∞. This is an elementary fact of algebraic number theory. In particular, if $\mathfrak{q}'_1, \ldots, \mathfrak{q}'_t$ in K' lie above t such primes $\mathfrak{q}_1, \ldots, \mathfrak{q}_t$, they will be distinct primes of K' splitting completely in K'_∞.

Theorem 5.1. *Let K_∞/K be a $\mathbf{Z}_p$-extension. Let $\mathfrak{q}_1, \ldots, \mathfrak{q}_t$ be prime ideals of K which split completely in K_∞. Let K' be a cyclic extension of K of degree l, in which $\mathfrak{q}_1, \ldots, \mathfrak{q}_t$ are totally ramified. Then*

$$\text{ord}_l |C(K'_n)| \geq (t - [K:\mathbf{Q}])p^n - 1.$$

Proof. This is merely a special case of Lemma 4.2.

A concrete example of the situation as in the theorem can be given once we have shown below how to construct a $\mathbf{Z}_p$-extension in which infinitely many primes of K split completely. Then we can take t arbitrarily large. We can then first lift the extension over the field $K(\boldsymbol{\mu}_l)$, and thus assume without loss of generality that K contains the l-th roots of unity. We then select an element α of K divisible exactly by the first power of $\mathfrak{q}_1, \ldots, \mathfrak{q}_t$ (and possibly other primes). We let

$$K' = K(\alpha^{1/l}).$$

Then the hypotheses of the theorem are satisfied.

Theorem 5.2. *Let K be a* CM *field. Then:*

(i) *There exists a $\mathbf{Z}_p$-extension K_∞ of K, Galois over K^+, such that if $\Gamma = \mathrm{Gal}(K_\infty/K)$, then*

$$\Gamma = \Gamma^-.$$

(ii) *For any such extension, let $\mathfrak{q}^+$ be a prime ideal of K^+ which does not divide p and remains prime in K, so $\mathfrak{q}$ is its unique extension in K. Then $\mathfrak{q}$ splits completely in K_∞. (Tchebotarev guarantees that there exist infinitely many such primes.)*

Proof. Let $M_p(K)$ be the maximal p-abelian p-ramified extension of K. By class field theory, e.g. Chapter 5, §5, there is a quasi-isomorphism

$$\mathrm{Gal}(M_p(K)/K) \sim U_p/\bar{E},$$

where U_p is the product of the local unit groups $U_\mathfrak{p}$ at the primes $\mathfrak{p}$ above p, and $\bar{E}$ is the closure of the global units in U_p. Each real prime $\mathfrak{p}^+$ either remains prime in K or splits into two primes $\mathfrak{p}_1, \mathfrak{p}_2$ in K. In either case, the semilocal component

$$\prod_{\mathfrak{p}|\mathfrak{p}^+} U_\mathfrak{p}$$

of U_p contains a subgroup of finite index isomorphic under the exponential map with

$$\prod_{\mathfrak{p}|\mathfrak{p}^+} p^m \mathfrak{o}_\mathfrak{p},$$

for m sufficiently large. Here $\mathfrak{o}_\mathfrak{p}$ denotes the local ring of integers at $\mathfrak{p}$. From this it is clear that U_p^- has $\mathbf{Z}_p$-rank ≥ 1. Furthermore, E contains a subgroup of finite index which is real, so $\bar{E}$ contains a subgroup of finite index which is fixed under complex conjugation. Put

$$\mathcal{G} = \mathrm{Gal}(M_p(K)/K),$$

and for simplicity of language, assume that p is odd. Then

$$\mathcal{G} = \mathcal{G}^+ \times \mathcal{G}^-,$$

and the $\mathbf{Z}_p$-rank of $\mathcal{G}^-$ is ≥ 1. Hence there exists a factor group Γ of $\mathcal{G}$ such that $\Gamma = \Gamma^-$, and Γ is isomorphic to $\mathbf{Z}_p$ as compact group. By Galois theory, Γ is the Galois group of a $\mathbf{Z}_p$-extension K_∞ of K, which is normal over K^+.

Let $\mathfrak{q}^+$ be a prime ideal of K^+ which remains prime in K. Let D be the decomposition group of $\mathfrak{q}^+$ in the group

$$\mathcal{G} = \mathrm{Gal}(K_\infty/K^+).$$

Then D is (topologically) cyclic because $\mathfrak{q}$ is unramified in K_∞. But $\mathcal{G}$ is "dihedral," in other words, $\mathcal{G}$ is generated by complex conjugation and Γ satisfying the relations

$$\rho\gamma\rho = \gamma^{-1},$$

for $\gamma \in \Gamma$. Since $\mathfrak{q}^+$ remains prime in K, we cannot have $D \subset \Gamma$. Then D contains an element $\rho\gamma$, and any such element has order 2. It is then immediately verified that D is cyclic of order 2 and that its intersection with Γ is 1. This proves that $\mathfrak{q}$ splits completely in K_∞, and concludes the proof.

Remark. If $p = 2$, then one has to take a factor group of $\mathcal{G}$ by $\mathcal{G}^{1+\rho}$ to obtain the minus part, and the argument is essentially the same.

§6. A Lemma of Kummer

Theorem 6.1. *Let p be an odd prime. Let u be a unit in $\mathbf{Q}(\boldsymbol{\mu}_p)$. Suppose there exists an integer $a \in \mathbf{Z}$ such that $u \equiv a \pmod{p}$. If p does not divide the class number h_p, then u is a p-th power in $\mathbf{Q}(\boldsymbol{\mu}_p)$.*

Proof. Let $K = \mathbf{Q}(\boldsymbol{\mu}_p)$. By class field theory, it suffices to show that the extension $K(u^{1/p})$ is unramified, because the hypothesis then implies that $K(u^{1/p}) = K$, so u is a p-th power in K. Raising u to the $(p-1)$-th power allows us to assume without loss of generality that $u \equiv 1 \pmod{p}$.

We contend that

$$u \equiv 1 \pmod{\pi p},$$

where $\pi = \zeta - 1$ and ζ is a primitive p-th root of unity. Otherwise, we have

$$u \equiv 1 + px \bmod \pi p$$

with x equal to a p-unit in $\mathbf{Z}_p^*$. But u is a global unit, so

$$N_{K/\mathbf{Q}}(u) = \pm 1 \equiv (1 + px)^{p-1} \equiv 1 - px \pmod{\pi p}.$$

In both cases where the norm is 1 or -1 we get a contradiction.

Let $\alpha = u^{1/p}$ be any p-th root of u. Then

$$\frac{1 - \alpha}{\pi} \quad \text{is a root of} \quad \frac{(\pi X - 1)^p + u}{\pi^p},$$

and this polynomial has p-integral coefficients. Its other roots are

$$\frac{1 - \zeta^i \alpha}{\pi},$$

and the different is a product of terms

$$\frac{(\zeta^i - \zeta^j)\alpha}{\pi},$$

each of which is a p-unit. Hence $K(\alpha) = K(u^{1/p})$ is unramified at p, and is trivially unramified at all other primes. This concludes the proof.

14 p-adic Preliminaries

The first section introduces the p-adic gamma function of Morita. After that, we deal with a topic which can be viewed temporarily as independent, the Artin–Hasse power series, and the Dwork power series closely related to it. The latter allows us to obtain an analytic representation of p-th roots of unity, which reappear later in the context of gauss sums, occurring as eigenvalues of p-adic completely continuous operators. Cf. Dwork's papers in the bibliography.

§1. The p-adic Gamma Function

In this section, we give Morita's definition of the p-adic gamma function. To start, we let f be the function defined for positive integers n by the formula

$$f(n) = (-1)^n \prod_{\substack{j=1 \\ (p,j)=1}}^{n} j.$$

We wish to show that there exists a continuous function on $\mathbf{Z}_p$ which restricts to f on the positive integers $\mathbf{Z}^+$. For this purpose, it suffices to prove the following lemma.

Lemma 1.1. *For any positive integers N, n, k we have*

$$f(n + p^N k) \equiv f(n) \pmod{p^N}.$$

Proof. Let $G = \mathbf{Z}(p^N)^*$. Pairing an element and its inverse in G we find:

$$\prod_{j \in G} j \equiv \begin{cases} -1 & \bmod p^N \text{ if } p \text{ is odd} \\ 1 & \bmod p^N \text{ if } p = 2. \end{cases}$$

If p is odd, then

$$\frac{f(n+p^N k)}{f(n)} = (-1)^{p^N k} \prod_{\substack{n+1 \le j \le n+p^N k \\ (j,p)=1}} j$$

$$\equiv (-1)^k \Big(\prod_{j \in G} j\Big)^k \equiv 1 \bmod p^N,$$

as desired. The proof is similar for $p = 2$ and is left to the reader.

By the lemma, we can extend f by continuity to all of $\mathbf{Z}_p$, and since $f(n)$ is a p-adic unit for n equal to a positive integer, it follows that

$$f: \mathbf{Z}_p \to \mathbf{Z}_p^*$$

is a map of $\mathbf{Z}_p$ into $\mathbf{Z}_p^*$.

We define the **p-adic gamma function**

$$\Gamma_p(x) = -f(x-1).$$

Thus on integers $n \ge 2$ we have

$$\boxed{\Gamma_p(n) = (-1)^n \prod_{\substack{j=1 \\ (p,j)=1}}^{n-1} j.}$$

We shall write Γ instead of Γ_p when p is fixed, and in particular, for the rest of this section.

Let us now calculate a few values, especially of negative integers. If $u(x)$, $v(x)$ are continuous functions of a p-adic variable x, we define:

$$\begin{cases} u(x) \\ v(x) \end{cases} \text{ is the function such that } \quad x \mapsto \begin{cases} u(x) & \text{if } x \in \mathbf{Z}_p^* \\ v(x) & \text{if } x \in p\mathbf{Z}_p. \end{cases}$$

With this notation, we have

$$\Gamma(n+1) = \begin{cases} -n\Gamma(n) \\ -\Gamma(n), \end{cases}$$

if n is an integer ≥ 2, and consequently by continuity,

$$\Gamma(x+1) = \begin{cases} -x\Gamma(x) \\ -\Gamma(x) \end{cases} \quad \text{for all } x \in \mathbf{Z}_p.$$

We call this the **functional equation**.

From $\Gamma(2) = \Gamma(1 + 1) = -\Gamma(1)$, we get

$$\Gamma(1) = -1.$$

From $\Gamma(1) = \Gamma(1 + 0) = -\Gamma(0)$, we get

$$\Gamma(0) = 1.$$

Theorem 1.2. *For any integer $n \geq 1$, we have*

$$\Gamma(n)\Gamma(1 - n) = (-1)^{n+[(n-1)/p]},$$

where the bracket, as usual, is the greatest integer function.

Proof. The theorem is true for $n = 1$ by the above. We can then proceed inductively, using $\Gamma(1 - n) = \{^{n}_{-1}\,\Gamma(-n)$, to get:

$$\Gamma(0) = \Bigg(\prod_{\substack{j=1 \\ (j,\,p)=1}}^{n-1} j\Bigg)(-1)^{\delta(n-1)}\Gamma(1 - n),$$

where $\delta(n - 1) =$ number of elements among $1, \ldots, n - 1$ which are divisible by p. This is immediate from the functional equation. Since $\delta(n - 1) = [(n - 1)/p]$, the right-hand side is equal to

$$(-1)^n\Gamma(n)(-1)^{[(n-1)/p]}\Gamma(1 - n).$$

The formula of the theorem then follows at once.

Let $x \in \mathbf{Z}_p$. We denote *for p odd*:

$$R(x) = \text{representative of } x \bmod p \text{ in the set } \{1, \ldots, p\}.$$

If p must be in the notation, we would write $R_p(x)$.

Theorem 1.3. *If $p \neq 2$ then*

$$\Gamma(x)\Gamma(1 - x) = (-1)^{R(x)}.$$

If $p = 2$, then

$$\Gamma(x)\Gamma(1 - x) = \varepsilon(x),$$

where

$$\varepsilon(x) = \begin{cases} -1 & \text{if } x \equiv 0, 1 \bmod 4 \\ 1 & \text{if } x \equiv 2, 3 \bmod 4. \end{cases}$$

Proof. By continuity, it suffices to prove the theorem when x is an integer $n \geq 1$. Write

$$n = a_0 + a_1 p + a_2 p^2 + \cdots + a_r p^r,$$

with $a_0 \in \{1, \ldots, p\}$ and $a_i \in \{0, \ldots, p-1\}$ for $i \geq 1$. Then

$$\left[\frac{n-1}{p}\right] = a_1 + \cdots + a_r p^{r-1}.$$

$$n + \left[\frac{n-1}{p}\right] = a_0 + a_1(1+p) + a_2(p+p^2) + \cdots + a_r(p^{r-1} + p^r).$$

Hence for p odd,

$$n + \left[\frac{n-1}{p}\right] \equiv a_0 = R(n) \bmod 2.$$

This proves the theorem in the present case.

If $p = 2$, then the parity of $n + [(n-1)/2]$ does not change for the elements in the residue class of n mod 4. We then determine explicitly the values of this number for $n = 1, 2, 3, 4$ to get the desired answer.

If $p = 2$, then it is convenient to define

$$R_2(x) = R(x) = \text{representative of } x \bmod 2 \text{ in } \{0, 1\}.$$

In both cases, p odd or even, we define

$$R'(x) = \frac{x - R(x)}{p}.$$

If $p \neq 2$ note that for a positive integer n,

$$R'(n) = \left[\frac{n-1}{p}\right].$$

This follows from the p-adic expansion in the proof of Theorem 1.3.

Theorem 1.4 (Distribution relation). *Let N be an integer ≥ 2 and prime to p. Then*

$$\prod_{i=0}^{N-1} \Gamma\left(\frac{x+i}{N}\right) = \Gamma(x) \prod_{i=1}^{N-1} \Gamma\left(\frac{i}{N}\right) g_N(x)^{-1},$$

where $g_N(x)$ is given by the formulas:

$$g_N(x) = N^{R(x)-1}N^{(p-1)R'(x)} \quad \text{if } p \neq 2$$
$$g_N(x) = N^{R'(x)} \quad \text{if } p = 2.$$

Remark. Both $R(x)$ and $R'(x)$ are continuous functions of $x \in \mathbf{Z}_p$. Since $R(x)$ is a positive integer, the exponentiation $N^{R(x)}$ is well defined. Since $N^{p-1} \equiv 1 \pmod p$, its exponentiation with $R'(x)$ is also well defined. When $x = n$ is a positive integer, then we can simplify the formula for $g_N(n)$, using

$$R(n) + pR'(n) = n.$$

For instance, for p odd, $g_N(n) = N^{n-1-R'(n)}$.

Proof. Define $g_N(x)$ by using the relation to be proved, so

$$g_N(x) = \prod_{i=1}^{N-1} \Gamma\left(\frac{i}{N}\right) \prod_{i=0}^{N-1} \Gamma\left(\frac{x+i}{N}\right)^{-1} \Gamma(x).$$

By continuity, it suffices to prove the theorem when x is a positive integer. We have

$$g_N(0) = 1$$

$$\frac{g_N(x+1)}{g_N(x)} = \frac{\Gamma(x+1)}{\Gamma(x)} \frac{\Gamma\left(\frac{x}{N}\right)}{\Gamma\left(\frac{x}{N}+1\right)}.$$

From the functional equation, we get

$$g_N(x+1) = \frac{\begin{cases} -x \\ -1 \end{cases}}{\begin{cases} -x/N \\ -1 \end{cases}} g_N(x) = \begin{cases} N \\ 1 \end{cases} g_N(x).$$

Hence for every positive integer n we find

$$g_N(n) = N^{n-1-\delta(n-1)} = N^{n-1-[(n-1)/p]}.$$

If p is odd, then $[(n-1)/p] = R'(n)$ and the assertion follows. The same argument also works for $p = 2$ and is left to the reader.

The next result shows that the fudge product occurring in the distribution relation is a fourth root of unity.

Proposition 1.5.

$$\prod_{i=1}^{N-1} \Gamma\left(\frac{i}{N}\right) = \begin{cases} \pm 1, & \textit{if } N \textit{ is odd} \\ \pm 1, \pm\sqrt{-1} & \textit{if } N \textit{ is even.} \end{cases}$$

Proof. Suppose N is odd. We write

$$\prod_{i=1}^{N-1} \Gamma\left(\frac{i}{N}\right) = \prod_{i=1}^{(N-1)/2} \Gamma\left(\frac{i}{N}\right)\Gamma\left(1 - \frac{i}{N}\right),$$

and each factor on the right is ± 1 by Theorem 1.3. If N is even, then

$$\prod_{i=1}^{N-1} \Gamma\left(\frac{i}{N}\right) = \Gamma\left(\frac{1}{2}\right) \prod_{i \leq (N-1)/2} \Gamma\left(\frac{i}{N}\right)\Gamma\left(1 - \frac{i}{N}\right).$$

Furthermore,

$$\Gamma(\tfrac{1}{2})\Gamma(\tfrac{1}{2}) = \pm 1,$$

and the proposition follows.

§2. The Artin–Hasse Power Series

Let p be a prime number. Define the **Artin–Hasse power series** by

$$\mathrm{AH}(X) = \exp\left(\sum_{n=0}^{\infty} \frac{X^{p^n}}{p^n}\right).$$

As usual, exp is the standard power series for the exponential function. Then $\mathrm{AH}(X)$ has rational coefficients.

Theorem 2.1. *We have* $\mathrm{AH}(X) \in \mathbf{Z}_p[[X]]$.

Lemma (Dieudonné–Dwork). *Let* $f(X) \in 1 + X\mathbf{Q}_p[[X]]$. *Then*

$$f(X) \in 1 + X\mathbf{Z}_p[[X]] \quad \Leftrightarrow \quad \frac{f(X)^p}{f(X^p)} \in 1 + pX\mathbf{Z}_p[[X]].$$

Proof. The left-hand side obviously implies the right-hand side in the equivalence to be proved. So assume the right-hand side. If R is any ring, and

$$f(X) \in 1 + XR[[X]],$$

then a simple recursion shows that $f(X)$ has an infinite product expression

$$f(X) = \prod_{n=1}^{\infty} (1 - a_n X^n) \quad \text{with } a_n \in R.$$

Furthermore, the elements a_n are uniquely determined.

Assume that $f(X)^p/f(X^p) \in 1 + pX\mathbf{Z}_p[[X]]$. Suppose that some coefficient a_n is not p-integral. Without loss of generality, we may assume that

$$f(X) = \prod_{n=r}^{\infty} (1 - a_n X^n) = 1 - a_r X^r + \text{higher terms}$$

and $a_r \notin \mathbf{Z}_p$. Then

$$\frac{f(X)^p}{f(X^p)} = \frac{1 - pa_r X^r + \cdots}{1 - a_r X^{rp} + \cdots} = 1 - pa_r X^r + \text{higher terms}.$$

By assumption, we must have $pa_r \in p\mathbf{Z}_p$ so $a_r \in \mathbf{Z}_p$, which proves the lemma.

We apply the test of the lemma to the Artin–Hasse series. We thus find

$$\frac{\mathrm{AH}(X)^p}{\mathrm{AH}(X^p)} = \frac{\exp(p \sum X^{p^n}/p^n)}{\exp(\sum X^{p^{n+1}}/p^n)} = \exp(pX) = \sum \frac{p^n}{n!} X^n.$$

To apply the lemma it suffices that $\operatorname{ord} p^n/n! \geq 1$ for all n, which is the case. This concludes the proof of the theorem.

Let us write

$$\exp\left(x + \frac{x^p}{p}\right) = \sum_{k=0}^{\infty} c_k x^k.$$

We wish to give estimates for the coefficients c_k. Let

$$\operatorname{ord} = \operatorname{ord}_p.$$

We shall prove:

(1) $$\operatorname{ord} c_k \geq -\frac{k}{p^2}\left(2 + \frac{1}{p-1}\right).$$

Proof. We can write

$$\exp\left(x + \frac{x^p}{p}\right) = \mathrm{AH}(x) \prod_{n \geq 2} \exp\left(-\frac{x^{p^n}}{p^n}\right).$$

Write

$$\exp\left(-\frac{x^{p^n}}{p^n}\right) = \sum_{k=0}^{\infty} d_k^{(n)} x^k = \sum_{m=0}^{\infty} \frac{x^{p^n m}}{p^{nm} m!}(-1)^m.$$

We prove:

$$\text{(2)} \qquad \operatorname{ord} d_k^{(n)} \geq -\frac{k}{p^n}\left(n + \frac{1}{p-1}\right).$$

Indeed, for any positive integer m, we recall that

$$\operatorname{ord} m! = \frac{m - s(m)}{p - 1},$$

where $s(m) = s_p(m)$ is the sum of the coefficients in the standard p-adic expansion of m. Hence

$$\operatorname{ord} d_{mp^n}^{(n)} = -nm - \frac{m - s(m)}{p - 1} \geq -nm - \frac{m}{p - 1}.$$

Factoring out $k = mp^n$ immediately implies (2).

Since

$$\min_{n \geq 2} \frac{-1}{p^n}\left(n + \frac{1}{p-1}\right) = \frac{-1}{p^2}\left(2 + \frac{1}{p-1}\right),$$

it follows that

$$\text{(3)} \qquad \operatorname{ord} d_k^{(n)} \geq \frac{-k}{p^2}\left(2 + \frac{1}{p-1}\right).$$

Hence the coefficients in the power series expansion of the product

$$\mathrm{AH}(x) \prod_{n=2}^{\infty} \exp\left(-\frac{x^{p^n}}{p^n}\right) = \exp\left(x + \frac{x^p}{p}\right)$$

satisfy an estimate like (3), and therefore the coefficients c_k satisfy

$$\operatorname{ord} c_k \geq -\frac{k}{p^2}\left(2 + \frac{1}{p-1}\right).$$

This proves the desired estimate (1).

For each element π such that $\pi^{p-1} = -p$, we define **Dwork's power series**

$$E_\pi(X) = \exp(\pi X - \pi X^p).$$

This can also be written

$$E_\pi(X) = \exp\left(\pi X + \frac{(\pi X)^p}{p}\right) = \sum e_n X^n.$$

The coefficients e_n lie in $\mathbf{Q}_p(\pi)$. In particular

$$\operatorname{ord} e_n \in \frac{1}{p-1}\mathbf{Z}.$$

As usual, $\operatorname{ord} = \operatorname{ord}_p$. If we want ord_π, then we shall specify π in the notation. Of course,

$$\operatorname{ord}_\pi = (p-1)\operatorname{ord}_p.$$

Lemma 2.2.

(i) *We have* $\operatorname{ord} e_n \geq n(p-1)/p^2$, *and* e_n *is* p*-integral.*
(ii) *If* $n \geq 2$ *then* $\operatorname{ord}_\pi e_n \geq 2$.

Proof. By (1), letting $X \mapsto \pi X$, we find at once that

$$\operatorname{ord} e_n \geq n\left(\frac{p-1}{p^2}\right).$$

To prove (ii), i.e. to prove that

$$\operatorname{ord} e_n \geq \frac{2}{p-1} \quad \text{for } n \geq 2,$$

it suffices to show that

$$\operatorname{ord} e_n > \frac{1}{p-1}$$

because ord e_n is a fraction whose denominator is $p - 1$. The inequality for ord e_n follows if

$$n\left(\frac{p-1}{p^2}\right) > \frac{1}{p-1}, \quad \text{or equivalently,} \quad n > \left(\frac{p}{p-1}\right)^2.$$

For $p \geq 5$ or $n \geq 5$ there is no problem. For $p = 3$ and $n \geq 3$, there is also no problem. For $p = 3$ and $n = 2$, we get directly

$$\exp(\pi X - \pi X^p) = 1 + \pi X + \tfrac{1}{2}\pi^2 X^2 + \text{higher terms},$$

so we get the right lower bound for ord e_2. For $p = 2$, we just compute explicitly the coefficients of X^2, X^3, X^4 and find that they are divisible by 4, thus settling the final cases, and proving the lemma.

Remark. From the first part of the lemma, we know that the coefficients of $E_\pi(X)$ tend to 0. In particular, we can evaluate $E_\pi(X)$ by substituting any p-adic number for X in the power series if this p-adic number has absolute value ≤ 1. However this value cannot usually be found by substituting the number directly in the expression $\exp(\pi X - \pi X^p)$. We shall see an example of this in the next section.

§3. Analytic Representation of Roots of Unity

Let p be a prime number. Throughout this section, we let $\pi \in \mathbf{C}_p$ be an element such that

$$\pi^{p-1} = -p.$$

Lemma 3.1. *For each element $\pi \in \mathbf{C}_p$ such that $\pi^{p-1} = -p$, there exists a unique p-th root of unity ζ_π such that*

$$\zeta_\pi \equiv 1 + \pi \pmod{\pi^2}.$$

The correspondence $\pi \mapsto \zeta_\pi$ establishes a bijection between p-th roots of unity $\neq 1$, and elements π as above.

Proof. If ζ is a non-trivial p-th root of unity, then $(\zeta - 1)^{p-1} \sim p$ (where $x \sim y$ means that x/y is a p-unit). If ζ_1, ζ_2 are two primitive p-th roots of unity which are both $\equiv 1 + \pi \pmod{\pi^2}$, then $\zeta_1 - \zeta_2 \equiv 0 \pmod{\pi^2}$, whence $\zeta_1 = \zeta_2$. Since there are exactly $p - 1$ non-trivial p-th roots of unity, and $p - 1$ possible elements π in $\mathbf{Q}_p(\pi) = \mathbf{Q}_p(\zeta)$, and since any element $\xi \equiv 1 \pmod{\pi}$ but $\xi \not\equiv 1 \pmod{\pi^2}$ can be written in the form

$$\xi \equiv 1 + \gamma\pi \pmod{\pi^2}$$

for some $(p - 1)$th root of unity γ, both existence and the bijection follow.

Remark. Each root of unity ζ gives rise to a character

$$\psi : \mathbf{Z}_p \to \boldsymbol{\mu}_p$$

such that $\psi(1) = \zeta$. The unique character whose value at 1 is ζ_π will be denoted by ψ_π, so by definition we have the formula

$$\psi_\pi(1) = \zeta_\pi,$$

and $\psi_\pi(1)$ is the unique root of unity $\equiv 1 + \pi \bmod \pi^2$.

Theorem 3.2 (Dwork). *We have $E_\pi(1) = \zeta_\pi$, so $E_\pi(1)$ is the unique p-th root of unity $\equiv 1 + \pi \bmod \pi^2$. For any $c \in \mathbf{Z}_p$ such that $c^p = c$, we have*

$$E_\pi(c) = E_\pi(1)^c.$$

Proof. First we observe that $E_\pi(1)$ is defined by substituting 1 for X in the power series for $E_\pi(X)$, which converges in light of the lower bound for the orders of the coefficients. Then note that

$$E_\pi(X)^p = \exp(p\pi X - p\pi X^p) = \exp(p\pi X)\exp(-p\pi X^p)$$

because the factor p inside the exponent makes the two series on the right-hand side converge. Substituting 1 for X can now be done to see at once that $E_\pi(1)^p = 1$. To conclude the proof of the first assertion, it suffices to show that

$$E_\pi(1) \equiv 1 + \pi (\operatorname{mod} \pi^2).$$

This is clear from Lemma 2.2(ii). Similarly,

$$E_\pi(c)^p = \exp(p\pi(c - c^p)) = \exp(0) = 1,$$

and

$$E_\pi(c) \equiv 1 + c\pi (\operatorname{mod} \pi^2),$$

so $E_\pi(c) = E_\pi(1)^c$, thus proving the theorem.

Similarly, let $\zeta \in \boldsymbol{\mu}_{q-1}$ where $q = p^r$ is a power of p. Let $\mathbf{T}$ denote the absolute trace to $\mathbf{F}_p$, and let

$$\psi_{\pi,q} = \psi_\pi \circ \mathbf{T}.$$

Then

$$\psi_{\pi,q} : \mathbf{F}_q \to \boldsymbol{\mu}_p$$

is a character on the additive group $\mathbf{F}_q$, and we let

$$E_{\pi,q}(x) = \exp(\pi x - \pi x^q) = E_\pi(x)E_\pi(x^p)\cdots E_\pi(x^{p^{r-1}}).$$

Theorem 3.3 (Dwork). *We have*

$$E_{\pi,q}(\zeta) = \psi_{\pi,q}(\zeta \bmod p),$$

and this is the unique p-th *root of unity* $\equiv 1 + T(\zeta)\pi \bmod \pi^2$.

Proof. From Lemma 2.2(ii), we know that $E_\pi(\zeta) \equiv 1 + \zeta\pi \bmod \pi^2$. Hence

$$\begin{aligned} E_{\pi,q}(\zeta) &= E_\pi(\zeta)E_\pi(\zeta^2)\cdots E_\pi(\zeta^{p^{r-1}}) \\ &\equiv 1 + T(\zeta)\pi \bmod \pi^2. \end{aligned}$$

The same argument as in Theorem 3.2 shows that $E_{\pi,q}(\zeta)$ is the unique p-th root of unity satisfying the above congruence, thus proving the theorem.

Appendix: Barsky's Existence Proof for the p-adic Gamma Function

We include this appendix to illustrate techniques which might be useful in similar contexts, say for p-adic differential equations, where an ad hoc argument as in §1 cannot be given. No use will be made of this appendix elsewhere in the book.

We wish to show the existence of a continuous function of $\mathbf{Z}_p$ into itself which takes on values related to the factorials at the positive integers. This is an interpolation problem, and thus we begin with a criterion for the existence of a continuous function taking given values. As in Chapter 4, we consider the space of power series

$$g(x) = \sum b_n \frac{x^n}{n!}, \qquad b_n \in \mathbf{C}_p,$$

with $\lim b_n = 0$ (the limit is of course the p-adic limit). This space is called the Leopoldt space. It is in fact a Banach algebra, under the sup norm of the coefficients, $\|g\|_{\mathscr{L}} = \max|b_n|$. If g, h are in the space, so is the product gh and

$$\|gh\|_{\mathscr{L}} \leq \|g\|_{\mathscr{L}}\|h\|_{\mathscr{L}}.$$

These properties are trivially verified. Note that a power series

$$\sum b'_n x^n$$

with coefficients b'_n which are p-integral is in the Leopoldt space, as one sees by writing $b_n = b'_n n!$.

Theorem. *Let $\{a_n\}$ be a sequence in $\mathbf{C}_p$. There exists a continuous function*

$$f: \mathbf{Z}_p \to \mathbf{C}_p$$

such that $f(n) = a_n$ for all integers $n \geq 0$ if and only if the following condition is satisfied. Let

$$e^{-x} \sum_{n=0}^{\infty} a_n \frac{x^n}{n!} = \sum_{n=0}^{\infty} b_n \frac{x^n}{n!}.$$

Then $\lim b_n = 0$ *(p-adically, of course).*

Proof. We use Mahler's theorem (Chapter 4, Theorem 1.3). Any continuous function f has an expansion

$$f(x) = \sum_{n=0}^{\infty} b_n \binom{x}{n} \quad \text{with } b_n \to 0,$$

and

$$b_n = (\Delta^n f)(0).$$

Conversely, if $f: \mathbf{Z}^+ \to \mathbf{C}_p$ is a function such that $\Delta^n f(0) \to 0$ as $n \to \infty$, then f can be extended to a continuous function on $\mathbf{Z}_p$. (We denote by $\mathbf{Z}^+$ the set of integers ≥ 0.) Note that

$$\Delta^n f(0) = \sum_{i=0}^{n} (-1)^{n-i} \binom{n}{i} f(i).$$

The theorem is then obvious in view of the identity

$$e^{-x} \sum f(n) \frac{x^n}{n!} = \sum \Delta^n f(0) \frac{x^n}{n!}.$$

We apply the theorem to the function f defined by

$$f(0) = 1, \qquad f(n) = (-1)^n \prod_{\substack{1 \leq j \leq n \\ (j,p)=1}} j.$$

Except for the power of -1, $f(n)$ is equal to $n!$ from which all the factors divisible by p have been deleted. Thus

$$f: \mathbf{Z}^+ \to \mathbf{Z}_p^*$$

takes its values in p-adic units.

This function has a continuous extension to $\mathbf{Z}_p$.

Proof. We have

$$\begin{aligned}
\sum f(n) \frac{x^n}{n!} &= \sum_{i=0}^{p-1} \sum_{n=0}^{\infty} f(pn+i) \frac{x^{pn+i}}{(pn+i)!} \\
&= \sum_{i=0}^{p-1} \sum_{n=0}^{\infty} (-1)^{pn+i} \frac{(pn+i)!}{p^n n!} \frac{x^{pn+i}}{(pn+i)!} \\
&= \exp\left(\frac{(-x)^p}{p}\right) \sum_{i=0}^{p-1} (-x)^i.
\end{aligned}$$

Multiplying by e^{-x}, we obtain

$$\begin{aligned}
e^{-x} \sum f(n) \frac{x^n}{n!} &= \exp\left(-x + \frac{(-x)^p}{p}\right) \sum_{i=0}^{p-1} (-x)^i \\
&= \sum b_n \frac{x^n}{n!},
\end{aligned}$$

with some coefficients b_n, which we must show tend to 0. The set of power series

$$\sum c_n' \frac{x^n}{n!} \quad \text{with } c_n' \to 0$$

is closed under multiplication. Hence after replacing x by $-x$, it suffices to prove that the coefficients c_k' of the series

$$\exp\left(x + \frac{x^p}{p}\right) = \sum c_k' \frac{x^k}{k!}$$

tend to 0. But by (1) of §2, we know that

$$\begin{aligned}
\operatorname{ord} c_k' &\geq \operatorname{ord} k! - \frac{k}{p^2}\left(2 + \frac{1}{p-1}\right) \\
&= \frac{k - s(k)}{p-1} - \frac{k}{p^2}\left(2 + \frac{1}{p-1}\right).
\end{aligned}$$

Since

$$s(k) \leq (p-1)(1+\log_p k),$$

(where $\log_p$ = high-school-log-to-the-base-p), we obtain

$$\operatorname{ord} c'_k \geq k\left(\frac{1}{p-1} - \frac{1}{p^2}\left(2 + \frac{1}{p-1}\right)\right) - 1 - \log_p k,$$

from which it is clear that $c'_k \to 0$. This concludes the proof.

15 The Gamma Function and Gauss Sums

The history of Gauss sums as eigenvalues of Frobenius on Fermat or Artin–Schreier curves goes back to Davenport–Hasse [Da–H] and Hasse [Ha 3]. However, Ron Evans has pointed out to me that Howard Mitchell [Mi] in 1916 considered Jacobi sums in connection with the number of points of the Fermat curve in arbitrary finite fields, and proved the first Davenport–Hasse relation between Jacobi sums in a finite field and in a finite extension.

Dwork introduced for the first time spaces of power series with p-adic coefficients tending to zero like a geometric series (he denotes such spaces by $L(b)$). He represents Frobenius on factor spaces $L(b)/DL(b)$, where D is an appropriate differential operator, and gets the Gauss sums as eigenvalues in this context, using a simple p-adic trace formula.

Washnitzer–Monsky saw the advantage of taking the union

$$\bigcup_{b>0} L(b),$$

to obtain a more functorial theory. They also introduced in addition certain affine rings over the p-adic numbers, lifting affine rings of hypersurfaces in characteristic p.

Several years ago, Honda looked at Gauss sums again in connection with the Jacobian of the Fermat curve [Ho], and conjectured an expression as limit of certain factorials. This was proved by Katz by taking the action of Frobenius on rather fancy cohomology (unpublished letter to Honda). More recently, Gross–Koblitz recognized this limit as being precisely the value of the p-adic gamma function [Gr–Ko]. In a course (1978), Katz showed that by using the Washnitzer–Monsky spaces and other techniques of Dwork (e.g. his special power series, and estimates of growths of certain coefficients), he could give an elementary proof that the eigenvalue of Frobenius acting on the space associated with the curve

$$y^p - y = x^N$$

was equal to the appropriate expression involving the p-adic gamma function. To establish the equality of this eigenvalue with the Gauss sum, he still referred to the rather extensive theory of Washnitzer–Monsky and their "trace formula," without going through the rather elaborate proofs.

On the other hand, Dwork in [Bo] showed that he could recover the Gross–Koblitz formula by working entirely with his spaces, and by using the completely elementary trace formula which he had proved already in his first paper on the zeta function of a hypersurface [Dw 1]. This paper, and a subsequent one, contained a concrete instance of what Serre [Se 3] recognized as p-adic Fredholm theory, giving a self-contained treatment which systematized Dwork's proofs in those respects which could be viewed as abstract nonsense.

Thus finally it was possible to give a completely elementary proof of the Gross–Koblitz formula. The exposition given here follows Katz in first obtaining the eigenvalue of Frobenius in terms of the gamma function. The second part, getting the Gauss sums, was worked out in collaboration with Dwork, mixing in what seemed the simplest way his spaces and those of Washnitzer–Monsky.

It also turned out that the use of the Artin–Schreier curve was unnecessary for the derivation of the formula, so the connection with that curve is postponed to the next chapter. I am much indebted to Katz for a number of illuminating comments. For instance, although for the special Artin-Schreier curve $y^p - y = x^N$ we can work ad hoc, Katz in his thesis [Ka 2] worked out completely in the general case the relations between Dwork cohomology and Washnitzer–Monsky cohomology. My purpose here was to derive the Gross–Koblitz formula essentially as simply as possible, and to avoid such general theories.

§1. The Basic Spaces

Let p be an odd prime, N a positive integer prime to p, and $q = p^r$ such that $q \equiv 1 \bmod N$. We let

$$0 \leq j \leq N - 1 \quad \text{and} \quad a = j\frac{q-1}{N}.$$

We let

$$\omega_q : \mathbf{F}_q^* \to \boldsymbol{\mu}_{q-1}$$

be the Teichmuller character, and we let

$$\psi_{\pi,q} = \psi_\pi \circ \mathrm{Tr}$$

be the additive character formed with the absolute trace Tr and the character ψ_π defined at the end of the last chapter. We let

$$\tau(\omega_q^a, \psi_{\pi,q}) = -S(\omega_q^a, \psi_{\pi,q})$$

be the negative of the Gauss sum, where

$$S(\chi, \psi) = \sum_{x \in \mathbf{F}_q} \chi(x)\psi(x).$$

We let R be a discrete valuation ring of characteristic 0, complete, such that the maximal ideal contains the prime p and also contains π. We let K be the quotient field of R.

We shall work with the ring of power series $R[[x]]_K$ whose coefficients have bounded denominators. This is also given by

$$R[[x]]_K = R[[x]][1/p].$$

We want a vector space

$$\mathscr{A}_{j,\pi} \subset R[[x]]_K$$

such that, if we put

$$H_{j,\pi} = \mathscr{A}_{j,\pi} \frac{dx}{x} \Big/ d\mathscr{A}_{j,\pi},$$

then $\dim H_{j,\pi} = 1$ and $H_{j,\pi}$ is a representation space for the Frobenius map Φ_q^* having $\tau(\omega_q^a, \psi_{\pi,q})$ as eigenvalue. We shall construct such a space.

We shall use an **embedding**

$$\mathrm{emb}_j\colon R[[x]] \to R[[x]]$$

given by

$$\varphi(x) \mapsto x^j \exp(-\pi x^N)\varphi(x^N).$$

We have a corresponding embedding into differential forms

$$\mathrm{emb}_j^*: R[[x]] \to R[[x]]\, dx$$

given by

$$\varphi(x) \mapsto x^j \exp(-\pi x^N)\varphi(x^N)\,\frac{dx}{x}.$$

(We use here the assumption that $j \geq 1$ so that no negative powers of x occur in the right-hand side.)

We have a commutative diagram

DW 1.

$$\begin{array}{ccc} R[[x]] & \xrightarrow{\text{emb}_j} & R[[x]] \\ \downarrow{\scriptstyle ND_j} & & \downarrow{\scriptstyle d} \\ R[[x]] & \xrightarrow[\text{emb}_j^*]{} & R[[x]]\,dx \end{array}$$

where D_j is the **differential operator** given by

$$D_j = x\frac{d}{dx} - \pi x + \frac{j}{N}.$$

This is immediate from the rule giving the derivative of a product.

Remark. For any α not equal to an integer ≤ 0, one verifies at once that the differential operator

$$D = x\frac{d}{dx} - \pi x + \alpha$$

is injective on power series, by looking at its effect on the term of lowest degree.

For each positive number δ we define

$L(\delta) = K$-vector space of power series $\sum a_n x^n$ in $R[[x]]_K$ such that

$$\operatorname{ord} a_n - \delta n \to \infty.$$

This condition could also be written $\operatorname{ord} a_n = \delta n + \infty(n)$.

Remark. This is a variation of Dwork's definition, and one may think of elements of $L(\delta)$ as functions holomorphic on the *closed* disc of "radius" p^δ. As usual, we take

$$\operatorname{ord} = \operatorname{ord}_p.$$

We define

$$L(0+) = \bigcup_{\delta} L(\delta).$$

We let

$$\begin{aligned}\mathscr{A}_{j,\pi} &= \text{Image of } L(0+) \text{ under emb}_j \\ &= \text{space of power series of the form } x^j \exp(-\pi x^N)\varphi(x^N) \text{ with } \\ &\quad \varphi(x) \in L(0+).\end{aligned}$$

This is the space used to define $H_{j,\pi}$ by the formula above. Thus we have an isomorphism

$$\boxed{L(0+)/D_j L(0+) \approx H_{j,\pi} = \mathscr{A}_{j,\pi} \frac{dx}{x} \Big/ d\mathscr{A}_{j,\pi}}$$

coming from commutative diagram **DW 1**, and induced by emb_j.

Lemma 1.1. *Let $\alpha \in \mathbf{Z}_p$ and suppose that α is not an integer ≤ 0. Let*

$$D = x\frac{d}{dx} - \pi x + \alpha.$$

(i) *We have a direct sum decomposition*

$$L(0+) = K \oplus DL(0+).$$

(ii) *If $\delta \geq 1/(p-1)$, then we have a direct sum decomposition*

$$L(\delta) = K \oplus DL(\delta).$$

Proof. We wish first to write an arbitrary series $\varphi(x)$ as some constant plus Dg for some g. We first solve this problem for powers of x. We have for integers $m \geq 0$:

$$Dx^m = (m + \alpha)x^m - \pi x^{m+1},$$

so that

$$x^{m+1} = \frac{1}{\pi}(m + \alpha)x^m - D\left(\frac{1}{\pi}x^m\right).$$

Recursively, we obtain

$$x^{m+1} = \frac{1}{\pi^{m+1}}(m+\alpha)(m-1+\alpha)\cdots(\alpha)$$
$$-D\left[\frac{1}{\pi}x^m + \frac{1}{\pi^2}(m+\alpha)x^{m-1} + \cdots + \frac{1}{\pi^{i+1}}(m+\alpha)\cdots(m-i+1+\alpha)x^{m-i} + \cdots\right].$$

Hence we have proved what we wanted for powers of x, of degree ≥ 1.
Next, we see what happens for a series in $L(0+)$:

$$(*) \quad \sum_{m=0}^{\infty} b_{m+1}x^{m+1} = \sum_{m=0}^{\infty} b_{m+1}\frac{1}{\pi^{m+1}}(m+\alpha)\cdots(\alpha)$$
$$-D\sum_{n=0}^{\infty} x^n\left[\sum_{m=n}^{\infty} b_{m+1}\frac{1}{\pi\pi^{m-n}}(m+\alpha)\cdots(n+1+\alpha)\right].$$

This gives a formal solution, and we want to see that the right-hand side lies in

$$K + DL(0+).$$

For this we need a lemma giving estimates for binomial coefficients.

Lemma 1.2. *Let $\alpha \in \mathbf{Z}_p$, and let n be a positive integer. Then*

$$\operatorname{ord} \alpha(\alpha-1)\cdots(\alpha-n+1) \geq -\frac{s(n)}{p-1} \geq -1-\log_p n.$$

Proof. Obvious, from

$$\frac{\alpha(\alpha-1)\cdots(\alpha-n+1)}{\pi^n} = \frac{n!}{\pi^n}\binom{\alpha}{n},$$

from the fact that the binomial coefficient is p-integral, and from

$$\operatorname{ord} n! = \frac{n-s(n)}{p-1}, \qquad \frac{s(n)}{p-1} \leq 1 + \log_p n.$$

Since the coefficients b_{m+1} tend to 0 like a geometric progression, it is clear that the first series giving the constant term in the formal solution is actually an element of K. As for the second, there is some positive δ such that

$$\text{ord}\, b_{m+1} \geq \delta(m+1) + \infty(m).$$

Hence by Lemma 1.2, for $m \geq n$ we get

$$(**) \quad \text{ord}\, b_{m+1} \frac{1}{\pi\pi^{m-n}} (m+\alpha)\cdots(n+1+\alpha)$$

$$\geq \delta(m+1) - \frac{1}{p-1} - \frac{s(m-n)}{p-1} + \infty(m).$$

This proves that the coefficients of x^n tend to 0 like a geometric series, and hence proves that

$$L(0+) = K + DL(0+).$$

There remains to prove that this sum is direct, in other words that $1 \neq Dg$ for $g \in L(0+)$. This is a special case of the following lemma.

Lemma 1.3. *The equation $Dg = 1$ has a unique solution*

$$g(x) = \sum b_n x^n \in K[[x]],$$

and the coefficients satisfy

$$\text{ord}\, b_n \leq 1 - \frac{1}{p-1} + \log_p (n+1).$$

Proof. Write

$$1 = \left(x\frac{d}{dx} + \alpha - \pi x\right)\left(\sum_{n=0}^{\infty} b_n x^n\right).$$

Then

$$1 + \sum \pi b_n x^{n+1} = \sum (nb_n + \alpha b_n)x^n,$$

so that

$$1 = \alpha b_0 \quad \text{and} \quad \pi b_{n-1} = (n+\alpha)b_n.$$

Hence

$$b_n = \frac{\pi^n}{(n+\alpha)\cdots(\alpha)} = \frac{\pi^n}{(n+1)!\binom{n+\alpha}{n+1}}.$$

The same estimate as before, using Lemma 1.2 shows that

$$\operatorname{ord} b_n \leq 1 + \frac{1}{p-1} + \log_p(n+1),$$

which proves the lemma.

From the lemma we see that the formal solution of $Dg = 1$ cannot lie in $L(0+)$. This concludes the proof of Lemma 1.1(i).

For the second part we disregard the factorials completely (they are p-integral), and use the more trivial estimate

$$\operatorname{ord} b_{m+1} \geq \delta(m+1) + \infty(m).$$

Then (**) is now estimated by

$$\begin{aligned}\operatorname{ord} b_{m+1} \frac{1}{\pi^{m-n+1}}(m+\alpha)\cdots(n+1+\alpha) &\geq \delta(m+1) - \frac{m-n}{p-1} + \infty(m)\\ &\geq \delta n + \infty(n)\end{aligned}$$

because $\delta \geq 1/(p-1)$. This proves the lemma, because the rest of the proof is identical with what we did before.

Theorem 1.4. *We have isomorphisms for* $\delta \geq 1/(p-1)$:

$$L(\delta)/D_j L(\delta) \approx L(0+)/D_j L(0+) \approx H_{j,\pi},$$

and these spaces are 1*-dimensional.*

Proof. Immediate from Lemma 1.1.

§2. The Frobenius Endomorphism

On the power series ring $K[[x]]$, we have the Frobenius endomorphism

$$\Phi_p : K[[x]] \to K[[x]]$$

such that $x \mapsto x^p$, and similarly $\Phi_q = \Phi_p^r$. Then for $h(x) \in K[[x]]$ we have

$$\Phi_q^*(h(x)\,dx) = h(x^q)\,dx^q = h(x^q)qx^q\,\frac{dx}{x}.$$

First we determine the mapping corresponding to the Frobenius under the embedding emb_j^*. Thus we **define**

$$B_{j,p} = x^{(pj-j')/N}E_\pi(x)\Phi_p,$$

where j' is the integer satisfying

$$1 \leq j' \leq N-1 \quad \text{and} \quad j' \equiv pj \bmod N.$$

Then we have commutative diagrams

FR 1.

$$\begin{array}{ccc} R[[x]] & \xrightarrow{\mathrm{emb}_j} & R[[x]] \\ \downarrow{\scriptstyle B_{j,p}} & & \downarrow{\scriptstyle \Phi_p} \\ R[[x]] & \xrightarrow[\mathrm{emb}_j]{} & R[[x]] \end{array}$$

and

$$\begin{array}{ccc} R[[x]] & \xrightarrow{\mathrm{emb}_j^*} & R[[x]]\,dx \\ \downarrow{\scriptstyle pB_{j,p}} & & \downarrow{\scriptstyle \Phi_p^*} \\ R[[x]] & \xrightarrow[\mathrm{emb}_j^*]{} & R[[x]]\,dx. \end{array}$$

The commutativity is done by direct verification, and is immediate, starting with a power series $\varphi(x)$ in the upper left-hand corner, and using the definitions of the various mappings going around one way, and then the other way.

A direct verification also shows that

$$\Phi_p : \mathscr{A}_{j,\pi} \to \mathscr{A}_{j',\pi}$$

maps $\mathscr{A}_{j,\pi}$ into $\mathscr{A}_{j',\pi}$, and furthermore

$$\Phi_p^* \circ d = d \circ \Phi_p.$$

Consequently we obtain a homomorphism

$$\Phi_p^* : H_{j,\pi} \to H_{j',\pi},$$

while the diagram **FR 1** yields the corresponding diagram

FR 2.

$$\begin{array}{ccc} L(0+)/D_j\, L(0+) & \xrightarrow{\text{emb}_j^*} & H_{j,\pi} \\ {\scriptstyle p\bar{B}_{j,p}}\downarrow & & \downarrow{\scriptstyle \Phi_p^*} \\ L(0+)/D_j\, L(0+) & \xrightarrow[\text{emb}_{j'}^*]{} & H_{j',\pi}. \end{array}$$

For our purposes, we also want to look at a composite Frobenius endomorphism, so we let

$$B_{j,q} = x^a E_{\pi,q}(x)\Phi_q.$$

Then we have the commutative diagram

FR 3.

$$\begin{array}{ccc} L(0+)/D_j L(0+) & \xrightarrow{\text{emb}_j^*} & H_{j,\pi} \\ {\scriptstyle q\bar{B}_{j,q}}\downarrow & & \downarrow{\scriptstyle \Phi_q^*} \\ L(0+)/D_j L(0+) & \xrightarrow[\text{emb}_j^*]{} & H_{j,\pi}. \end{array}$$

Thus $q\bar{B}_{j,q}$ is an endomorphism of $L(0+)/D_jL(0+)$, corresponding to the endomorphism Φ_q^* under the embedding emb_j^*.

Inverse of the Frobenius Endomorphism

We also want to tabulate formulas about the inverse of the Frobenius endomorphism, actually one-sided. We define it on power series by

$$\Psi_q : \sum a_n x^n \mapsto \sum_{q|n} a_n x^{n/q},$$

and on differential forms by

$$\Psi_q^*\left(x^n \frac{dx}{x}\right) = \begin{cases} \dfrac{1}{q} x^{n/q} \dfrac{dx}{x} & \text{if } q \mid n \\ 0 & \text{otherwise.} \end{cases}$$

Then

$$\Psi_q \circ \Phi_q = \text{id}.$$

Again we have the commutation rule (trivially verified)

DW 2. $$\Psi_q^* \circ d = d \circ \Psi_q.$$

Let

$$A_j = A_{j,q} = \Psi_q \circ (x^{-a} E_{\pi,q}(x)^{-1}).$$

This means that A_j is the composite of multiplication by $x^{-a}E_{\pi,q}(x)^{-1}$ followed by Ψ_q. Since $0 < a < q$, it follows that A_j maps x-integral power series into x-integral power series, and we have

$$A_{j,q} \circ B_{j,q} = \text{id}.$$

We shall need the commutation rule

FR 4. $$A_j \circ D_j = qD_j \circ A_j.$$

This can be proved either directly, or by the same pattern as for the corresponding fact for B_j, using **DW 2** and **FR 3**, together with the fact that emb_j is injective. This gives rise to the commutative diagrams:

FR 5.
$$\begin{array}{ccc} R[[x]] & \xrightarrow{\text{emb}_j^*} & R[[x]]\dfrac{dx}{x} \\ {\scriptstyle A_{j,q}}\downarrow & & \downarrow{\scriptstyle q\Psi_q^*} \\ R[[x]] & \xrightarrow[\text{emb}_j^*]{} & R[[x]]\dfrac{dx}{x} \end{array} \quad \text{and} \quad \begin{array}{ccc} R[[x]] & \xrightarrow{\text{emb}_j} & R[[x]] \\ {\scriptstyle A_{j,q}}\downarrow & & \downarrow{\scriptstyle \Psi_q} \\ R[[x]] & \xrightarrow[\text{emb}_j]{} & R[[x]]. \end{array}$$

We now consider the effect of the inverse of the Frobenius endomorphism on the Dwork spaces. We see directly from the definitions that

$$\Psi_q : L(\delta) \to L(q\delta)$$

maps $L(\delta)$ into $L(q\delta)$.

Furthermore, let $g \in L(\delta)$. Then multiplication by g,

$$f \mapsto gf$$

maps $L(\delta)$ into itself.

Let δ' satisfy $\delta' < \delta < q\delta'$. Let $g \in L(\delta')$. Then

$$\Psi_q \circ g : L(\delta) \to L(\delta)$$

maps $L(\delta)$ into itself. Indeed, this map is obtained as a composite map:

$$L(\delta) \xrightarrow{\text{inc.}} L(\delta') \xrightarrow{g} L(\delta') \xrightarrow{\Psi_q} L(q\delta') \xrightarrow{\text{inc.}} L(\delta).$$

We shall apply this to the case when $g(x) = E_\pi(x)$. Since $p \geq 3$ we can select

$$\frac{1}{p-1} \leq \delta < \frac{p-1}{p},$$

and then select $\delta' < (p-1)/p^{r+1}$ but close to $(p-1)/p^{r+1}$ such that

$$\delta' < \delta < q\delta'.$$

Then

$$E_{\pi,q}(x) \in L(\delta').$$

Lemma 2.1. *Under the above conditions on δ and δ',*

$$A_{j,q}: L(\delta) \to L(\delta)$$

maps $L(\delta)$ into itself, and induces a homomorphism

$$\bar{A}_{j,q}: L(\delta)/D_j L(\delta) \to L(\delta)/D_j L(\delta).$$

Proof. This is a special case of the preceding discussion, except that the negative power x^{-a} occurs inside the operator A_j. However, since $a < q - 1$, we have already seen that Ψ_q annihilates such negative powers, so $A_{j,q}$ maps $L(\delta)$ into itself. The commutation rule **FR 4** shows that $A_{j,q}$ induces a homomorphism on the factor space mod $D_j L(\delta)$, as desired.

By Lemma 1.1 we know that this factor space is one-dimensional. Consequently, the operator $\bar{A}_{j,q}$ has an eigenvalue on this space, which we denote by $\lambda'_{j,q} = \lambda'_j$. By definition, we have the relation

$$A_j(1) = \lambda'_j + D_j h_j$$

where h_j is a uniquely determined power series, which lies in $L(\delta)$. Our object is to determine λ'_j. We shall show that λ'_j is a Gauss sum.

We have a commutative diagram

FR 6.

$$\begin{array}{ccc} L(0+)/D_j L(0+) & \longrightarrow & L(\delta)/D_j L(\delta) \\ {\scriptstyle B_j}\downarrow & & \uparrow{\scriptstyle \bar{A}_j} \\ L(0+)/D_j L(0+) & \longrightarrow & L(\delta)/D_j L(\delta). \end{array}$$

The horizontal maps are the inverses of those obtained from the natural inclusion $L(\delta) \subset L(0+)$, see Theorem 1.4. We know that

$$A_j \circ B_j = \mathrm{id},$$

so we conclude that the eigenvalue of $\bar{B}_j$ is the inverse of the eigenvalue of $\bar{A}_j$.

§3. The Dwork Trace Formula and Gauss Sums

Consider first the complete space $R\langle\langle x\rangle\rangle_p$, consisting of the power series whose coefficients tend to 0. Each such power series converges on the (closed) unit disc. We may view the powers of x, namely

$$1, x, x^2, \ldots$$

as forming a "basis" for this space. If

$$u: R\langle\langle x\rangle\rangle_p \to R\langle\langle x\rangle\rangle_p$$

is an endomorphism, let us work formally, and represent u by an infinite matrix with respect to this basis. We define the **trace** to be the sum of the diagonal elements. For this to make sense, we must have appropriate conditions of convergence, and in §5 we justify the formal computations which we shall make here in light of the p-adic Banach Fredholm theory of completely continuous operators.

Theorem 3.1. *Let $g(x) \in R\langle\langle x\rangle\rangle_p$. Let $1 \leq a \leq q - 1$. Then*

$$(q - 1)\,\mathrm{tr}(\Psi_q \circ x^{-a}g(x)) = \sum_{\zeta} \zeta^{-a}g(\zeta),$$

where the sum is taken over $\zeta \in \boldsymbol{\mu}_{q-1}$.

Proof. Write

$$h(x) = x^{-a}g(x) = \sum a_n x^n.$$

Then $a_n \to 0$ (p-adically). Let

$$\Psi_q(x^i h(x)) = \sum a_{ij} x^j.$$

Then a_{ij} is the coefficient of x^{qj} in $x^i h(x)$, and so

$$a_{ij} = a_{qj-i}.$$

Hence $a_{ii} = a_{(q-1)i}$, and so

$$(q-1)\sum a_{ii} = (q-1)\sum a_{(q-1)i}.$$

But

$$\sum_{\zeta} h(\zeta) = \sum_{\zeta, n} a_n \zeta^n,$$

and in this sum, the n-th term is equal to 0 unless n is divisible by $q-1$. The remaining terms give precisely the desired value stated in the theorem.

Let

$$\omega_q : \mathbf{F}(q)^* \to \boldsymbol{\mu}_{q-1}$$

be the Teichmuller character, such that

$$\omega_q(c) \equiv c \bmod p.$$

The **Gauss sum** is defined by

$$S_q(\chi, \psi) = \sum_{c \in \mathbf{F}(q)^*} \chi(c)\psi(c).$$

We apply the preceding discussion with $g(x) = E_{\pi,q}(x)^{-1}$. Then Theorem 3.3 of Chapter 14 yields:

Theorem 3.2. *Let $a = j(q-1)/N$ and $\psi_{\pi,q} = \psi_\pi \circ \mathbf{T}$, where $\mathbf{T}$ is the absolute trace to the prime field. Then*

$$(q-1)\operatorname{tr} A_{j,q} = S_q(\omega_q^{-a}, \psi_{\pi,q}^{-1}).$$

Note that the additive character is determined in the usual canonical way (composing with the trace) from the character ψ_π on the prime field. Hence we shall usually omit the additive character from the notation.

Remark. In the next theorem, we assume that the formalism of determinants works in the present situation. This is justified in the last section of the chapter. Specifically, the operator $A_{j,q}$ is viewed as an endomorphism of $L(\delta)$. A Banach basis consists of appropriate scalar multiples of the powers x^n, and the argument used in Theorem 3.1 applies to this case to yield Theorem 3.2. Cf. §5, Example from Dwork Theory.

In the next theorem, we follow usual notation, and let

$$\tau_q = -S_q$$

be the negative of the Gauss sum.

Theorem 3.3. *Let* $1 \leq j \leq N - 1$, *and* $a = j(q-1)/N$, *where* N *divides* $q - 1$. *Let* δ *be a rational number such that*

$$\frac{1}{p-1} \leq \delta < \frac{p-1}{p}.$$

Let $W = L(\delta)/D_j L(\delta)$. *Let* λ'_j *be the eigenvalue of* $\bar{A}_j$ *on* W. *Then*

$$\lambda'_j = \tau_q(\omega_q^{-a}, \psi_{\pi,q}^{-1}).$$

Proof. By **FR 4** of §2, we have a commutative diagram with exact rows:

$$\begin{array}{ccccccccc} 0 & \longrightarrow & L(\delta) & \xrightarrow{D_j} & L(\delta) & \longrightarrow & W & \longrightarrow & 0 \\ & & \downarrow{\scriptstyle qA_j} & & \downarrow{\scriptstyle A_j} & & \downarrow{\scriptstyle \bar{A}_j} & & \\ 0 & \longrightarrow & L(\delta) & \xrightarrow[D_j]{} & L(\delta) & \longrightarrow & W & \longrightarrow & 0. \end{array}$$

By the additivity of the trace (cf. Proposition 5.6 below), we have

$$\operatorname{tr} A_j = \operatorname{tr} qA_j + \operatorname{tr} \bar{A}_j.$$

Since W is 1-dimensional, $\operatorname{tr} \bar{A}_j$ is the eigenvalue of $\bar{A}_j$. The theorem is now a direct consequence of Theorem 3.2.

§4. Eigenvalues of the Frobenius Endomorphism and the p-adic Gamma Function

By Lemma 1.1(i) we know that

$$\omega_j = x^j \exp(-\pi x^N) \frac{dx}{x}$$

represents a basis of $H_{j,\pi}$. Consequently there is some element $\lambda(j, \pi, N)$ such that

$$\Phi_p^*(\omega_j) = \lambda(j, \pi, N)\omega_{j'}, \quad \text{in } H_{j',\pi}. \tag{1}$$

We shall relate this element λ to the p-adic gamma function.

Note that Φ_q^* is an endomorphism of $H_{j,\pi}$, so its eigenvalue λ_q is given by

$$\lambda_q = \lambda(j, \pi, N)\lambda(j', \pi, N) \cdots \lambda(j^{(r-1)}, \pi, N). \tag{2}$$

Theorem 4.1. *We have*

$$\lambda(j, \pi, N) = \frac{-p\Gamma_p\left(1 - \frac{j'}{N}\right)}{\pi^{(pj-j')/N}}.$$

Proof. (Following Katz.) We have

$$\Phi_p^*\left(x^j \exp(-\pi x^N)\frac{dx}{x}\right) = \lambda x^{j'} \exp(-\pi x^N)\frac{dx}{x} + dg$$

with some $g \in R[[x]]_K$, that is g is a power series with bounded denominators. Expanding out, we find

$$p\sum \frac{(-\pi)^n}{n!} x^{npN+pj}\frac{dx}{x} = \lambda(j, \pi, N)\sum \frac{(-\pi)^m}{m!} x^{mN+j'}\frac{dx}{x} + dg.$$

Equating coefficients, using $npN + pj = mN + j'$, we obtain

$$p\frac{(-\pi)^n}{n!} = \lambda(j, \pi, N)\frac{(-\pi)^m}{m!} + O(npN + pj). \tag{3}$$

Note that $O(npN + pj) = O(n + j/N)$, and

$$m = np + \frac{pj - j'}{N}.$$

This yields

$$p = \lambda(j, \pi, N)(-\pi)^{m-n}\frac{n!}{m!} + O\left(n + \frac{j}{N}\right)\frac{n!}{\pi^n}.$$

On the other hand, we have

$$\Gamma(1 + m) = (-1)^{1+m}\frac{m!}{n!p^n}. \tag{4}$$

This is immediate from the definition of the p-adic gamma function, and the fact that a positive integer $i \leq m$ divisible by p can be written in the form

$$i = pi_0 \quad \text{with} \quad i_0 \leq n + \frac{pj - j'}{pN},$$

and therefore $i_0 \leq n$. Hence the denominator $n!p^n$ is exactly the product of such i divisible by p.

Substituting (4) in (3), using the fact that $\pi^{p-1} = -p$, and therefore $\pi^{n(p-1)} = (-p)^n$, we find

$$(5) \qquad -p = \frac{\lambda(j, N, \pi)}{\Gamma(1+m)} \pi^{(pj-j')/N} + O\left(n + \frac{j}{N}\right) \frac{n!}{\pi^n}.$$

The following lemma then concludes the proof.

Lemma. *There exists a sequence of positive integers n such that*

(i) $n \to -j/N$ *p-adically, and hence* $m \underset{p}{\to} -j'/N$;
(ii) $\operatorname{ord}(n + j/N) + \operatorname{ord}(n!/\pi^n) \to \infty$.

Proof. First we simplify slightly the exponent in (ii). We have:

$$\begin{aligned} \operatorname{ord}\left(n + \frac{j}{N}\right) + \operatorname{ord}\frac{n!}{\pi^n} &= \operatorname{ord}\left(n + \frac{j}{N}\right) + \frac{n - s(n)}{p-1} - \frac{n}{p-1} \\ &= \operatorname{ord}\left(n + \frac{j}{N}\right) - \frac{s(n)}{p-1}. \end{aligned}$$

Let $q = p^r$ be such that N divides $q - 1$. Write the expansion

$$-\frac{j}{N} = \frac{j(q-1)}{N(1-q)} = \frac{j(q-1)}{N}(1 + q + q^2 + \cdots).$$

There is also a finite expansion

$$\frac{j(q-1)}{N} = a_0 + a_1 p + \cdots + a_{r-1} p^{r-1}$$

with standard integral representatives $0 \leq a_i \leq p - 1$. Let

$$n = \frac{j(q-1)}{N}(1 + q + \cdots + q^{M-1}).$$

For M tending to infinity, we see that n approaches $-j/N$ to satisfy (i). Furthermore

$$\operatorname{ord}\left(n + \frac{j}{N}\right) \geq M \cdot \operatorname{ord} q = Mr.$$

On the other hand, since $j(q-1)/N < q - 1$, we have

$$s(n) \leq M((p-1)r - 1).$$

The second condition of the lemma is then obviously satisfied also. This concludes the proof of the lemma, and of Theorem 4.1.

Theorem 4.2. *Let λ_q be the eigenvalue of Frobenius Φ_q^* on $H_{j,\pi}$. Then*

$$\lambda_q = \tau_q(\omega_q^a, \psi_{\pi,q}).$$

Proof. This follows from Theorem 3.3, the remark at the end of §2, **FR 3**, and the fact that

$$\tau(\chi, \psi)\tau(\chi^{-1}, \psi^{-1}) = q.$$

Let a be a positive integer, with $1 \leq a < q - 1$ and $q = p^r$. Write

$$a = a_0 + a_1 p + \cdots a_{r-1} p^{r-1}$$

with integers a_i such that $0 \leq a_i \leq p - 1$. As usual, define

$$s(a) = \sum a_i.$$

Theorem 4.3 (Gross–Koblitz formula).

$$\tau_q(\omega_q^a, \psi_{\pi,q}) = (-1)^r q \pi^{-s(a)} \prod_{i=0}^{r-1} \Gamma_p\left(1 - \left\langle \frac{p^i a}{q-1} \right\rangle\right).$$

Proof. We merely put together Theorem 4.2, and the expression for the eigenvalue obtained in Theorem 4.1. The only quantity remaining to be worked out is the power of π appearing on the right-hand side, and this follows from the following lemma.

Lemma 4.4. *Let $a = j(q - 1)/N$. Then*

$$\sum_{i \bmod r} \frac{pj^{(i)} - j^{(i+1)}}{N} = s\left(j \frac{q-1}{N}\right) = s(a).$$

Proof. This is essentially the easy Lemma 1 of Chapter 1, §2. Indeed,

$$\frac{j^{(i)}}{N} = \frac{a^{(i)}}{q-1}$$

where $a^{(i)}$ is defined by

$$1 \leq a^{(i)} < q - 1 \quad \text{and} \quad \frac{a^{(i)}}{q-1} = \left\langle \frac{p^i a}{q-1} \right\rangle.$$

With this change of variables from j to the corresponding a, the present lemma is identical with Lemma 1, loc. cit.

Let $\mathfrak{P}$ be the prime in the algebraic closure of $\mathbf{Q}_p$. In the next theorem we show how Stickelberger's result on the Gauss sum mod $\mathfrak{P}$ follows from the Gross–Koblitz formula.

Theorem 4.5 (Stickelberger). *Let* $\gamma(a) = \prod a_i!$. *Then*

$$\pi^{-s(a)}\tau(\omega_q^{-a}, \psi_{\pi,q}) \equiv \frac{1}{\gamma(a)} \mod \mathfrak{P}.$$

Proof. From the formula $\tau\bar{\tau} = q$, and Theorem 4.3 we obtain

$$(*) \qquad \frac{\tau(\omega_q^{-a}, \psi_{\pi,q}^{-1})}{\pi^{s(a)}} = \frac{(-1)^r}{G(a)},$$

where

$$G(a) = \prod_i \Gamma_p\left(1 - \left\langle \frac{p^i a}{q-1} \right\rangle\right) = \prod_i \Gamma_p\left(1 - \left\langle \frac{p^{r-i} a}{q-1} \right\rangle\right).$$

Replacing π by $-\pi$ in the left hand side of (*) yields

$$\pi^{-s(a)}\Gamma(\omega_q^{-a}, \psi_{\pi,q})(-1)^{s(a)}.$$

The theorem then follows from the next lemma.

Lemma 4.6.

$$\Gamma_p\left(1 - \left\langle \frac{p^{r-i} a}{q-1} \right\rangle\right) \equiv (-1)^{1+a_i} a_i! \mod p.$$

Proof. Note that

$$\Gamma_p(1 + a_i) = (-1)^{1+a_i} a_i!$$

This is true by definition if $a_i \geq 2$ because $a_i \leq p - 1$. It is also true when $a_i = 1$ and $a_i = 0$ by the direct computations of Chapter 14, §1. On the other hand, by Lemma 1.1 of Chapter 14, we know that

$$\Gamma_p(1 + x) \equiv \Gamma_p(1 + y) \quad \text{if} \quad x \equiv y \mod p.$$

Thus it suffices to prove that

$$-\left\langle \frac{p^{r-i}a}{q-1} \right\rangle \equiv a_i \bmod p.$$

As usual, we write cyclically

$$\begin{aligned} a &= a_0 + a_1 p + \cdots + a_{r-1}p^{r-1} \\ pa &\equiv a_{r-1} + a_0 p + \cdots + a_{r-2}p^{r-1} \bmod q-1 \\ &\vdots \end{aligned}$$

The desired congruence follows at once, thus proving the lemma and the theorem.

§5. p-adic Banach Spaces

The purpose of this section is to do as rapidly as possible the linear algebra in p-adic Banach spaces justifying the use of determinants and traces in this chapter. Thus we cover only part of Serre's paper [Se 3], whose exposition leaves nothing to be desired.

Let R be a complete discrete valuation ring, with quotient field K, maximal ideal $\mathfrak{m}$, residue class field k, and prime element π.

By a **Banach space** E over K, we mean a normed space which is complete, and such that the norm $x \mapsto |x|$ satisfies

$$|cx| = |c||x| \quad \text{and} \quad |x + y| \leq \max\{|x|, |y|\}$$

for $c \in K$ and $x, y \in E$. We also assume that the value group $|x|$ (for $x \in E$, $x \neq 0$) is the same group of positive real numbers as the value group of K^*. So for each $x \in E$, there exists $c \in K$ such that $|x| = |c|$.

Let E, F be Banach spaces over K. Let $L(E, F)$ be the vector space of continuous linear maps

$$u: E \to F.$$

As usual, we can define a **norm** on $L(E, F)$ by

$$|u| = \sup_{x \neq 0} \frac{|ux|}{|x|}.$$

Also as usual, we have

$$|u| = \sup_{|x| \leq 1} |ux|.$$

By an **isomorphism**, we mean a norm-preserving continuous linear map having an inverse.

Example. Let I be a set of indices, and let $C(I, K)$ be the set of families $x = (x_i)_{i \in I}$, $x_i \in K$, such that x_i tends to 0 as $i \to \infty$. By this we mean that given ε there exists a finite set of indices S such that for $i \notin S$, we have $|x_i| < \varepsilon$. The reader may as well think of the positive integers where $i = 1, 2, 3, \ldots$. We define

$$|x| = \sup|x_i|.$$

Then $C(I, K)$ is a Banach space.

More generally, if F is a Banach space, we may form families (x_i) with $x_i \in F$ such that $x_i \to 0$. Such families with the sup norm form a Banach space $C(I, F)$.

A family $\{e_i\}$ in a Banach space E will be called a **Banach basis** if every element $x \in E$ can be written uniquely as a series

$$x = \sum_{i \in I} x_i e_i \quad \text{with } x_i \to 0$$

and $|x| = \sup|x_i|$. It is clear that the space $C(I, K)$ has such a basis, with $e_i(i) = 1$ and $e_i(i') = 0$ if $i' \neq i$.

Let E be a Banach space and let E_0 denote the subset of elements $x \in E$ such that $|x| \leq 1$. If $\{e_i\}$ is a Banach basis, then E_0 consists of the set of all elements

$$\sum x_i e_i$$

such that $|x_i| \leq 1$ for all i.

Lemma 5.1. *Let $\mathfrak{m}$ be the maximal ideal of R. Let $\bar{E} = E_0/\mathfrak{m}E_0$. A family $\{e_i\}$ in E is a Banach basis if and only if all $e_i \in E_0$, and their images $\bar{e}_i$ in $\bar{E}$ form an algebraic basis of $\bar{E}$ as vector space over k.*

Proof. Suppose that $\{e_i\}$ is a Banach basis for E. Then first it is clear that every element of $\bar{E}$ can be written as a linear combination of the $\bar{e}_i$ with coefficients in k, and such a combination is unique, as one sees by lifting back to E_0.

Conversely, suppose $\{\bar{e}_i\}$ is an algebraic basis for $\bar{E}$. Any element $x \in E_0$ can be written

$$x = \sum x_i^{(1)} e_i + \pi x^{(1)} \quad \text{with } x^{(1)} \in E_0.$$

Iterating, and expressing $x^{(1)}$ in a similar way, we get an expression

$$x = \sum x_i e_i \quad \text{with } x_i \in R \text{ and } x_i \to 0.$$

Such an expression is unique, for if we have a relation

$$\sum x_i e_i = 0, \quad \text{not all } x_i = 0,$$

then first we may divide by the highest power of π dividing all x_i, so that not all x_i are divisible by π, and then we may reduce mod $\mathfrak{m}$ to get a contradiction. Finally if $|x| = 1$, we have

$$|x| = \sup |x_i|,$$

and the same relation holds for any x by multiplying with an appropriate power of π. This proves the lemma.

Proposition 5.2. *Every Banach space over K is isomorphic with a space $C(I, K)$. Equivalently, every Banach space has a Banach basis.*

Proof. Immediate from the lemma, by lifting an algebraic basis from $\bar{E}$ back to E_0.

Corollary. *Every closed subspace F of a Banach space E has a complementary closed subspace F' such that $E \approx F \times F'$.*

Proof. Let F be a closed subspace of E. By the theorem, there exists a Banach basis $\{e_i'\}$ of the factor space E/F. Let e_i be a representative of e_i' in E such that $|e_i| \leq 1$. The map

$$u : e_i' \mapsto e_i$$

extends to a continuous linear map $E/F \to E$, of norm ≤ 1. Then the map

$$E/F \times F \to E$$

given by

$$(\bar{x}, y) \mapsto u\bar{x} + y$$

is an isomorphism, as is immediately verified. This proves the corollary.

Let F be a Banach space. Let I be a set of indices, and let $B(I, F)$ be the set of bounded maps of I into F, i.e., the set of bounded families $(f_i)_{i \in I}$ with the sup norm. Then $B(I, F)$ is a Banach space.

Proposition 5.3. *Let $E = C(I, K)$ and let $\{e_i\}$ be a Banach basis. Then we have an isomorphism*

$$L(E, F) \to B(I, F)$$

given by

$$u \mapsto (ue_i)_{i \in I}.$$

Proof. Let φ be the map from $L(E, F)$ to $B(I, F)$ as given in the statement of the proposition. Conversely, if (f_i) is a bounded family in F, define a map

$$\psi : B(I, F) \to L(E, F)$$

by associating to (f_i) the map u such that

$$u(\sum x_i e_i) = \sum x_i f_i.$$

It is then immediate that φ, ψ are continuous linear, inverse to each other, and have norms ≤ 1, so that they are isomorphisms. This proves the proposition.

The proposition will be used especially when F has a Banach basis, so that $F = C(J, K)$ for some set of indices J. Each element ue_i can then be written uniquely

$$ue_i = \sum_j c_{ij} e'_j$$

where (e'_j) is the natural Banach basis for $C(J, K)$. Thus u has an associated matrix (c_{ij}), relative to the bases (e_i) and (e'_j). By the definitions, and Proposition 5.3, we have

$$|u| = \sup_{i,j} |c_{ij}|.$$

Example. Let $F = K$ so that $L(E, K)$ is the dual space, denoted by E^*. If we let $ue_i = c_i$, then

$$|u| = \sup_i |c_i|.$$

Let $u \in L(E, F)$. We say that u is **completely continuous** if u is a limit of elements of $L(E, F)$ with finite dimensional image. (The limit of course taken with respect to the norm on $L(E, F)$.) We denote by $CC(E, F)$ the space of such completely continuous linear maps. It is obviously a Banach subspace of $L(E, F)$.

Let E, E', E'' be Banach spaces, and let

$$E \xrightarrow{u} E' \xrightarrow{v} E''$$

be continuous linear maps. If u or v is completely continuous, then so is $v \circ u$. This is immediate from the definition. In particular, $CC(E, E)$ is a two-sided ideal in $L(E, E)$.

Suppose now that $F = C(J, K)$ with Banach basis (f_j). Let $u \in L(E, F)$. For each $x \in E$ we can write

$$ux = \sum u_j(x) f_j$$

where $u_j \in E^*$ is an element of the dual of E, i.e. a functional. We define

$$r_j(u) = |u_j|.$$

Then

$$|u| = \sup_j r_j(u) = \sup_j |u_j|,$$

directly from the definitions.

Let $E = C(I, K)$, and let (c_{ij}) be the matrix associated with u, relative to the bases (e_i) and (f_j). Then

$$|u_j| = \sup_i |c_{ij}|.$$

This is clear from the example following Proposition 5.3.

Proposition 5.4. *Let $F = C(J, K)$, with Banach basis (f_j). The map*

$$u \mapsto (u_j)$$

gives an isomorphism

$$CC(E, F) \approx C(J, E^*)$$

between $CC(E, F)$ and the space of families (u_j) of elements in E^* such that $u_j \to 0$. In particular, an element $u \in L(E, F)$ is completely continuous if and only if $u_j \to 0$.

Proof. First suppose u has finite dimensional image, so without loss of generality, we may assume the image is one-dimensional. Then $u(x) = v(x)f$ for some $f \in F$, and v is a functional. Then $u_j = f_j v$, where f_j is the j-th coordinate of f, so clearly $u_j \to 0$. Next let $u^{(n)} \in L(E, F)$ approach u, and have finite dimensional images. Then

$$u_j^{(n)} \to u_j$$

uniformly, so $u_j \to 0$ also in this general case.

Finally, let (v_j) be a family of elements in the dual E^* such that $v_j \to 0$. Define u by

$$ux = \sum v_j(x) f_j.$$

Then u is certainly continuous linear. The partial sums u_S defined for every finite set S of indices by

$$u_S(x) = \sum_{j \in S} v_j(x) f_j$$

approach u in the norm of $L(E, F)$, and have finite rank. Hence the correspondence between $CC(E, F)$ and $C(J, E^*)$ is bijective. It is immediately verified to be norm preserving. This concludes the proof.

Example from Dwork theory. Let $E = L(\delta)$, with the norm defined as follows. If

$$f(x) = \sum a_n x^n$$

is in $L(\delta)$, we define

$$|f|_\delta = \sup \left| \frac{a_n}{p^{n\delta}} \right|.$$

Then E is a Banach space. We assume that δ is rational, and that the constant field K has been extended by a finite extension if necessary so that p^δ lies in K. The powers

$$(p^\delta x)^n, \qquad n = 0, 1, 2, \ldots$$

form a Banach basis for E. If $\delta' < \delta$, then the inclusion

$$L(\delta) \to L(\delta')$$

is completely continuous. This is obvious by looking at the associated matrix, and using Proposition 5.4.

On the other hand, if $g \in L(\delta')$, then the map $f \mapsto fg$ is a continuous linear map of $L(\delta')$ into itself. Furthermore, the Dwork map Ψ_q is a continuous linear map of $L(\delta')$ into $L(q\delta')$. It follows that the Dwork operator $A_{j,q}$ is completely continuous on the space $L(\delta)$.

We now come to the results on Fredholm determinants which were used in the context of Dwork theory. First we make some remarks about determinants in finite dimensional spaces.

Let L be a free module over a commutative ring, and let u be an endomorphism of L such that $u(L)$ is contained in a finitely generated submodule. Let M be a submodule of L which is finitely generated, free, and contains $u(L)$. Let

$$u_M : M \to M$$

be the restriction of u to M. The polynomial

$$\det(I - tu_M)$$

is well defined, and it is easy to verify that it does not depend on the choice of M. We may therefore denote it by

$$\det(I - tu) = 1 + c_1 t + \cdots + c_m t^m + \cdots.$$

The coefficients c_m can be expressed in terms of u as follows.

Let (e_i) be a basis of L. Let S be a finite subset of the set of indices I, with card $S = m$. Let $A = (a_{ij})$ be the matrix of u with respect to this basis, and let

$$\det_S(A) = \text{determinant of the submatrix of } a_{ij} \text{ with } i, j \in S.$$

Then

$$\boxed{c_m = (-1)^m \sum_{|S|=m} \det_S(A),}$$

where the sum is taken over subsets S having m elements.

This formula actually relates to a finite square matrix. It suffices to prove it when the matrix consists of algebraically independent elements in characteristic 0 (by specializing afterward). In that case, the matrix has distinct eigenvalues, and the formula is obvious if the matrix is diagonal. To get the formula in general, all we need is an invariant characterization of the terms in the sum over S. This is provided by the second expression for the coefficients:

$$c_m = (-1)^m \operatorname{tr} \textstyle\bigwedge^m u,$$

where $\bigwedge^m u$ is the m-th exterior power of u, acting on $\bigwedge^m M$, and M may be assumed to be a finite dimensional vector space. But in that case, one finds at once that

$$\operatorname{tr} \bigwedge^m u = \sum_{|S|=m} \det_S(A),$$

by looking at the trace with respect to the basis

$$\{e_{i_1} \wedge \cdots \wedge e_{i_m}\}.$$

This proves that the coefficients c_m have the value as stated. In particular, $c_1 = -\operatorname{tr} u$ (sum of the diagonal elements), so

$$\det(I - tu) = 1 - (\operatorname{tr} u)t + \cdots.$$

Now suppose E is a Banach space with Banach basis (e_i), $i \in I$. We wish to define $\det(I - tu)$ for any $u \in CC(E, E)$. Suppose first that $|u| \leq 1$. Let E_0 be as before, the subspace of elements x in E such that $|x| \leq 1$. Then $u(E_0) \subset E_0$. For each positive integer n, the endomorphism u defines an endomorphism

$$u(\pi^n): E_0/\pi^n E_0 \to E_0/\pi^n E_0.$$

We denote $E_0/\pi^n E_0 = E_0(\pi^n)$. If (a_{ij}) is the matrix of u with respect to the basis (e_i), then we know that

$$|u_j| = \sup_i |a_{ij}| \to 0.$$

Hence there exists a finite subset S of indices j such that for all i and all $j \notin S$, the components a_{ij} lie in the ideal (π^n). Hence the polynomial

$$\det(I - tu(\pi^n))$$

is defined, as a polynomial in $R/\pi^n R$. As n tends to infinity, these polynomials form a projective system, and we define

$$\det(I - tu)$$

to be their limit. It is a formal power series, in $R[[t]]$.

If u is arbitrary in $CC(E, E)$, we can find an element $c \in R$, $c \neq 0$, such that $|cu| \leq 1$. Then $\det(I - tcu)$ is defined, as a power series $D(t)$, and we define $\det(I - tu) = D(tc^{-1})$. This is independent of the choice of c.

Proposition 5.5. *Let u be a completely continuous endomorphism of the Banach space E. Let $(e_i)_{i \in I}$ be a Banach basis, and let $A = (a_{ij})$ be the matrix of u with respect to this basis. Let*

$$\det(I - tu) = \sum_{m=0}^{\infty} c_m t^m.$$

Then:

(i)
$$c_m = (-1)^m \sum_S \det_S(A).$$

where S ranges over subsets of I with m elements.

(ii) *Given a positive number r, we have $|c_m| \ll r^m$ for all m sufficiently large.*

(iii) *If $\{u_n\}$ is a sequence in $CC(E, E)$ and $u_n \to u$, then*

$$\det(I - tu_n) \to \det(I - tu),$$

where the convergence is the simple convergence of coefficients.

(iv) *If u has finite rank, then* $\det(I - tu)$ *is the polynomial defined from elementary linear algebra.*

Proof. Suppose first that $|u| \le 1$. Then the formula for c_m is valid for each $u(\pi^n)$, and so remains valid in the limit. The general case follows by using $c \neq 0$ such that $|cu| \le 1$.

For (ii), let S be a finite subset of m elements. Each product in the expansion for the determinant $\det_S(A)$ contains m elements, indexed by m distinct indices $j_1, \ldots, j_m$. Let

$$r_j = |u_j| \quad \text{so} \quad r_j \to 0.$$

Then any product of m elements as above is bounded in absolute value by the product $r_1 \cdots r_m$. Hence

$$|\det_S(A)| \le r_1 \cdots r_m \quad \text{for each } S,$$

so

$$|c_m| \le r_1 \cdots r_m.$$

Since $r_j \to 0$ it follows that $r_j \le r$ for all j sufficiently large, and (ii) follows.

The third assertion follows trivially from (i).

Finally, suppose u has finite dimensional image. After replacing u by a scalar multiple, we may assume that $|u| \leq 1$. Let F be a finite dimensional subspace of E containing $u(E)$. Let

$$F_0 = F \cap E_0 \quad \text{and} \quad F(\pi^n) = F_0/\pi^n F_0.$$

Then F_0 is a direct summand of E_0, and hence $F_0(\pi^n)$ is a direct summand of $E_0(\pi^n)$, and is a free submodule. Hence the determinant

$$\det(I - tu(\pi^n))$$

can be computed in $F_0(\pi^n)$. Since

$$\det(I - tu) = \lim \det(I - tu(\pi^n)),$$

it follows that

$$\det(I - tu) = \det(I - tu_F),$$

where u_F is the restriction of u to F. Then (iv) follows, and the proposition is proved.

By the estimate of (ii) in the proposition, we conclude that the power series $\det(I - tu)$ represents an entire function of the variable t for values in K, or in any complete extension of K. In particular, $\det(I + u)$ is defined for all $u \in CC(E, E)$.

Corollary 1. *Let $u, v \in CC(E, E)$. Then*

$$\det((I - tu)(I - tv)) = \det(I - tu)\det(I - tv).$$

Proof. After multiplying u, v by appropriate scalars, we may assume without loss of generality that $|u| \leq 1$ and $|v| \leq 1$. In that case, we view u, v as acting on E_0, and reduce mod π^n, in which case the formula is true when u, v are replaced by $u(\pi^n)$ and $v(\pi^n)$ respectively. Taking the limit yields the corollary.

Corollary 2. *Let $u \in CC(E, F)$ and $v \in L(F, E)$. Then*

$$\det(I - tu \circ v) = \det(I - tv \circ u).$$

Proof. By (iii) and (iv) we may assume that u, v have finite rank, in which case the assertion is standard by the linear algebra of finite dimensional spaces.

We define the **trace** $\mathrm{tr}(u)$ of an element $u \in CC(E, E)$ as the coefficient of $-t$ in $\det(I - tu)$. If (a_{ij}) is the matrix associated with u with respect to a Banach basis, then as usual,

$$\mathrm{tr}(u) = \sum a_{ii}.$$

This formula shows that $|\mathrm{tr}(u)| \leq |u|$.

Corollary 3. *Assume that K has characteristic* 0. *Then*

$$\det(I - tu) = \exp\left(-\sum_{m=1}^{\infty} \mathrm{tr}(u^m)\frac{t^m}{m!}\right).$$

Proof. The assertion is true if u has finite rank by ordinary linear algebra. Let $\{u_n\}$ be a sequence of such endomorphisms converging to u. Then both the right-hand side and left-hand side with u replaced by u_n converge to the corresponding expressions of the corollary, thus establishing their equality.

Note. The above three corollaries are included to show how to apply Proposition 5.5. They were not needed in the applications of the preceding sections. Only the following proposition was needed.

Proposition 5.6. *Let*

$$0 \to E' \to E \to E'' \to 0$$

be an exact sequence of Banach spaces, and let u', u, u'' be continuous linear maps making the diagram commutative:

$$\begin{array}{ccccccccc} 0 & \longrightarrow & E' & \longrightarrow & E & \longrightarrow & E'' & \longrightarrow & 0 \\ & & \downarrow{\scriptstyle u'} & & \downarrow{\scriptstyle u} & & \downarrow{\scriptstyle u''} & & \\ 0 & \longrightarrow & E' & \longrightarrow & E & \longrightarrow & E'' & \longrightarrow & 0 \end{array}$$

If u is completely continuous, so are u' and u'', and we have

$$\det(I - tu) = \det(I - tu')\det(I - tu''),$$

$$\mathrm{tr}\, u = \mathrm{tr}\, u' + \mathrm{tr}\, u''.$$

Proof. We may view E' as a subspace of E, and u' as the restriction of u to E'. Thus u'' is the map induced on E'' as factor space E/E'. By the Corollary of Proposition 5.2 there exists a Banach basis $\{e_i'\}$ for E' and a Banach basis $\{e_j''\}$ for a complementary subspace such that the images of $\{e_j''\}$ in E/E' form a Banach basis. Then $\{e_i', e_j''\}$ forms a Banach basis for E. From this it is clear that if u is completely continuous, so are u' and u''. Reducing mod π^n yields the desired identity for $u(\pi^n)$, $u'(\pi^n)$, and $u''(\pi^n)$, whence the identity as in the proposition for u, u', u''. The identity for the trace follows from that of the determinant since

$$\det(1 - tu) = 1 - (\operatorname{tr} u)t + \cdots.$$

This concludes the list of properties of the determinant which has been used in the Dwork theory.

16 Gauss Sums and the Artin–Schreier Curve

In this chapter we establish the connection between the spaces used to represent the Frobenius endomorphism, and eigenspaces for a Galois group of automorphisms of the Artin–Schreier curve. This connects Dwork theory, the Washnitzer–Monsky theory, and the fact realized by Monsky that the series $E_\pi(x)$ can be used to construct explicitly such eigenspaces. The first section lays the foundations for the special type of ring under consideration. After that we study the Artin–Schreier equation and the Frobenius endomorphism.

§1. Power Series with Growth Conditions

We let:

R = discrete valuation ring of characteristic 0, complete, with prime element π dividing the prime number p.
K = quotient field of R.
$k = R/\pi$ = residue class field, which has characteristic p.

$R\langle\langle x\rangle\rangle$ = set of power series

$$\varphi(x) = \sum_{n=0}^{\infty} a_n x^n \quad \text{with } a_n \in R \tag{1}$$

such that there exists $\delta > 0$ for which

$$\operatorname{ord}_\pi a_n \geq \delta n \quad \text{for all } n \text{ sufficiently large.}$$

It is clear that $R\langle\langle x\rangle\rangle$ is a ring, which will be called the **Washnitzer–Monsky ring**. Furthermore, the elements of $R\langle\langle x\rangle\rangle$ are precisely the power series which converge (absolutely) on a disc of radius > 1. Of course, the radius depends on the power series, there is no uniformity required. Later, we shall deal with the Dwork spaces, taking the uniformity into account. Indeed, it will also be useful to consider power series such that a finite number of coefficients may lie in K.

Pick δ rational > 0, and let π^δ denote any rational power of π, well defined up to a root of unity in the integral closure of R. If $\varphi(x)$ as above is an element of $R\langle\langle x\rangle\rangle$, then we may write

$$\varphi(x) = \sum b_n(\pi^\delta x)^n \tag{2}$$

with coefficients b_n (possibly algebraic over R) which are π-integral for all but a finite number of n. Conversely, if a power series $\varphi(x)$ in $R[[x]]$ has a representation as in (2) with coefficients b_n whose denominators are bounded, then it is clear that $\varphi(x) \in R\langle\langle x\rangle\rangle$.

We let $R\langle\langle x\rangle\rangle_p$ be the p-adic completion of $R\langle\langle x\rangle\rangle$. Then $R\langle\langle x\rangle\rangle_p$ is the ring of power series

$$\sum a_n x^n, \qquad a_n \in R,$$

such that $a_n \to 0$, under the sup norm of coefficients. To prove that this is the completion, we observe that this ring $R\langle\langle x\rangle\rangle_p$ is p-adically complete, and that $R\langle\langle x\rangle\rangle$ is dense in it. Furthermore, the polynomial ring $R[x]$ is dense, so that

$$R\langle\langle x\rangle\rangle_p = R[x]_p$$

is also the p-adic completion of $R[x]$.

The ring $R\langle\langle x\rangle\rangle_p$ still has the advantage that

$$R\langle\langle x\rangle\rangle_p \bmod \pi = k[x],$$

in other words, its reduction mod π is the polynomial ring.

The following property is immediately verified.

Let $w \in R\langle\langle x\rangle\rangle$ (respectively $R\langle\langle x\rangle\rangle_p$) be such that

$$w \equiv 0 \bmod \pi x.$$

Then the geometric series gives an inverse for $1 - w$ in $R\langle\langle x\rangle\rangle$ (respectively $R\langle\langle x\rangle\rangle_p$).

Lemma 1.1. *The ring $R\langle\langle x\rangle\rangle_p$ is integrally closed.*

Proof. An element $\varphi(x)$ of $R\langle\!\langle x\rangle\!\rangle_p$ not divisible by π can be written

$$\varphi(x) = b_0 + \cdots + b_d x^d + \cdots$$

such that b_d is a unit, and $b_n \equiv 0 \bmod \pi$ for $n \geq d + 1$. Then the Manin proof of the Weierstrass preparation theorem given for instance in Chapter 5, §2 applies to show that

$$\varphi(x) = P(x)u(x),$$

where $P(x) = x^d + \cdots + c_0 \in R[x]$ is a polynomial, and $u(x)$ is a unit in $R\langle\!\langle x\rangle\!\rangle_p$. (In that reference, the crucial step occurs when we assert that $\tau(f)$ is invertible, which is true in $R\langle\!\langle x\rangle\!\rangle_p$, by the remark made before the lemma.) This implies that the zeros of $P(x)$ all lie in the unit disc in the algebraic closure of the quotient field of R.

If the quotient $\varphi(x)/\psi(x)$ of elements in $R\langle\!\langle x\rangle\!\rangle_p$ is integral over $R\langle\!\langle x\rangle\!\rangle_p$, then we use the Weierstrass preparation theorem to write

$$\frac{\varphi(x)}{\psi(x)} = \pi^r \frac{P(x)}{Q(x)} u(x),$$

where $P(x)$, $Q(x)$ are polynomials, and $u(x)$ is a unit in $R\langle\!\langle x\rangle\!\rangle_p$. Then without loss of generality, we may assume that $u(x) = 1$. Since elements of $R\langle\!\langle x\rangle\!\rangle_p$ converge on the unit disc, if $Q(x)$ does not divide $P(x)$, we may evaluate the right-hand side at a zero of $Q(x)$ which is not a zero of $P(x)$ to make the right-hand side infinity. An elementary criterion for integrality then shows that $\varphi(x)/\psi(x)$ cannot be integral over $R\langle\!\langle x\rangle\!\rangle_p$, which proves the lemma.

Lemma 1.2. *The ring $R\langle\!\langle x\rangle\!\rangle$ is algebraically closed in $R\langle\!\langle x\rangle\!\rangle_p$, and therefore it is integrally closed (in its quotient field).*

Proof. As pointed out to me by Dwork, this lemma can be viewed as a special case of a result in Dwork–Robba [Dw–Ro], Theorem 3.1.6. The proof given below was derived in collaboration with Dwork.

Let $A = R\langle\!\langle x\rangle\!\rangle$ and $A_p = R\langle\!\langle x\rangle\!\rangle_p$. Let $y \in A_p$ be algebraic over A, satisfying a polynomial equation

$$F(y) = 0$$

with coefficients in A. By the x-topology, we mean the topology of formal power series (high powers of x are close to zero). Given $\alpha \in A_p$, if α' is x-close to α, then α' is also p-close to α.

Let y_0 be a polynomial, in $R[x]$, which is x-close to y. Let $\|\ \ \|$ be the **Gauss norm** on power series (sup norm of the coefficients) extended to the quotient field $K(A)$. Then $F'(y_0) \in A$, and in particular is holomorphic on a disc of radius > 1. Then for y_0 sufficiently close to y, we have

$$\left\| \frac{F(y_0)}{F'(y_0)^2} \right\| < 1.$$

Hence the Newton sequence

$$y_{n+1} = y_n - \frac{F(y_n)}{F'(y_n)}$$

converges in the Gauss norm in the completion of $K(A)$, and in fact to y itself since A_p is complete, if we pick y_0 sufficiently close to y (closer than any other root of F in the completion of $K(A)$). Cf. for instance my paper, "On quasi algebraic closure," *Ann. of Math.* (1952) pp. 373–392; or also my *Algebraic Number Theory*, Chapter II.

We shall now use another norm to see that the sequence converges to a holomorphic function on a disc of radius $\geq 1 + \varepsilon$ for some ε, from which relatively small sets have been deleted.

To avoid introducing a new letter, let us view K as a subfield of $\mathbf{C}_p$. If t is a real number > 0, we let $D(t, 0)$ be the closed disc of radius t around the origin in $\mathbf{C}_p$. Let $\alpha_1, \ldots, \alpha_m$ be elements of this disc, and let $r_1, \ldots, r_m > 0$. We let

$B = B(t; r, \alpha) =$ set obtained by deleting from $D(t, 0)$ the union of the balls $D(r_j, \alpha_j)$ of radius r_j, centered at α_j.

In the sequel, we assume that $t > 1$ and $r_j < 1$,

If $H(x)$ is a rational function with no poles in a set B as above, then from its factorization into linear factors, we see that the norm

$$\|H(x)\|_B = \sup_{x \in B} |H(x)|$$

is defined.

Lemma 1.2.1. *For any rational function H holomorphic on B, we have*

$$\|H\| \leq \|H\|_B.$$

Proof. We factor the rational function into a product of a constant factor and linear factors of type

$$x - a \quad \text{and} \quad \frac{1}{x - a}, \quad \text{with } a \in \mathbf{C}_p.$$

The Gauss norm of $x - a$ is $\max(1, |a|)$. We pick $x \in B$ to be a unit which is not congruent to any a mod $\mathfrak{m}_p$. For such x we have

$$|H(x)| \leq \|H\|_B,$$

and the lemma is then obvious.

Lemma 1.2.2. *Given a rational function $H(x)$, with $\|H\| < 1$, there exists a set $B(t; \alpha, r)$ with $t > 1$ and $r_j < 1$, such that*

$$\|H\|_B < 1.$$

Proof. Factor

$$H(x) = c \prod (1 - a_i x)^{m_i} \prod (x - b_j)^{n_j}$$

where $|a_i| < 1$ and $|b_j| \leq 1$. Then $|c| < 1$ because $\|H\| < 1$. Let N be the number of linear factors, counting multiplicities. Let $s > 1$ be such that $s^N c < 1$. It will suffice to find B such that each factor of H has B-norm $\leq s$. We consider factors of four types:

$$1 - ax, \quad (1 - ax)^{-1}, \quad x - b, \quad (x - b)^{-1}$$

with $|a| < 1$ and $|b| \leq 1$. For factors of the first three types, it suffices to select the radius t of B to be $< s$ and sufficiently small > 1. For the factor of last type, namely $(x - b)^{-1}$, it suffices to delete from B a disc of radius $1/t'$, where $t' < t$ and t' is very close to t. This proves the sublemma.

We now return to the proof of Lemma 1.2. We define:

Hol(B) = completion of the ring of rational functions having no poles in B, under the B-norm.

Since the Gauss norm is bounded by the B-norm, every Cauchy sequence for the B-norm is a Cauchy sequence for the Gauss norm. Consequently there is a natural injection of the completions

$$\text{Hol}(B) \to K(x)_{\text{Gauss}}.$$

Observe that $A_p = R\langle\langle x \rangle\rangle_p$ is contained in the Gauss completion.

Given our element $y \in A_p$ algebraic over A, we first select a polynomial $y_0 \in R[x]$ as in the beginning of the proof, so that

$$\left\| \frac{F(y_0)}{F'(y_0)^2} \right\|$$

is small, and in particular is < 1. We let $H(x) = F(y_0)/F'(y_0)^2$ be this rational function. We then select B as in Lemma 1.2.2 so that $\|H\|_B < 1$. Then the Newton sequence also converges to an element of Hol(B). This implies that y is the power series expansion on the *closed* disc of radius 1 of a holomorphic function on B. It is now a matter of foundations of p-adic analytic functions that the power series for y also represents the analytic function on some disc of radius > 1. This implies that the coefficients of the power series $y = \varphi(x)$ tend to 0 like some geometric series, thereby proving Lemma 1.2.

For a proof of the foundational fact we have just used, see for instance Amice [Am]. Note that the power series for y may converge only on a disc of radius smaller than the radius of the original set B. For instance, the holomorphic function may have a finite number of poles near but outside the circle of radius 1. The power series will converge only on a disc which does not contain these poles. The proof requires the p-adic analogue of the Mittag-Leffler theorem, in lieu of analytic continuation over the complex numbers.

Let $f_0(Y)$ be an irreducible polynomial of degree d, with coefficients in $k[x]$, leading coefficient 1. By a **lifting** of f_0 we mean a polynomial $f(Y)$ in $R\langle\langle x\rangle\rangle[Y]$ of the same degree, leading coefficient 1, such that

$$f_0 = f \mod \pi.$$

Then f is necessarily irreducible over $R\langle\langle x\rangle\rangle_p$. Indeed, the coefficients in a factor are integral over $R\langle\langle x\rangle\rangle$, so in $R\langle\langle x\rangle\rangle$ by Lemmas 1.1 and 1.2. Such a factor then reduces to a factor of f_0.

Let y_0 be a root of f_0 and let y be a root of f. Let

$$\mathscr{A} = R\langle\langle x\rangle\rangle[y], \quad \text{and} \quad \mathscr{A}_0 = k[x, y_0].$$

Then there is a unique homomorphism

$$\mathscr{A} \to k[x, y_0] = \mathscr{A}_0$$

reducing $R\langle\langle x\rangle\rangle$ mod π, and sending y to y_0. This is a standard fact of elementary field theory. Thus the ideal (π) in $R\langle\langle x\rangle\rangle$ extends uniquely to a prime ideal of $\mathscr{A}$, and

$$\mathscr{A}_0 = \mathscr{A} \mod \pi.$$

We view f as defining an affine curve V and f_0 as defining its reduction mod π. Thus we shall write

$$\mathscr{A} = \mathscr{A}(V) \quad \text{and} \quad \mathscr{A}/\pi = \mathscr{A}_0(V_0).$$

Then $\mathscr{A}_0(V_0)$ is the ordinary affine ring of V_0 over the field $k = R/\pi$, but $\mathscr{A} = \mathscr{A}(V)$ is a more complicated ring, arising from the work of Washnitzer and Monsky, following Dwork. We shall also say that V is a **lifting** of V_0, corresponding to the lifting f of f_0.

Lemma 1.3. *Let $w \in \mathscr{A}$ and assume $w \equiv 0 \bmod \pi$. Then the geometric series*

$$1 + w + w^2 + \cdots$$

converges to an inverse of $1 - w$ in $\mathscr{A}$.

Proof. For convenience, assume that the lifted polynomial $f(Y)$ has coefficients in $R[x]$. This is all that we need in the applications, and the proof is slightly easier in this case. We write

$$w = \pi \sum_{i=0}^{d-1} g_i(x) y^i$$

with $g_i(x) \in R\langle\langle x \rangle\rangle$. Then for some $\delta > 0$ we also have

$$w = \sum_{i=0}^{d-1} h_i(\pi^\delta x)(\pi^\delta y)^i,$$

where h_i are power series with coefficients in the algebraic closure of R, and all of these coefficients are divisible by the small power π^δ. Furthermore, for each positive integer n, we can write

$$y^n = \sum_{i=0}^{d-1} \varphi_{n,i}(x) y^i$$

where $\varphi_{n,i}$ is a polynomial of degree $\ll n$ ($\leq$ some constant times n, the constant depending only on f). It then follows at once that there exists ε such that if we write

$$w^n = \sum_{i=0}^{d-1} \sum_{j=0}^{\infty} c_{j,i}^{(n)} x^j y^i$$

then $\operatorname{ord} c_{j,i}^{(n)} \geq \varepsilon(j + n)$. This proves the lemma.

We shall say that V_0 or f_0 is **special** if $f_0'(y_0)^{-1} \in k[x, y_0]$. It then follows that

$$f'(y)^{-1} \in \mathscr{A}.$$

Indeed, let $z \in \mathscr{A}$ be such that $z \equiv f_0'(y_0)^{-1} \bmod \pi$. Then

$$zf'(y) = 1 + w,$$

where $w \in \mathscr{A}$ and $w \equiv 0 \bmod \pi$. Applying Lemma 1.3 proves our assertion.

Lemma 1.4. *Assume that f_0 is special. Then $\mathscr{A}$ is integrally closed in $R\langle\langle x\rangle\rangle_p[y]$, and also in the quotient field K of $\mathscr{A}$.*

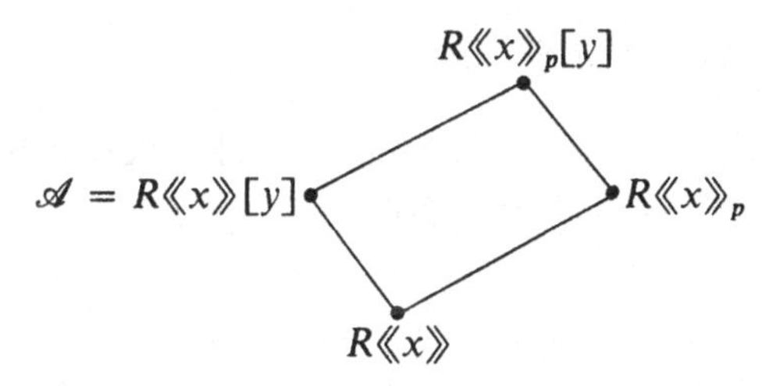

Proof. The powers y^j ($j = 0, \ldots, d - 1$) form a basis of K over the quotient field of $R\langle\langle x\rangle\rangle$. The dual basis with respect to the trace is contained in $f'(y)^{-1}\mathscr{A}$, and hence in $\mathscr{A}$. If an element $z \in R\langle\langle x\rangle\rangle_p[y]$ is integral over $\mathscr{A}$, and we write

$$z = \sum_{i=0}^{d-1} g_i(x)y^i \quad \text{with } g_i(x) \in R\langle\langle x\rangle\rangle_p,$$

then the coefficients $g_i(x)$ can be expressed as traces,

$$g_i(x) = \mathrm{Tr}(zy_i'),$$

where $\{y_1', \ldots, y_d'\}$ is the dual basis. Since each zy_i' and its conjugates are integral over $R\langle\langle x\rangle\rangle$, so is the trace, which lies in $R\langle\langle x\rangle\rangle$ by Lemma 1.2. Hence $z \in R\langle\langle x\rangle\rangle[y]$. The same argument shows that $\mathscr{A}$ is integrally closed. (All of this is standard elementary theory of fields and integral closure, as in algebraic number theory.)

We shall need to lift roots of polynomials, as expressed in the next lemma. We let Q be the quotient field of $R\langle\langle x\rangle\rangle$.

Lemma 1.5. *Let V be a lifting of V_0 as above, and assume that V_0 is special. Let*

$$\mathscr{A} = R\langle\langle x\rangle\rangle[y] \to k[x, y_0]$$

be reduction mod π. *To each root* $\bar{z}$ *of* f_0 *in* $k[x, y_0]$ *there exists a unique root* z *of* f *in* $\mathscr{A}$ *such that*

$$z \bmod \pi = \bar{z}.$$

If in addition $k(x, y_0)$ *is Galois over* $k(x)$ *with group* G_0, *then* $Q(y)$ *is Galois over* Q. *Say* G *is its Galois group. Then the map*

$$\sigma \mapsto \sigma \bmod \pi$$

gives an isomorphism of G *with* G_0.

Proof. Let $z_1 \in \mathscr{A}$ reduce to $\bar{z}$ mod π. Such z_1 exists because reduction mod π gives a surjective homomorphism of $\mathscr{A}$ onto $k[x, y_0]$. We can find $w \in \mathscr{A}$ such that

$$wf'(z_1) \equiv 1 \bmod \pi.$$

We then define the usual sequence

$$z_{n+1} = z_n - wf(z_n).$$

We obtain

$$f(z_{n+1}) = f(z_n) - wf(z_n)f'(z_n) \bmod f(z_n)^2.$$

It follows by induction that

$$f(z_n) \equiv 0 \bmod \pi^n \quad \text{and} \quad z_n \equiv z_1 \bmod \pi.$$

The limit z of the sequence $\{z_n\}$ then a priori lies in the ring

$$R\langle\langle x \rangle\rangle_p[y],$$

and is a root of f in that ring. By Lemma 1.4, we conclude that the root actually lies in $R\langle\langle x \rangle\rangle[y] = \mathscr{A}$, as desired.

Under the additional assumption of Galois extensions as stated, this proves that $\mathscr{A}$ is Galois over $R\langle\langle x \rangle\rangle$. Since the prime π remains a prime ideal in $\mathscr{A}$, by standard decomposition group arguments we see that G is the decomposition group of the prime, and we get the isomorphism of G with G_0 by reduction mod π. Cf. Proposition 14, Chapter I, §5 of my *Algebraic Number Theory*.

The last lemma is only a special case of a more general situation having to do with the possibility of lifting morphisms. What we use in connection with the Frobenius morphism is summarized in the next lemma.

Lemma 1.6. *Let W_0, V_0 be two special affine curves, defined over k by polynomials g_0, f_0 respectively. Let W, V be liftings, defined by g, f. Let*

$$\varphi_0 : \mathscr{A}_0(V_0) \to \mathscr{A}_0(W_0)$$

be a homomorphism of the affine rings, such that

$$\varphi_0(x) \in xk[x].$$

Then there exists a lifting homomorphism

$$\varphi : \mathscr{A}(V) \to \mathscr{A}(W)$$

of φ_0. If $\varphi(x) \in R\langle\langle x \rangle\rangle$ is a lifting of $\varphi_0(x)$, then there is a unique choice of $\varphi(y)$ lifting $\varphi_0(y)$ and making φ a p-adically continuous homomorphism.

Proof. Suppose $\varphi(x)$ lifts $\varphi_0(x)$. Then we can define a unique p-adically continuous homomorphism $R\langle\langle x \rangle\rangle \to R\langle\langle x \rangle\rangle$ by

$$\sum a_n x^n \mapsto \sum a_n \varphi(x)^n.$$

Let φf be the polynomial obtained by applying φ to the coefficients of f, and let $(\varphi f)_0$ be its reduction mod π. Then $(\varphi f)_0$ has a root $\bar{z}$ in $\mathscr{A}_0(W_0)$, namely $\bar{z} = \varphi_0(y_0)$. Let $z_1 \in \mathscr{A}(W)$ reduce to $\bar{z}$ mod π. As in the proof of Lemma 1.5 we then find w in $\mathscr{A}$ such that

$$w(\varphi f)'(z_1) \equiv 1 \pmod{\pi},$$

and the same argument as in Lemma 1.5 then shows that z_1 can be uniquely refined to a root of φf in $\mathscr{A}(W)$. This concludes the proof.

§2. The Artin–Schreier Equation

The example we shall consider is given by the equation

$$Y^p - Y = x^N$$

where N is a positive integer prime to p. If k is a field of characteristic p, then $Y^p - Y - x^N$ is irreducible. (For instance, a root is ramified of order p over $x = \infty$.) Furthermore, putting $f_0(Y) = Y^p - Y - x^N$ (as a polynomial over k), we have

$$f_0'(Y) = pY^{p-1} - 1 = -1.$$

Viewing now

$$f(Y) = Y^p - Y - x^N$$

as a polynomial in characteristic 0, i.e. with coefficients in R, then for the root y of $f(Y)$ we have

$$f'(y) = py^{p-1} - 1.$$

We see that f_0 (or the affine curve V_0) is special as we defined it in the last section, and that it satisfies all the hypotheses of the lemmas, which are therefore applicable. We call V_0 the **Artin–Schreier curve**.

The equation $Y^p - Y - x^N = 0$ has a formal solution both in characteristic p and characteristic 0. Indeed, it has the approximate solution $Y = 0 \bmod x$, and since $f'(0) = -1$, the Newton sequence of approximate solutions converges in the formal power series to a unique solution $h(x)$ such that $h(0) = 0$. In fact,

$$h(x) = -x^N + \cdots.$$

In characteristic p, it has the group of automorphisms $G_0(p)$ isomorphic *to* $\mathbf{Z}/p\mathbf{Z}$, sending

$$y_0 \mapsto y_0 + \alpha, \quad \text{for } \alpha \in \mathbf{Z}(p).$$

By Lemma 1.5, there is a unique automorphism σ_α of $\mathscr{A}$ leaving $R\langle\langle x\rangle\rangle$ fixed such that

$$\sigma_\alpha(y) \equiv y + \alpha \,(\text{mod } \pi).$$

We let $G(p)$ be the lifting of $G_0(p)$ as a group of automorphisms of $\mathscr{A}$, as in Lemma 1.6. The map $\alpha \mapsto \sigma_\alpha$ is an isomorphism.

On the other hand, we have a lattice of rings:

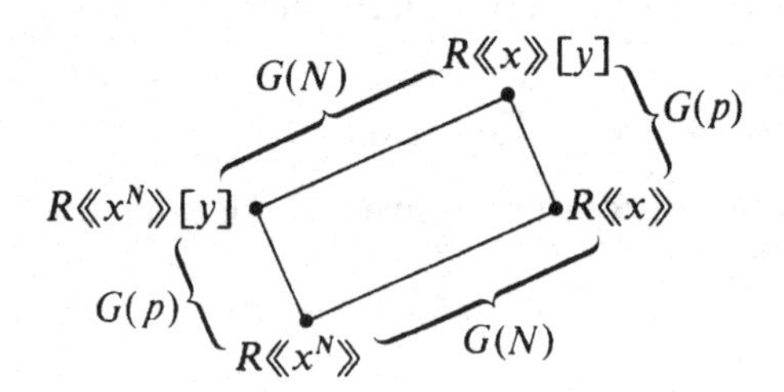

From now on, assume that R contains the N-th roots of unity. Then $R\langle\langle x\rangle\rangle$ is Galois over $R\langle\langle x^N\rangle\rangle$ with group $G(N)$ isomorphic to a cyclic group of order N, sending

$$x \mapsto \zeta x \quad \text{for } \zeta \in \boldsymbol{\mu}_N.$$

The two extensions corresponding to the two bottom sides of the parallelogram are linearly disjoint (being of relatively prime degree), and so opposite sides of the parallelogram have isomorphic Galois groups. In particular, the ring

$$\mathscr{A} = R\langle\langle x \rangle\rangle[y]$$

admits the group

$$G = G(p) \times G(N)$$

as a group of automorphisms, whose fixed ring is $R\langle\langle x^N \rangle\rangle$. The fixed rings of $G(p)$ and $G(N)$ are $R\langle\langle x \rangle\rangle$ and $R\langle\langle x^N \rangle\rangle[y]$ respectively.

The character group $\hat{G}$ consists of character (ψ, χ), where ψ is a character on $\mathbf{Z}(p)$ and χ is a character which may be identified with a character on roots of unity. We write

$$\chi = \chi_j$$

if $\chi(\zeta) = \zeta^j$, with $0 \leq j \leq N - 1$.

If (ψ, χ) is a character of G, and M is a G-module, we let $M(\psi, \chi)$ be the eigenspace corresponding to this character.

Lemma 2.1. *We have* $\mathscr{A}(1, \chi_j) = x^j R\langle\langle x^N \rangle\rangle$, *for* $0 \leq j \leq N - 1$.

Proof. Obvious.

We now let $\pi^{p-1} = -p$ and

$$R = \mathbf{Z}_p[\pi, \boldsymbol{\mu}_N].$$

In the previous chapter, §3, we had associated with each π such that $\pi^{p-1} = -p$ a character ψ_π of $\mathbf{Z}(p)$, and the Dwork power series

$$E_\pi(X) = \exp(\pi X - \pi X^p).$$

It follows from Lemma 2.2 of the preceding chapter that

$$E_\pi(y) \in R\langle\langle y \rangle\rangle \subset \mathscr{A}.$$

Lemma 2.2. *The element $E_\pi(y)$ is an eigenvector of $G(p)$ with eigencharacter ψ_π. In other words, for $\alpha \in \mathbf{Z}(p)$,*

$$\sigma_\alpha(E_\pi(y)) = \psi_\pi(\alpha) E_\pi(y).$$

Furthermore, $E_\pi(y)$ is a unit in $R\langle\langle y \rangle\rangle$.

Proof. For p odd, the inverse of $E_\pi(y)$ is $E_\pi(-y)$, which is thus also in $R\langle\langle y\rangle\rangle$. If $p = 2$, one has to apply Lemma 1.3. Next we verify that $\sigma_\alpha E_\pi(y)$ and $E_\pi(y)$ differ by a p-th root of unity. We have:

$$\begin{aligned}(\sigma_\alpha E_\pi(y))^p &= \sigma_\alpha(E_\pi(y)^p)\\ &= \sigma_\alpha \exp(p\pi y - p\pi y^p)\\ &= \sigma_\alpha \exp(-\pi p x^N)\\ &= \exp(-\pi p \sigma_\alpha x^N) = \exp(-p\pi x^N) = E_\pi(y)^p.\end{aligned}$$

Hence

$$\frac{\sigma_\alpha E_\pi(y)}{E_\pi(y)} \in \boldsymbol{\mu}_p.$$

The Galois theory of Lemma 1.5 applied to

$$R\langle\langle x^N\rangle\rangle[y] = R\langle\langle y\rangle\rangle$$

shows that σ_α is an automorphism of $R\langle\langle y\rangle\rangle$. Hence $\sigma_\alpha E_\pi(y)$ is a power series in y, and in fact,

$$\begin{aligned}\sigma_\alpha E_\pi(y) = E_\pi(\sigma_\alpha y) &\equiv 1 + \pi\sigma_\alpha y \pmod{\pi^2}\\ &\equiv 1 + \pi y + \pi\alpha \pmod{\pi^2}.\end{aligned}$$

On the other hand,

$$E_\pi(y) \equiv 1 + \pi y \pmod{\pi^2}.$$

Taking the quotient shows that

$$\frac{\sigma_\alpha E_\pi(y)}{E_\pi(y)} \equiv 1 + \alpha\pi \pmod{\pi^2}.$$

This proves that the quotient on the left-hand side is equal to $\psi_\pi(\alpha)$, as desired.

The units $\sigma_\alpha E_\pi(y)$ will allow us to determine an eigenspace decomposition for $\mathscr{A}$, except that we need p in the denominator of the orthogonal idempotents of the group ring $G(p)$. Hence we let K be the quotient field of R, so that

$$K = R[1/p].$$

Then we shall abbreviate by $\mathscr{A}_K$ the ring

$$\mathscr{A}_K = \mathscr{A}[1/p] \approx \mathscr{A} \otimes K.$$

For each $\alpha \in \mathbf{Z}(p)$ we suppose given a unit $E_\alpha \in \mathscr{A}$ such that E_α is an eigenvector for $G(p)$ with eigenvalue $\psi_\pi(\alpha)$. The family of units $\sigma_\alpha E_\pi(y)$ forms one example, but there is another, equally natural, namely the family of units

$$E_\pi(y)^i \quad \text{with } i = 0, \ldots, p-1.$$

Since $E_\pi(y)$ has eigenvalue $\psi_\pi(1)$, it follows that $E_\pi(y)^i$ has eigenvalue $\psi_\pi(1)^i$.

Theorem 2.3. *We have the eigenspace decomposition*

$$\mathscr{A}_K = \bigoplus_{\alpha \in \mathbf{Z}(p)} E_\alpha \cdot R\langle\!\langle x \rangle\!\rangle_K$$

or also

$$\mathscr{A}_K = R\langle\!\langle x \rangle\!\rangle_K \oplus \bigoplus_{i=1}^{p-1} E_\pi(y)^i R\langle\!\langle x \rangle\!\rangle_K.$$

Proof. Note that $\mathscr{A}_K$ is free of dimension p over $R\langle\!\langle x \rangle\!\rangle_K$. Each eigenvector provides for a one-dimensional subspace, and hence their sum (necessarily direct) is the whole space $\mathscr{A}_K$.

The fact that $R\langle\!\langle x \rangle\!\rangle$ is the fixed subring of $G(p)$ implies that the eigenspace for the trivial character is precisely

$$R\langle\!\langle x \rangle\!\rangle_K.$$

Recall that not only $G(p)$ acts on $\mathscr{A}$ but also $G(N)$, and so the group

$$G = G(p) \times G(N).$$

A character of $G(N)$ can be viewed as χ_j with $0 \leq j \leq N-1$, and χ_j is nontrivial if and only if $1 \leq j \leq N-1$.

Theorem 2.4. *Let $0 \leq i \leq p-1$ and $0 \leq j \leq N-1$. Then*

$$\mathscr{A}_K(\psi_\pi^i, \chi_j) = x^j E_\pi(y)^i R\langle\!\langle x^N \rangle\!\rangle_K.$$

Proof. Obvious from Lemma 2.1 and Lemma 2.2, and the fact that $E_\pi(y)$ is fixed under $G(N)$.

Remark. The use of the Dwork power series $E_\pi(X)$ to obtain eigenspace decompositions on spaces associated with the Artin–Schreier curve is due to Monsky, cf. [Dw 5], last section of the paper.

§3. Washnitzer–Monsky Cohomology

Cohomology of differential forms with growth conditions on the coefficients was considered long ago by Dwork [Dw 1], [Dw 2]. The particular cohomology considered here is due to Washnitzer–Monsky [Wa–M], [M 1], [M 2], following the work of Dwork.

The K-algebra $\mathscr{A}_K$ is finite separable over $R\langle\langle x\rangle\rangle_K$. We have the ordinary differentiation d/dx on the power series in x. This differentiation extends in a unique way to $\mathscr{A}$, because from the relation $f(x, y) = 0$ with coefficients in R, we get

$$D_1 f(x, y)\, dx + D_2 f(x, y)\, dy = 0,$$

so

$$dy = -\frac{D_1 f(x, y)}{D_2 f(x, y)}\, dx.$$

Here, if $z \in \mathscr{A}_K$ then dz denotes the functional on derivations arising from the pairing

$$(z, D) \mapsto Dz.$$

Since we are dealing with a special curve, we know that $D_2 f(x, y)$ is invertible in $\mathscr{A}$. Consequently the above formula expressing dy in terms of dx is valid over $\mathscr{A}$. (For an elementary discussion of the foundations of the theory of derivations, cf. my *Algebra*, Chapter X, §7.)

We let

$$\Omega(\mathscr{A}) = \mathscr{A}\, dx \quad \text{and} \quad \Omega(\mathscr{A}_K) = \Omega_K(\mathscr{A}) = \mathscr{A}_K\, dx$$

be the spaces of 1-forms. We let the **Washnitzer–Monsky cohomology group** be

$$H_K^1(\mathscr{A}) = H^1(\mathscr{A}_K) = \Omega_K/d\mathscr{A}_K.$$

Note that the differential d acts like 0 on K, so

$$d(cz) = c\, dz \quad \text{for } z \in \mathscr{A},\ c \in K.$$

We are interested in the eigenspace decomposition of $H_K^1(\mathscr{A})$ with respect to $G = G(p) \times G(N)$. We note that the differential

$$\frac{dx}{x}$$

is invariant under G. The theorems of the last section give an eigenspace decomposition into three pieces:

$$\mathscr{A}_K = R\langle\langle x\rangle\rangle_K \oplus \bigoplus_{i=1}^{p-1} E_\pi(y)^i R\langle\langle x^N\rangle\rangle_K \oplus \bigoplus_{i=1}^{p-1}\bigoplus_{j=1}^{N-1} E_\pi(y)^i x^j R\langle\langle x^N\rangle\rangle_K.$$

These three pieces correspond to:

(i) trivial action by $G(p)$;
(ii) trivial action by $G(N)$;
(iii) direct sum of $\mathscr{A}_K(\psi_\pi^i, \chi_j)$ with both ψ_π^i and χ_j non-trivial.

In abbreviated notation, this direct sum can be written

$$\mathscr{A}_K = \mathscr{A}_K^{G(p)} \oplus \mathscr{A}_{K,0}^{G(N)} \oplus \bigoplus_{\substack{\psi \neq 1\\ \chi \neq 1}} \mathscr{A}_K(\psi, \chi),$$

where $\mathscr{A}_{K,0}^{G(N)}$ is the second piece in the above direct sum. Note that

$$\mathscr{A}_K^{G(p)} \cap \mathscr{A}_K^{G(N)} = R\langle\langle x^N\rangle\rangle_K,$$

whence the need for a subdivision of $\mathscr{A}_K^{G(N)}$ into two smaller subspaces to be able to write the direct sum as above.

We now want a similar decomposition for Ω_K.

Lemma 3.1. *We have:*

(i) $\Omega_K^{G(p)} = R\langle\langle x\rangle\rangle_K\, dx = \mathscr{A}_K^{G(p)}\, dx$;

(ii) $\Omega_K^{G(N)} = \mathscr{A}_K^{G(N)} \dfrac{dx}{x} \cap \Omega_K$;

(iii) *for non-trivial ψ, χ we have*

$$\Omega_K(\psi, \chi) = \mathscr{A}_K(\psi, \chi)\frac{dx}{x}.$$

Proof. Write a differential form as

$$\sum_{i=0}^{p-1} g_i(x) y^i\, dx.$$

Invariance under $G(p)$ implies that $g_i = 0$ if $i \geq 1$, and conversely, so (i) is clear. If ζ is a primitive N-th root of unity, then $d(\zeta x) = \zeta dx$. Hence invariance under $G(N)$ implies that

$$\zeta g_i(\zeta x) = g_i(x) \quad \text{for } i = 0, \ldots, p-1,$$

or equivalently

$$g_i(\zeta x) = \zeta^{-1} g_i(x).$$

If $N = 1$ then assertion (ii) is clear. If $N > 1$, then this last relation implies that $g_i(x)$ has no constant term, and that in the power series expansion, all terms are zero except those involving x^n with $n \equiv -1 \pmod N$. Then (ii) is also clear in this case. For (iii), we write an element of $\Omega_K(\psi, \chi)$ in the form

$$g(x, y)\frac{dx}{x}$$

with $g(x, y) \in \mathscr{A}_K$. Since dx/x is invariant under G. it follows that

$$g(x, y) \in \mathscr{A}_K(\psi, \chi),$$

and $g(x, y)$ is divisible by x because $j \geq 1$ if $\chi = \chi_j$, so the lemma is proved.

Lemma 3.2. *The differential operator d maps:*

$$d: \mathscr{A}_K^{G(p)} \to \Omega_K^{G(p)}$$

$$d: \mathscr{A}_K^{G(N)} \to \Omega_K^{G(N)}$$

$$d: \mathscr{A}_K(\psi, \chi) \to \mathscr{A}_K(\psi, \chi)\frac{dx}{x} = \Omega_K(\psi, \chi) \quad \textit{for } \psi \neq 1, \chi \neq 1.$$

Proof. The first inclusion is clear. For the other two, we have

$$\frac{dE_\pi(y)}{E_\pi(y)} = d(\pi y - \pi y^p) = d(-\pi x^N) = -\pi N x^{N-1}\,dx.$$

Since $\varphi \mapsto d\varphi/\varphi$ is homomorphic, we get

$$\frac{dE_\pi(y)^i}{E_\pi(y)^i} = -i\pi N x^N \frac{dx}{x}. \tag{1}$$

Using the rule for the derivative of a product, we then also easily find for any $\varphi(x) \in R\langle\langle x \rangle\rangle$:

$$d(E_\pi(y)^i x^j \varphi(x^N)) = E_\pi(y)^i x^j (D_{j,i}\varphi)(x^N) N \frac{dx}{x} \tag{2}$$

where $D_{j,i}$ is the differential operator on a power series $\varphi(x)$ given by

$$D_{j,i} = x\frac{d}{dx} - i\pi x + \frac{j}{N}.$$

We shall work with $i = 1$, and we thus let

$$\boxed{D_j = x\frac{d}{dx} - \pi x + \frac{j}{N}.}$$

This proves the lemma, and in addition gives explicit formulas for the differentials, summarized by the following commutative diagram.

$$\begin{array}{ccc} R\langle\langle x\rangle\rangle_K & \longrightarrow & x^j E_\pi(y) R\langle\langle x^N \rangle\rangle_K \\ {\scriptstyle ND_j}\downarrow & & \downarrow{\scriptstyle d} \\ R\langle\langle x\rangle\rangle_K & \longrightarrow & x^j E_\pi(y) R\langle\langle x^N \rangle\rangle_K \dfrac{dx}{x} \end{array}$$

The horizontal map on top sends

$$\varphi(x) \mapsto x^j E_\pi(y)\varphi(x^N),$$

and similarly on the bottom, with dx/x on the right-hand side.

Recall from Theorem 2.4 that for $1 \leq j \leq N - 1$, we have

$$\mathscr{A}_K(\psi_\pi, \chi_j) = x^j E_\pi(y) R\langle\langle x^N \rangle\rangle_K.$$

Theorem 3.3.

(i) *For $\psi \neq 1$ and $\chi \neq 1$ we have*

$$H_K^1(\mathscr{A})(\psi, \chi) = \mathscr{A}_K(\psi, \chi)\frac{dx}{x}\bigg/ d(\mathscr{A}_K(\psi, \chi)).$$

(ii) *For $1 \leq j \leq N - 1$, we have an isomorphism*

$$R\langle\langle x\rangle\rangle_K / D_j R\langle\langle x\rangle\rangle_K \xrightarrow{\approx} H^1(\psi_\pi, \chi_j)$$

given by

$$\varphi(x) \mapsto x^j E_\pi(y)\varphi(x^N)\frac{dx}{x}.$$

Proof. The first assertion is clear from Lemma 3.2. The second comes from the above diagram and Lemma 3.2.

The affine ring $\mathbf{Z}[x, y]$ has a natural embedding in the power series ring,

$$\mathbf{Z}[x, y] \to \mathbf{Z}[[x]],$$

mapping y on a uniquely determined power series $y = h(x)$, such that $h(0) = 0$. We have already mentioned this formal solution, and the fact that

$$h(x) = -x^N + \cdots.$$

Since $f(Y)$ is irreducible over $R\langle\langle x\rangle\rangle$ and its quotient field (for instance by the Galois theory), the root $h(x)$ of $f(Y)$ in $R[[x]]$ gives rise to the embedding

$$\mathscr{A} = R\langle\langle x\rangle\rangle[y] \to R[[x]],$$

sending y on $h(x)$. This gives rise to a natural homomorphism

$$\begin{array}{ccc} H^1(\mathscr{A}_K) & \longrightarrow & H^1(R[[x]]_K) \\ \approx\big\downarrow & & \big\downarrow\approx \\ \mathscr{A}_K\, dx/d\mathscr{A}_K & & R[[x]]_K\, dx/dR[[x]]_K. \end{array}$$

Theorem 3.4. *We have an isomorphism*

$$H^1(\mathscr{A}_K)(\psi, \chi_j) \xrightarrow{\approx} H_{j,\pi},$$

where $H_{j,\pi}$ is the representation space of the last chapter.

Proof. Clear from Theorem 3.3.

§4. The Frobenius Endomorphism

If φ is a lifting of a morphism φ_0 of special varieties, then we let $\varphi^* = H_K^1(\varphi)$ be the induced homomorphism on the cohomology groups. It can be shown that φ^* is independent of the lifting, but we shall not need this here.

We shall deal especially with automorphisms of the Artin–Schreier curve

$$\sigma = (\sigma_\alpha, \sigma_\zeta) \quad \text{with } \alpha \in \mathbf{Z}(p) \text{ and } \zeta \in \boldsymbol{\mu}_N.$$

Such $\sigma \in G = G(p) \times G(N)$ is an automorphism of $\mathscr{A} = R\langle\langle x\rangle\rangle[y]$. We have

$$\sigma^* = (\sigma_\alpha^*, \sigma_\zeta^*).$$

We also have the Frobenius endomorphism

$$F_0 : \mathscr{A}_0(V_0) \to \mathscr{A}_0(V_0)$$

such that $F_0(x, y_0) = (x^p, y_0^p)$ in characteristic p. By Lemma 1.6 we know that it can be lifted uniquely to an endomorphism

$$F : \mathscr{A} \to \mathscr{A} \quad \text{such that } F(x) = x^p,$$

extending by K-linearity to an endomorphism of $\mathscr{A}_K$.

In characteristic p, that is on V_0, we obviously have

$$(1) \qquad F \circ (\sigma_\alpha, \sigma_\zeta) = (\sigma_\alpha, \sigma_\zeta^p) \circ F.$$

Indeed, in characteristic p, $(\sigma_\alpha, \sigma_\zeta)(x, y_0) = (\zeta x, y_0 + \alpha)$, so

$$\begin{aligned} F_0 \circ (\sigma_\alpha, \sigma_\zeta)(x, y_0) &= (\zeta^p x^p, y_0^p + \alpha^p) \\ &= (\zeta^p x^p, y_0^p + \alpha) \\ &= (\sigma_\alpha, \sigma_\zeta^p)(x^p, y^p) \\ &= (\sigma_\alpha, \sigma_\zeta^p) \circ F_0(x, y_0). \end{aligned}$$

The commutation rule in characteristic zero follows by the uniqueness of liftings of homomorphisms when the x-value is prescribed.

From (1) we get on the cohomology

$$(1^*) \qquad (\sigma_\alpha, \sigma_\zeta)^* \circ F^* = F^* \circ (\sigma_\alpha, \sigma_\zeta^p)^*.$$

In particular, on an eigenspace we find that for $\psi \neq 1$, $\chi \neq 1$,

$$F^* : H_K^1(\psi, \chi) \to H_K^1(\psi, \chi^p).$$

Now we wish to see what happens to the Frobenius endomorphism under the embedding of $\mathscr{A}$ in $R[[x]]$. Mutatis mutandis, we know that $h(x^p)$ is the unique solution in power series with zero constant term of the equation

$$T^p - T = x^{pN}.$$

Theorem 3.4 now shows that the representation of F^* on the distinguished elements

$$x^j E_\pi(y) \frac{dx}{x}$$

corresponds to the representation on the elements

$$x^j \exp(-\pi x^N) \frac{dx}{x}$$

arising from the last chapter. Thus we find:

Theorem 4.1. *The eigenvalue of F_q^* on $H^1(\mathscr{A}_k)(\psi_\pi, \chi_j)$ is the same as the eigenvalue of Φ_q^* on $H_{j,\pi}$.*

Gauss Sums as Distributions 17

The Stickelberger theorem giving the factorization of Gauss sums, the Gross–Koblitz formula, and the Davenport–Hasse distribution relations will be combined to interpret Gauss sums as universal odd distributions (Yamamoto's theorem).

On the other hand, Diamond [Di 1] and Morita [Mo] gave a value of $L_p'(0, \chi)$ in terms of the p-adic gamma function. Ferrero–Greenberg gave a variation of Diamond's formula, and used it to show that $L_p'(0, \chi) \neq 0$ under the appropriate conditions. The value $L_p'(0, \chi)$ is essentially the generator for the χ-eigenspace of the Stickelberger distribution, but the proof of the non-vanishing requires the analogue of Baker's theorem (Brumer in the p-adic case), as in the proof of the non-vanishing of the p-adic regulator of cyclotomic fields. This is combined with the linear algebra of distribution relations, especially in the composite case. Cf. Kubert–Lang [KL 5].

The formula for $L_p'(0, \chi)$ will be derived by an elegant method of Washington [Wa 3], who gives the p-adic analogue of the partial zeta functions. Over the complex numbers, the coefficients in the expansion at $s = 1$ are themselves interesting functions (of gamma type), which appear thus in a natural way as homomorphic images of the partial zeta distributions. The same thing happens in the p-adic case.

§1. The Universal Distribution

We assume that the reader is acquainted with Chapter 2, §8, §9, §10. In particular, suppose that $N > 1$ is an integer. Let $(\mathbf{Q}/\mathbf{Z})_N = (1/N)\mathbf{Z}/\mathbf{Z}$, and let

$$g : (\mathbf{Q}/\mathbf{Z})_N \to A$$

be a function into some abelian group A. Such a function is called an **ordinary distribution**—**distribution** for short—if it satisfies the condition

$$\sum_{i=0}^{M-1} g\left(x + \frac{i}{M}\right) = g(Mx)$$

for every divisor M of N, and all $x \in (\mathbf{Q}/\mathbf{Z})_N$.

We let $\mathbf{F}(N)$ be the free abelian group generated by $(\mathbf{Q}/\mathbf{Z})_N$. We let $\mathbf{DR}(N)$ be the subgroup of distribution relations, that is, the subgroup generated by the elements of the form

$$\sum_{i=0}^{M-1} \left(x + \frac{i}{M}\right) - (Mx), \quad \text{for all } M \mid N.$$

We let $\mathbf{U}(N) = \mathbf{F}(N)/\mathbf{DR}(N)$ be the factor group, which we call the **universal distribution** of level N. The natural map

$$(\mathbf{Q}/\mathbf{Z})_N \to \mathbf{U}(N)$$

is then universal for distributions into abelian groups in the obvious sense. Given a distribution g on $(\mathbf{Q}/\mathbf{Z})_N$, there exists a unique homomorphism g_* making the diagram commutative:

$$\begin{array}{ccc} & & \mathbf{U}(N) \\ & \nearrow & \downarrow g_* \\ (\mathbf{Q}/\mathbf{Z})_N & & \\ & \searrow_{g} & A \end{array}$$

Kubert's theorem (Chapter 2, Theorem 9.2) asserts that $\mathbf{U}(N)$ is free on $\phi(N)$ generators.

Theorem 1.1. *Let g be a distribution as above. Let K be a field of characteristic 0. Assume that the distribution obtained by following g with the natural homomorphism*

$$A \to A \otimes K$$

has K-rank $\phi(N)$, in the sense that the dimension of the vector space generated by the image of $(\mathbf{Q}/\mathbf{Z})_N$ has dimension $\phi(N)$. Then g is the universal distribution.

Proof. The rank of the image is at most $\phi(N)$. If the vector space generated by the image has that rank, then the Kubert generators must remain free under g and the tensor product, so they must be linearly independent over $\mathbf{Z}$

in the abelian group generated by $g((\mathbf{Q}/\mathbf{Z})_N)$. Hence the canonical homomorphism from the universal distribution to g must be an isomorphism, as was to be shown.

Let

$$G(N) \approx \mathbf{Z}(N)^*$$

be a group isomorphic to $\mathbf{Z}(N)^*$, the isomorphism being denoted by

$$\sigma_c \leftrightarrow c.$$

We think of $G(N)$ as the Galois group of $\mathbf{Q}(\mu_N)$ over $\mathbf{Q}$. Let h be an ordinary distribution, and let as before the **Stickelberger distribution associated with** h be defined by

$$\mathrm{St}_h(x) = \mathrm{St}(x) = \sum_{c \in \mathbf{Z}(N)^*} h(xc)\sigma_c^{-1}.$$

If χ is a character of $G(N)$ with conductor m, we define

$$S(\chi, h) = S_m(\chi_m, h_m) = \sum_{c \in \mathbf{Z}(m)^*} \chi(c)h(c/m).$$

In fact, Theorem 1.1 can be made more precise, and again in Chapter 2, §8 we proved the following facts. Let $M \mid N$, and let χ be a character of $G(N)$. Let

$$e_\chi = \frac{1}{|G(N)|} \sum \bar{\chi}(c)\sigma_c$$

be the usual idempotent projecting on the χ-eigenspace.

ST 1. *If* cond χ *does not divide* M, *then*

$$\mathrm{St}_h\left(\frac{1}{M}\right)e_\chi = 0.$$

ST 2. *If* cond χ *divides* M *and has the same prime factors as* M, *then*

$$\mathrm{St}_h\left(\frac{1}{M}\right)e_\chi = \frac{|G(N)|}{|G(M)|} S(\bar{\chi}, h)e_\chi.$$

ST 3. *If* cond χ *divides* M, *and we let* $m =$ cond χ, *then*

$$\mathrm{St}_h\left(\frac{1}{M}\right)e_\chi = \frac{|G(N)|}{|G(M)|} \prod_{\substack{p \mid M \\ p \nmid m}} (1 - \bar{\chi}(p))S(\bar{\chi}, h)e_\chi.$$

An arbitrary value $\mathrm{St}_h(a/M)$ for a prime to M comes from the formula

$$\mathrm{St}_h\left(\frac{a}{M}\right) = \sigma_a \, \mathrm{St}_h\left(\frac{1}{M}\right),$$

so such values do not contain essentially more information on the image of the Stickelberger distribution than the normalized values $\mathrm{St}_h(1/M)$. We suppose that h takes its values in an algebraically closed field K of characteristic 0, which in the applications is $\mathbf{C}$ or $\mathbf{C}_p$ for some prime p. It would suffice to suppose that h takes values in a field containing enough roots of unity, but we can always extend scalars and still be able to use Theorem 1.1. The element $S(\chi, h)$ is then an element of K.

Theorem 1.2. *Let h be a distribution with values in K. Let V be the vector space generated by the image of the Stickelberger distribution. Then $e_\chi V$ is generated by the single element $S(\bar{\chi}, h)$, and in particular has dimension* 0 *or* 1 *according as that element is* 0 *or* $\neq 0$.

Proof. The proof of Theorem 8.2 in Chapter 2 in fact proves the statement as given here, although we stated previously only the corresponding dimension property.

Corollary. *If $S(\bar{\chi}, h) \neq 0$ for all χ, then St_h is universal.*

Proof. This follows just like Theorem 1.1, but we don't even need to tensor with K since the values of the distribution h are already in K, and the values of the Stickelberger distribution are already in a vector space over K.

A distribution h is called **odd** or **even** according as

$$h(-x) = -h(x) \quad \text{or} \quad h(-x) = h(x).$$

From now on we restrict ourselves to distributions whose values are in abelian groups without 2-torsion. This condition will not be repeated. Such a group may then be embedded in a group where multiplication by 2 is invertible. When that is the case, any distribution is uniquely expressible as an even distribution plus an odd distribution, in the usual manner.

Theorem 1.1 then remains valid for odd (respectively even) universal distributions, except that the rank is then $\phi(N)/2$ for $N \geq 3$, which we assume.

Likewise, in Theorem 1.2, if h is, say, an odd distribution, then St_h is also odd, and is universal odd if $S(\bar{\chi}, h) \neq 0$ for all odd characters χ.

Example. Let

$$h(x) = \mathbf{B}_1(\langle x \rangle) \quad \text{if } x \neq 0$$
$$h(0) = 0.$$

We call h the **first Bernoulli distribution**. As we saw in Chapter 2, Theorem 8.3, its associated Stickelberger distribution is the universal odd distribution. This comes back to the fact that

$$B_{1,\chi} = S(\chi, h) \neq 0$$

for odd characters χ.

§2. The Gauss Sums as Universal Distributions

In Chapter 2, §10 we already gave the Davenport–Hasse distribution relation. Another proof of this relation follows from the Gross-Koblitz formula, and is left as an exercise for the reader. Here we are concerned with showing to what extent the Gauss sums give the universal odd distribution.

We let $h(x)$ be the first Bernoulli distribution as mentioned at the end of the last section. We pick $N = q - 1$ where $q = p^r$ for some *odd prime* p, and we have by definition

$$\mathrm{St}(x) = \sum_c h(xc)\sigma_c^{-1}.$$

The sum is taken for $c \in \mathbf{Z}(q-1)^*$. Write

$$x = \frac{a}{q-1}.$$

Define

$$g(x) = \tau(\omega_q^{-a})/\tau(\omega_q^{(q-1)/2}).$$

Stickelberger's theorem gives the ideal factorization of $g(x)$, namely by Theorem 2.2 of Chapter 1, we know that

$$(g(x)) = \mathfrak{p}^{\mathrm{St}(x)}.$$

It is convenient typographically and otherwise to write this formally additively, and thus we call

$$\mathrm{St}(x) \cdot \mathfrak{p}$$

the **associated divisor** of $g(x)$. We view at first

$$g : (\mathbf{Q}/\mathbf{Z})_{q-1} \to \mathbf{C}_p^*$$

as a map into the multiplicative group of p-adic complex numbers. The distribution relation is then satisfied only with fudge factors. To get rid of them, we let

$P = \{\boldsymbol{\mu}, p^{1/\infty}\}$ be the group generated by all roots of unity and fractional powers of p.

We call P the **pure group** (with respect to the prime p). We may then compose g with the canonical homomorphism $\mathbf{C}_p^* \to \mathbf{C}_p^*/P$ to obtain a map

$$g_P : (\mathbf{Q}/\mathbf{Z})_{q-1} \to \mathbf{C}_p^*/P$$

into the factor group, which is a distribution by the Davenport–Hasse relation. We call g_P the **Davenport–Hasse** or **Gauss sum distribution**. Note that $\mathbf{C}_p^*/P$ is uniquely divisible by 2. Recall that the p-adic logarithm has kernel equal precisely to the pure group P. Consequently we have a natural isomorphism of distributions

GSD 1. $$\boxed{g_P \approx \log_p g.}$$

The map

$$\mathrm{St}(x) \mapsto \mathrm{St}(x) \cdot \mathfrak{p}$$

is a homomorphism of the Stickelberger distribution (which we know is universal odd). The decomposition group D_p of $\mathfrak{p}$ consists of the powers of p mod N. If we let $G(N) \approx \mathbf{Z}(N)^*$ under the notation

$$a \mapsto \sigma_a,$$

then the values of the homomorphic image above can be viewed as lying in the group ring

$$\mathbf{Q}[G(N)/D_p],$$

and

$$\mathrm{St}_h(x) \bmod D_p = \sum_{c \in \mathbf{Z}(N)^*/D_p} \left(\sum_{i=0}^{r-1} h(p^i xc) \right) \sigma_c^{-1}.$$

Lemma 2.1. *Let $\{m(x)\}$ be a family of integers and let*

$$\alpha = \prod_x g(x)^{m(x)}.$$

Then α is pure if and only if $\operatorname{div} \alpha = 0$, *and in that case α is a root of unity.*

Proof. The absolute value of $g(x)$ in the complex numbers is 1. Hence $|\alpha| = 1$. If α is pure, this implies that α is a root of unity. Conversely, assume that $\operatorname{div} \alpha = 0$, so α is a unit. The conjugates of $g(x)$ also have absolute value 1 (themselves being of the same form as $g(x)$), and it is standard that a unit all of whose conjugates have absolute value 1 must be a root of unity. This proves the lemma.

From the lemma, it follows that we have an isomorphism

GSD 2.
$$\boxed{g_P \approx \mathrm{St}_h \bmod D_p.}$$

Note. Factoring out by D_p corresponds to the obvious classical fact that the Gauss sums satisfy the relation

$$\tau(\chi) = \tau(\chi^p),$$

cf. Chapter 1, **GS 4**. In the present notation, this is written

$$g(x) = g(px).$$

Let χ be an odd character of conductor d prime to p. Since the Gauss sum distribution is also odd, it follows that the function

$$a \mapsto \chi(a) \log_p g\left(\frac{a}{d}\right)$$

is even, for $a \in \mathbf{Z}(d)^*$. Hence the function is defined on $\mathbf{Z}(d)^*/\pm 1$.

Theorem 2.2 (Ferrero–Greenberg). *Let χ be an odd character of conductor d. Assume that $\chi(p) = 1$. Then*

$$\sum_{a \in \mathbf{Z}(d)^*} \chi(a) \log_p g\left(\frac{a}{d}\right) \neq 0.$$

The proof will be based on the following lemma.

Lemma 2.3. *There exists a family of integers $\{m(d')\}$, for divisors d' of d, $d' \neq d$, having the following property. Let*

$$\xi = \mathrm{St}_h\left(\frac{1}{d}\right) + \sum_{d'} m(d')\, \mathrm{St}_h\left(\frac{1}{d'}\right).$$

Let R be a set of representatives for cosets of the group generated by D_p and ± 1 in $\mathbf{Z}(d)^*$. Then the elements

$$\sigma_a \xi, \qquad a \in R$$

are linearly independent over $\mathbf{Q}$ *as elements of* $\mathbf{Q}[G(N)/D_p]$.

Proof. As in Theorem 1.2 we look at the ψ-eigenspace for odd characters ψ such that $\psi(p) = 1$ and cond ψ divides d. It suffices to prove that we can choose the family $\{m(d')\}$ such that

$$\xi e_\psi \neq 0 \quad \text{for all } \psi.$$

First we know from **ST 2** that if cond $\psi = d$ then

$$\mathrm{St}_h\left(\frac{1}{d}\right)e_\psi \neq 0$$

because $B_{1,\psi} \neq 0$. Let $d_1 > d_2 > \cdots$ be the other divisors of d, unequal to d. Pick a sequence of integers $m_1 < m_2 < \cdots$ which is rapidly increasing. Then for any ψ we have

$$\mathrm{St}_h\left(\frac{1}{d}\right)e_\psi + \sum_i m_i \mathrm{St}_h\left(\frac{1}{d_i}\right)e_\psi \neq 0.$$

Indeed, suppose d_s is the conductor of ψ. By **ST 1** we conclude that the i-th term in the sum is 0 if $i > s$. By **ST 2** and the fact that $B_{1,\psi} \neq 0$ we know that

$$\mathrm{St}_h\left(\frac{1}{d_s}\right)e_\psi \neq 0.$$

If the sequence is selected increasing sufficiently fast, then this s-term dominates in the sum, which is therefore also not equal to 0, thus proving the lemma.

We return to Theorem 2.2. Let

$$\alpha = g\left(\frac{1}{d}\right) \prod_{d' \neq d} g\left(\frac{1}{d'}\right)^{m(d')},$$

with the family $\{m(d')\}$ chosen as in the lemma. By **GSD 1**, **GSD 2** we conclude that the elements

$$\log_p \sigma_a \alpha \quad \text{with } a \in R$$

are linearly independent over the rational numbers. By Baker's theorem (in the p-adic case, Brumer) it follows that they are linearly independent over the algebraic numbers. Therefore

$$0 \neq \sum_{a \in \mathbf{Z}(d)^*/\{D_p, \pm 1\}} \chi(a) \log_p(\sigma_a \alpha) = \sum_a \chi(a) \log_p g\left(\frac{a}{d}\right),$$

because

$$\sum_a \chi(a) \log_p g\left(\frac{a}{d'}\right) = 0$$

for each $d' \neq d$ since the conductor of χ is d. This concludes the proof of Theorem 2.2.

§3. The L-function at $s = 0$

Let d be a positive integer prime to p (where p is an *odd prime*). Let χ be a primitive even Dirichlet character with conductor d or dp. We take the values of χ to be in $\mathbf{C}_p$. We let $\omega = \omega_p$ be the Teichmuller character, and we define

$$\chi_n = \chi\omega^{-n}.$$

We know that

$$L_p(1 - n, \chi) = -(1 - \chi_n(p)p^{n-1}) \frac{1}{n} B_{n, \chi_n}.$$

For $n = 1$, we obtain

$$L_p(0, \chi) = -(1 - \chi_1(p))B_{1, \chi_1}.$$

Since χ_1 is odd, we know that $B_{1, \chi_1} \neq 0$. Hence:

$$L_p(0, \chi) = 0 \quad \text{if and only if} \quad \chi_1(p) = 1.$$

The next formula is a slight variation of a formula of Diamond [Di], but expressed in terms of the p-adic gamma function itself by Ferrero–Greenberg [Fe–Gr].

Theorem 3.1. *Let χ be an even character such that χ_1 has conductor d. Then*

$$L_p'(0, \chi) = \sum_{c=1}^{d} \chi_1(c) \log_p \Gamma_p\left(\frac{c}{d}\right) + (1 - \chi_1(p))B_{1, \chi_1} \log_p(d).$$

We shall prove this formula in §4. Here we derive consequences. Indeed, the next theorem amounts to a p-adic analogue of Stark's conjectures in a special case. Considerably more insight in this direction was provided by Gross [Gr].

As in the preceding section, let $x = a/(q-1)$ and let

$$\varphi(x) = \tau(\omega_q^{-a}), \qquad g(x) = \varphi(x)\varphi\left(\frac{q-1}{2}\right)^{-1}.$$

The formula of Theorem 3.1 involves the log of an analytic expression (the gamma function). We shall transform it so that it involves the log of an algebraic expression (a Gauss sum).

Theorem 3.2. *Let χ be an even character such that χ_1 has conductor d and such that $\chi_1(p) = 1$. Let D_p be the subgroup of $\mathbf{Z}(d)^*$ generated by the powers of p. Then*

$$L_p'(0, \chi) = \sum_{c \in \mathbf{Z}(d)^*/D_p} \chi_1(c) \log_p g\left(\frac{c}{d}\right) \neq 0.$$

Proof. By assumption the formula of Theorem 3.1 simplifies to

$$L_p'(0, \chi) = \sum_{c=1}^{d} \chi_1(c) \log_p \Gamma_p\left(\frac{c}{d}\right).$$

Let $d|(q-1)$, $q = p^r$ where r is the period of p mod d. Let $G = \mathbf{Z}(q-1)^*$. Then χ_1 is defined modulo D_p by assumption. Since

$$\Gamma_p(z)\Gamma_p(1-z) = \pm 1,$$

it follows at once from the Gross–Koblitz formula that

$$\log_p \varphi(x) = \sum_{i=1}^{r} \log_p \Gamma_p(\langle p^i x \rangle).$$

In particular,

$$\log_p \varphi(px) = \log_p \varphi(x).$$

The sum over $c = 1, \ldots, d$ in the formula for $L_p'(0, \chi)$ is then written in the form

$$\sum_{c \in \mathbf{Z}(d)^*/D_p} \chi_1(c) \sum_{i=0}^{r-1} \log_p \Gamma_p\left(\left\langle \frac{p^i c}{d} \right\rangle\right) = \sum_{c \in \mathbf{Z}(d)^*/D_p} \chi_1(c) \log_p g\left(\frac{c}{d}\right)$$

where

$$g(x) = \varphi(x)\varphi\left(\frac{q-1}{2}\right)^{-1}$$

is the odd distribution of the preceding section. That $L_p'(0, \chi) \neq 0$ is then the main result of §2.

§4. The p-adic Partial Zeta Function

In the complex case, the Hankel transform gives an analytic continuation of the partial zeta function to the whole plane. In the p-adic case, we shall give an analogue of this partial zeta function due to Washington [Wa 3], who also pointed out to me that his formula for the p-adic L-function immediately gives the value of the derivative at $s = 0$ in terms of the gamma-type functions.

Let $x \in \mathbf{C}_p^*$ be such that $x^{-1} \equiv 0 \bmod \mathfrak{m}_p$. Let $s \in \mathbf{Z}_p$. We define the **Hurwitz–Washington** function

$$H(s, x) = \sum_{j=0}^{\infty} \binom{1-s}{j} x^{-j} B_j,$$

where B_j is the j-th Bernoulli number. Since the Bernoulli numbers have bounded denominator at p (Kummer and von Staudt congruences), it is easily shown that $H(s, x)$ is holomorphic for $s \in \mathbf{Z}_p$. For an integer $k \geq 1$, we find

H 1. $$H(1-k, x) = x^{-k}\mathbf{B}_k(x),$$

where $\mathbf{B}_k$ is the k-th Bernoulli polynomial. This is immediate from the value of this polynomial,

$$\mathbf{B}_k(X) = \sum_{j=0}^{k} \binom{k}{j} X^{k-j} B_j$$

which comes directly from the definitions in terms of the generating power series, product of e^{tX} and $t/(e^t - 1)$.

Let N be a positive integer divisible by p. For $a \in \mathfrak{o}_p^*$ we therefore obtain

$$H\left(1-k, \frac{a}{N}\right) = N^k a^{-k} \mathbf{B}_k\left(\frac{a}{N}\right).$$

H 2. *If f is a function on $\mathbf{Z}(N)$, then*

$$\frac{1}{N}\sum_{\substack{a=1 \\ p\nmid a}}^{N-1} f(a)a^k H\left(1-k, \frac{a}{N}\right) = B_{k,f} - p^{k-1}B_{k,f\circ p}$$

where $(f \circ p)(x) = f(px)$ and

$$B_{k,f} = N^{k-1}\sum_{0}^{N-1} f(a)\mathbf{B}_k\left(\left\langle\frac{a}{N}\right\rangle\right).$$

This is immediate by taking the sum over all $a = 0, \ldots, N-1$ and subtracting the sum over $a = py$, with $0 \leq y \leq (N/p) - 1$. It is convenient to use the following notation. Put

$$M_p f(x) = f(px), \qquad f_k = f\omega^{-k}.$$

Then **H 2** can be written in the form

$$\textbf{H 3.}\quad \frac{1}{N}\sum_{\substack{a=1 \\ p\nmid a}}^{N-1} f(a)\langle a\rangle_p^k \frac{1}{k} H\left(1-k, \frac{a}{N}\right) = \int (1 - p^{k-1}M_p)f_k\, dE_k.$$

The formula expresses the Bernoulli distribution in terms of H, giving the possibility of analytic continuation.

If χ is a Dirichlet character whose conductor divides N, then the preceding formula reads

$$\textbf{H 4.}\quad \frac{1}{N}\sum_{\substack{a=1 \\ p\nmid a}}^{N-1} \chi(a)\langle a\rangle_p^k H\left(1-k, \frac{a}{N}\right) = (1 - \chi_k(p)p^{k-1})B_{k,\chi_k}.$$

We now define the **Hurwitz–Washington function** in three variables,

$$H(s; a, N) = -\frac{1}{1-s}\langle a\rangle_p^{1-s}\frac{1}{N} H\left(s, \frac{a}{N}\right),$$

for $a \in \mathbf{Z}_p^*$, $s \in \mathbf{Z}_p$, and N equal to a positive integer divisible by p. Then $H(s; a, N)$ is again holomorphic in s except at $s = 1$. It is the p-adic partial zeta function, cf. [Wa 3].

One could take the relation of the next theorem as the definition of the p-adic L-function, and thus make the present chapter independent of Chapter 12.

Theorem 4.1. *Let χ be a Dirichlet character, and let N be any multiple of the conductor of χ such that N is divisible by p. Then*

$$L_p(s, \chi) = \sum_{\substack{a=1 \\ p \nmid a}}^{N-1} \chi(a) H(s; a, N).$$

Proof. The left-hand side and the right-hand side have the same values at the negative integers, which are dense in $\mathbf{Z}_p$, and they are both holomorphic, hence they coincide.

We are here concerned with finding $L_p'(0, \chi)$. That we are dealing with a character χ is basically irrelevant, and so for any function on $\mathbf{Z}(N)$ we now define

$$L_p(s, f) = \sum_{\substack{a=1 \\ p \nmid a}}^{N-1} f(a) H(s; a, N).$$

In finding the expansion at $s = 0$, we shall meet the **Diamond function** defined by the formula

$$G_p(x) = (x - \tfrac{1}{2}) \log_p(x) - x + \sum_{j=2}^{\infty} \frac{B_j}{j(j-1)} x^{1-j},$$

cf. [Di 1]. This formula arises from the asymptotic expansion of the classical complex log gamma function. It converges p-adically for $|x| > 1$, so $G_p(x)$ is defined in that domain. We shall analyze later the relation between the Diamond function and the gamma function.

Theorem 4.2. *Let $p \neq 2$. Let N be a positive integer divisible by p and let f be a function on $\mathbf{Z}(N)$. Then*

$$L_p'(0, f) = \sum_{\substack{a=1 \\ p \nmid a}}^{N-1} f_1(a) G_p\left(\frac{a}{N}\right) + \sum_{\substack{a=1 \\ p \nmid a}}^{N-1} f_1(a) \mathbf{B}_1\left(\frac{a}{N}\right) \log_p(N).$$

If $f = \chi$ is a Dirichlet character, then

$$L_p'(0, \chi) = \sum_{\substack{a=1 \\ p \nmid a}}^{N-1} \chi_1(a) G_p\left(\frac{a}{N}\right) + (1 - \chi_1(p)) B_{1, \chi_1} \log_p(N).$$

Proof. The desired result is an immediate consequence of the next lemma.

Lemma 4.3. *Let $|a| > 1/p$ and let N be a positive integer divisible by p. Then the coefficient of s in $H(s; a, N)$ is equal to*

$$\omega(a)^{-1}G_p\left(\frac{a}{N}\right) + \omega(a)^{-1}\mathbf{B}_1\left(\frac{a}{N}\right)\log_p(N).$$

Proof. We have the expansions:

$$\frac{1}{1-s} = 1 + s + \cdots$$

$$\langle a\rangle^{1-s} = \langle a\rangle(1 - s\log_p\langle a\rangle + \cdots)$$

$$\text{If} \quad j \geq 2, \qquad \binom{1-s}{j} = \frac{(-1)^{j-1}}{j(j-1)^s} + \cdots.$$

Using the fact that $B_j = 0$ for j odd, $j > 1$, we find that the coefficient of s in $H(s; a, N)$ is

$$\begin{aligned} -\frac{\langle a\rangle}{N}&\left[1 - \log_p(a) + \frac{N}{2a}\log_p(a) - \sum_{j=2}^{\infty}\left(\frac{N}{a}\right)^j \frac{B_j}{j(j-1)}\right] \\ &= \omega(a)^{-1}\left[-\frac{a}{N} + \left(\frac{a}{N} - \frac{1}{2}\right)\log_p\left(\frac{a}{N}\right) + \sum_{j=2}^{\infty}\left(\frac{a}{N}\right)^{1-j}\frac{B_j}{j(j-1)}\right. \\ &\qquad\qquad \left. + \left(\frac{a}{N} - \frac{1}{2}\right)\log_p(N)\right] \\ &= \omega(a)^{-1}G_p\left(\frac{a}{N}\right) + \omega(a)^{-1}\left(\frac{a}{N} - \frac{1}{2}\right)\log_p(N). \end{aligned}$$

This proves the lemma.

Remark. In [Di 2], Diamond discusses the regularization of his function, giving rise to certain measures which are then related to the Bernoulli measures. See also Koblitz [Ko 1]. In the classical case, gamma-type functions appear as coefficients of partial zeta functions (Hurwitz functions), and we meet a similar phenomenon here.

Next, we derive some functional equations. First, for the Washington function, we get for p odd:

H 5. $$H(s; a, N) = H(s; N - a, N).$$

Proof. It suffices to prove the formula when $s = 1 - k$, $k \geq 2$ such that $k \equiv 0 \bmod p - 1$, because such integers are dense in $\mathbf{Z}_p$. But then the formula is immediate from the fact that k is even (p is assumed odd), and the property

$$\mathbf{B}_k(1 - X) = (-1)^k \mathbf{B}_k(X),$$

which follows directly from the generating function for Bernoulli polynomials.

Washington has also pointed out that one can give elegant proofs for the following properties of the Diamond function by using the formalism of the H-function. We assume p odd.

$\mathbf{G}_p$ 1. $$G_p(1 - x) + G_p(x) = 0.$$

Proof. In Lemma 4.3 we found the coefficient of s in $H(s; a, N)$. Using **H 5**, and replacing a by $N - a$ in this coefficient, we now see that

$$G_p\left(\frac{a}{N}\right) + G_p\left(1 - \frac{a}{N}\right) = 0.$$

This is true for any positive integer N divisible by p, and any p-unit a, thus proving the formula.

$\mathbf{G}_p$ 2. $$G_p(-x) + G_p(x) = -\log_p(x).$$

Proof. Immediate from the power series expansion.

$\mathbf{G}_p$ 3. $$G_p(1 + x) - G_p(x) = \log_p(x).$$

Proof. Immediate from the preceding two properties.

Theorem 4.4. *Extend $G_p(x)$ to $\mathbf{Q}_p$ by putting $G_p(x) = 0$ if $x \in \mathbf{Z}_p$. Then for all $x \in \mathbf{Z}_p$ we have*

$$\sum_{b=0}^{p-1} G_p\left(\frac{x + b}{p}\right) = \log_p \Gamma_p(x).$$

Proof. Both sides are continuous and satisfy the functional equation

$$f(x + 1) = f(x) + \delta(x)\log_p x,$$

where $\delta(x) = 0$ if $x \equiv 0 \bmod p$, and $\delta(x) = 1$ otherwise. This is true for $\log_p \Gamma_p(x)$ directly from the definition of Γ_p, and is true for the other side by $\mathbf{G}_p$ 3. Hence the two functions differ by a constant. Putting $x = 0$ gives 0 on the right-hand side. By $\mathbf{G}_p\,\mathbf{1}$ we conclude that the left-hand side is also equal to 0. This proves the theorem.

Theorem 4.5. *Let χ be a Dirichlet character such that the conductor d of χ_1 is not divisible by p. Then*

$$L_p'(0, \chi) = \sum_{c=1}^{d-1} \chi_1(c) \log_p \Gamma_p\left(\frac{c}{d}\right) + (1 - \chi_1(p))B_{1,\chi_1} \log_p(d).$$

Proof. Let $N = pd$. In Theorem 4.2, write

$$a = c + bd$$

with $1 \le c \le d - 1$ and $0 \le b \le p - 1$. Then $\chi_1(a) = \chi_1(c)$. If a is divisible by p then $a/N \in \mathbf{Z}_p$, so that $G_p(a/N) = 0$ in Theorem 4.4. The desired formula is then a direct consequence of Theorem 4.2 combined with the relation of Theorem 4.4, and the fact that $\log_p(p) = 0$.

The formula of Theorem 4.2 is due to Diamond. The argument used to derive the variation in Theorem 4.5 is in Ferrero–Greenberg [Fe–Gr].

For the record, we state the distribution relation for the Diamond function.

$\mathbf{G}_p\,\mathbf{4}$. *For any positive integer m and $|x| > 1$, we have*

$$\sum_{a=0}^{m-1} G_p\left(\frac{x+a}{m}\right) = G_p(x) - (x - \tfrac{1}{2}) \log_p(m).$$

In particular, if $m = p^r$ is a power of p, then:

$\mathbf{G}_p\,\mathbf{5}$.
$$\sum_{a=0}^{p^r-1} G_p\left(\frac{x+a}{p^r}\right) = G_p(x).$$

This is a special case of $\mathbf{G}_p\,\mathbf{4}$ because $\log_p(p) = 0$. The proof of $\mathbf{G}_p\,\mathbf{4}$ can easily be given following a similar argument to that of Theorem 4.4, using the analyticity of $G_p(x)$ for $|x| > 1$. Our intent was to deal mostly with the cyclotomic applications, and we don't go into a systematic treatment of these p-adic functions.

Appendix by Karl Rubin: The Main Conjecture

Introduction

In [Th], Thaine introduced a new method for studying ideal class groups of real cyclotomic fields using cyclotomic units. Recently, in [Kol], Kolyvagin developed a remarkable strengthening of this method, an inductive procedure which has Thaine's method as its first step. If p is an odd prime, Kolyvagin was able to determine the orders of the different eigenspaces of the p-part of the ideal class group of $\mathbf{Q}(\boldsymbol{\mu}_p)$, decomposed with respect to characters of $\mathrm{Gal}(\mathbf{Q}(\boldsymbol{\mu}_p)/\mathbf{Q})$ (see Theorems 4.2 and 8.8). These results were already known from the work of Mazur and Wiles [M–W], but Kolyvagin's proof is very much simpler.

In this appendix we give an exposition of Kolyvagin's results on ideal class groups and extend Kolyvagin's methods, using ideas from [Ru], to prove the full Mazur–Wiles theorem (the "main conjecture") for $\mathbf{Q}(\boldsymbol{\mu}_{p^\infty})$ (see §§5 and 8 for the precise statements). In §§1–7 we concentrate on cyclotomic units and ideal class groups of real cyclotomic fields. Then in §8 we combine these results with the Kummer duality of Chapter 6 to study "minus" ideal class groups. It is also possible, as in [Kol], to do a direct analysis of the minus class groups using Gauss sums instead of cyclotomic units.

The appendix relies only on Chapters 1 through 7.

§1. Setting and Notation

Fix a positive integer m and let $F = \mathbf{Q}(\boldsymbol{\mu}_m)^+$. Let $\mathscr{S}$ denote the set of positive squarefree integers divisible only by primes $l \equiv \pm 1 \pmod{m}$ (that is, by those l which split completely in $F/\mathbf{Q}$). For every $r \in \mathcal{S}$ write

$$G_r = \mathrm{Gal}(F(\boldsymbol{\mu}_r)/F) \approx \mathrm{Gal}(\mathbf{Q}(\boldsymbol{\mu}_r)/\mathbf{Q})$$

and write $\mathbf{N}_r$ for the norm operator

$$\mathbf{N}_r = \sum_{\tau \in G_r} \tau \in \mathbf{Z}[G_r].$$

We will often use additive notation for the multiplicative group $F(\boldsymbol{\mu}_r)^\times$. There is a natural isomorphism $G_r = \prod_{l|r} G_l$ (product over primes l dividing r), and

$$\mathbf{N}_r = \prod_{l|r} \mathbf{N}_l \in \mathbf{Z}[G_r].$$

If $l \equiv \pm 1 \pmod{m}$ and $l \nmid r$, then we will identify G_l with $\mathrm{Gal}(F(\boldsymbol{\mu}_{rl})/F(\boldsymbol{\mu}_r))$, and we will write Fr_l for the Frobenius of l in G_r, the automorphism which sends each r-th root of unity to its l-th power. For every prime $l \equiv \pm 1 \pmod{m}$ fix a generator σ_l of G_l (which is cyclic of order $l-1$) and define

$$\mathbf{D}_l = \sum_{i=1}^{l-2} i\sigma_l^i \in \mathbf{Z}[G_l].$$

This "operator" is constructed to satisfy the identity

$$(\sigma_l - 1)\mathbf{D}_l = (l-1) - \mathbf{N}_l \tag{1}$$

in $\mathbf{Z}[G_l]$. For $r \in \mathcal{S}$ define

$$\mathbf{D}_r = \prod_{l|r} \mathbf{D}_l \in \mathbf{Z}[G_r].$$

Fix a primitive m-th root of unity ζ_m and, for each prime $l \equiv \pm 1 \pmod{m}$, a primitive l-th root of unity ζ_l. For $r \in \mathcal{S}$ define

$$\xi_r = \left(\zeta_m \prod_{l|r} \zeta_l - 1\right)\left(\zeta_m^{-1} \prod_{l|r} \zeta_l - 1\right).$$

These algebraic integers satisfy, for $r \in \mathcal{S}$ and primes $l|r$:

ES 1. $\xi_r \in F(\boldsymbol{\mu}_r)^\times$.

ES 2. ξ_r *is a (cyclotomic) unit if* $r > 1$.

ES 3. $\mathbf{N}_l \xi_r = (\mathrm{Fr}_l - 1)\xi_{r/l}$.

ES 4. $\xi_r \equiv \xi_{r/l}$ *modulo every prime above* l.

Proof. The first two assertions are clear. **ES 3** is the distribution relation satisfied by the cyclotomic units (see **CU 2**, Chapter 6, §3) and **ES 4** is immediate from the fact that $\zeta_l \equiv 1$ modulo all primes above l.

Fix an odd integer M (later M will be a large power of some chosen prime p); we will work "modulo M". Let

$$\mathcal{S}_M = \{r \in \mathcal{S}: r \text{ is divisible only by primes } l \equiv 1 \pmod{M}\}.$$

§2. Properties of Kolyvagin's "Euler System"

Lemma 2.1. *If* $r \in \mathcal{S}_M$, *then* $\mathbf{D}_r\xi_r \in [F(\boldsymbol{\mu}_r)^\times/(F(\boldsymbol{\mu}_r)^\times)^M]^{G_r}$.

Proof. We prove this by induction on the number of primes dividing r. If $r = 1$ the statement is clear. If $l|r$, by **ES 3** and (1)

$$(\sigma_l - 1)\mathbf{D}_r\xi_r = ((l-1) - \mathbf{N}_l)\mathbf{D}_{r/l}\xi_r \equiv (1 - \mathrm{Fr}_l)\mathbf{D}_{r/l}\xi_{r/l} \pmod{(F(\boldsymbol{\mu}_r)^\times)^{l-1}}.$$

Since Fr_l fixes F, our induction hypothesis shows

$$(1 - \mathrm{Fr}_l)\mathbf{D}_{r/l}\xi_{r/l} \in (F(\boldsymbol{\mu}_{r/l})^\times)^M.$$

These σ_l generate G_r, so this proves the lemma.

Lemma 2.2. *For every* $r \in \mathcal{S}_M$ *there is a* $\kappa_r \in F^\times/(F^\times)^M$ *such that*

$$\kappa_r \equiv \mathbf{D}_r\xi_r \pmod{(F(\boldsymbol{\mu}_r)^\times)^M}.$$

Proof. First observe that since M is odd and F is real, $\boldsymbol{\mu}_M \cap F = \{1\}$. Since r is prime to M, a simple ramification argument shows that $\boldsymbol{\mu}_M \cap F(\boldsymbol{\mu}_r) = \{1\}$ as well.

Define a 1-cocycle c_r: $\mathrm{Gal}(F(\boldsymbol{\mu}_r)/F) \to F(\boldsymbol{\mu}_r)^\times$ by

$$c_r(\sigma) = [(\sigma - 1)\mathbf{D}_r\xi_r]^{1/M}.$$

By Lemma 2.1, $c_r(\sigma) \in F(\boldsymbol{\mu}_r)^\times$; it is uniquely defined because $F(\boldsymbol{\mu}_r)$ has no nontrivial M-th roots of unity. Since $H^1(F(\boldsymbol{\mu}_r)/F, F(\boldsymbol{\mu}_r)^\times) = 0$, there is a $\beta \in F(\boldsymbol{\mu}_r)^\times$ such that $(\sigma - 1)\beta = c_r(\sigma)$ for all σ. Define $\kappa_r = \mathbf{D}_r\xi_r/\beta^M$. Then $\kappa_r \in F^\times$, and since β is unique modulo $F^\times$, κ_r is well-defined modulo $(F^\times)^M$. This proves Lemma 2.2.

Remark. We have obtained κ_r from $\mathbf{D}_r\xi_r$ via the sequence of isomorphisms

$$[F(\boldsymbol{\mu}_r)^\times/(F(\boldsymbol{\mu}_r)^\times)^M]^{G_r} \approx H^1(\bar{F}/F(\boldsymbol{\mu}_r), \boldsymbol{\mu}_M)^{G_r} \approx H^1(\bar{F}/F, \boldsymbol{\mu}_M) \approx F^\times/(F^\times)^M.$$

Each κ_r gives a principal ideal of F (modulo M-th powers of ideals) which can be viewed as a relation in the ideal class group of F. These relations will be used to bound the size of the ideal class group. To do this, we must understand the prime factorizations of these ideals (Proposition 2.4 below) and also how to choose r so as to get useful relations (Theorem 3.1).

Let $\mathcal{O}_F$ denote the ring of integers of F, and write $\mathscr{I} = \bigoplus_{\lambda} \mathbf{Z}\lambda$ for the group of fractional ideals of F, written additively. For every rational prime l write $\mathscr{I}_l = \bigoplus_{\lambda | l} \mathbf{Z}\lambda$, so $\mathscr{I} = \bigoplus_{l} \mathscr{I}_l$, and if $y \in F^\times$ let $(y) \in \mathscr{I}$ denote the principal ideal generated by y and $(y)_l \in \mathscr{I}_l$, $[y] \in \mathscr{I}/M\mathscr{I}$ and $[y]_l \in \mathscr{I}_l/M\mathscr{I}_l$ the projections of (y). Note that $[y]$ and $[y]_l$ are also well defined for $y \in F^\times/(F^\times)^M$. Write $G = \mathrm{Gal}(F/\mathbf{Q})$.

Lemma 2.3. *Suppose l splits completely in F and $l \equiv 1 \pmod{M}$. There is a unique G-equivariant surjection*

$$\varphi_l: (\mathcal{O}_F/l\mathcal{O}_F)^\times \to \mathscr{I}_l/M\mathscr{I}_l$$

which makes the following diagram commute:

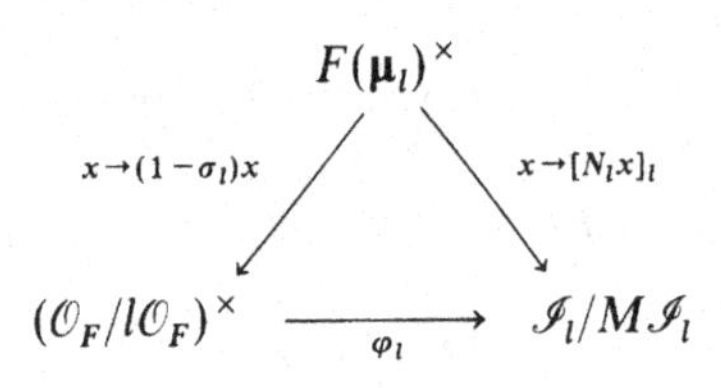

(For each λ of F above l and λ' of $F(\boldsymbol{\mu}_l)$ above λ, we have identified $\mathcal{O}_{F(\boldsymbol{\mu}_l)}/\lambda'$ with $\mathcal{O}_F/\lambda$.)

Proof. Since $[F(\boldsymbol{\mu}_l):F] = l - 1$ and all primes above l are totally, tamely ramified in $F(\boldsymbol{\mu}_l)/F$, the vertical maps are both surjective and the kernel of the left-hand map, namely the subgroup

$$\{x \in F(\boldsymbol{\mu}_l)^\times : x \text{ has order divisible by } l-1 \text{ at all primes above } l\},$$

is clearly contained in the kernel of the right-hand map. This proves the lemma.

For l as in Lemma 2.3 we will also write φ_l for the induced homomorphism

$$\varphi_l: \{y \in F^\times/(F^\times)^M : [y]_l = 0\} \to \mathscr{I}_l/M\mathscr{I}_l.$$

Proposition 2.4 (Kolyvagin). *Suppose $r \in \mathcal{S}_M$ and l is a rational prime.*

(i) *If $l \nmid r$, then $[\kappa_r]_l = 0$.*
(ii) *If $l \mid r$, then $[\kappa_r]_l = \varphi_l(\kappa_{r/l})$.*

Proof. By definition, $\kappa_r \equiv \mathbf{D}_r \xi_r \pmod{(F(\boldsymbol{\mu}_r)^\times)^M}$. If $l \nmid r$, then all primes above l are unramified in $F(\boldsymbol{\mu}_r)/F$, so (i) follows from **ES 2**. Suppose $l \mid r$, say $r = ls$. Recall that we can represent κ_r and κ_s by $\kappa_r = \mathbf{D}_r \xi_r / \beta_r^M$ and $\kappa_s = \mathbf{D}_s \xi_s / \beta_s^M$ where $\beta_r \in F(\boldsymbol{\mu}_r)^\times$ and $\beta_s \in F(\boldsymbol{\mu}_s)^\times$ satisfy

(2) $\quad (\sigma - 1)\beta_r = [(\sigma - 1)\mathbf{D}_r \xi_r]^{1/M} \quad$ and $\quad (\sigma - 1)\beta_s = [(\sigma - 1)\mathbf{D}_s \xi_s]^{1/M}.$

By (i) we may choose β_s prime to l.

In particular, the ideal of $F(\boldsymbol{\mu}_r)$ generated by β_r^M is the lift of an ideal of F, so (since all primes above l are unramified in $F(\boldsymbol{\mu}_r)/F(\boldsymbol{\mu}_l)$) there is a $\gamma \in F(\boldsymbol{\mu}_l)^\times$ such that $\beta_r \gamma^{(l-1)/M}$ is a unit at all primes above l. Then

$$[\mathbf{N}_l \gamma]_l = [\kappa_r]_l.$$

Modulo any prime above l, using (2), (1), **ES 3**, and **ES 4**,

$$(1 - \sigma_l)\gamma^{(l-1)/M} \equiv (\sigma_l - 1)\beta_r = [((l-1) - \mathbf{N}_l)\mathbf{D}_s \xi_r]^{1/M} = \frac{\mathbf{D}_s \xi_r^{(l-1)/M}}{[(\mathrm{Fr}_l - 1)\mathbf{D}_s \xi_s]^{1/M}}$$

$$\equiv \frac{\mathbf{D}_s \xi_s^{(l-1)/M}}{(\mathrm{Fr}_l - 1)\beta_s} \equiv (\mathbf{D}_s \xi_s / \beta_s^M)^{(l-1)/M}.$$

Therefore, applying the diagram of Lemma 2.3 with $\gamma \in F(\boldsymbol{\mu}_l)^\times$, we conclude that

$$[\kappa_r]_l = \varphi_l(\kappa_s).$$

§3. An Application of the Chebotarev Theorem

Fix a rational prime $p > 2$ and let C denote the p-part of the ideal class group of F. The next theorem together with Proposition 2.4 will enable us to construct all the relations we need in the ideal class group of F. In the simpler case where $p \nmid \#(G)$ it already appears in the work of Thaine ([Th], Proposition 4); the version below which we will need is essentially Theorem 5.5 of [Ru].

Theorem 3.1. *Suppose one is given $\mathfrak{c} \in C$, $M \in \mathbf{Z}$ a power of p, a finite G-submodule W of $F^\times/(F^\times)^M$, and a Galois-equivariant map*

$$\psi \colon W \to (\mathbf{Z}/M\mathbf{Z})[G].$$

Then there are infinitely many primes λ of F such that

(i) $\lambda \in \mathfrak{c}$.
(ii) $l \equiv 1 \pmod{M}$ *and l splits completely in $F/\mathbf{Q}$, where l is the rational prime below λ.*
(iii) $[w]_l = 0$ *for all $w \in W$, and there is a $u \in (\mathbf{Z}/M\mathbf{Z})^\times$ such that $\varphi_l(w) = u\psi(w)\lambda$ for all $w \in W$.*

Proof. Let H be the maximal unramified abelian p-extension of F, so that C is identified with $\mathrm{Gal}(H/F)$ by class field theory. Write $F' = F(\boldsymbol{\mu}_M)$. We have the diagram below.

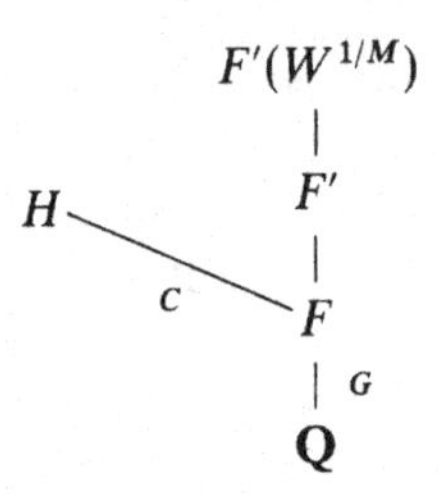

Step I. $F' \cap H = F$.
The inertia group of p in $\mathrm{Gal}(F(\boldsymbol{\mu}_M)/F)$ has index either 1 or 2, so there is no nontrivial unramified p-extension of F in F'.

Step II. $F'(W^{1/M}) \cap H = F$.
As in Chapter 6, Kummer theory gives a nondegenerate $\mathrm{Gal}(F'/\mathbf{Q})$-equivariant pairing

$$\mathrm{Gal}(F'(W^{1/M})/F') \times W/W' \to \boldsymbol{\mu}_M,$$

where W' is the kernel of the map from W into $(F')^\times/[(F')^\times]^M$. Let τ denote complex conjugation in $\mathrm{Gal}(F'/F)$. Then τ acts trivially on W and by -1 on $\boldsymbol{\mu}_M$, so τ acts by -1 on $\mathrm{Gal}(F'(W^{1/M})/F')$. Since H is abelian over F, τ acts trivially on $\mathrm{Gal}(H/F) \approx \mathrm{Gal}(HF'/F')$. Therefore τ acts on $\mathrm{Gal}(F'(W^{1/M}) \cap HF'/F')$ by both 1 and -1, so $F'(W^{1/M}) \cap HF' = F'$ and step II follows from step I.

Step III. *$F^\times/(F^\times)^M \to (F')^\times/((F')^\times)^M$ is injective, so $W' = 0$.*
By the inflation–restriction sequence of Galois cohomology,

$$\begin{aligned}\ker(F^\times/(F^\times)^M \to (F')^\times/((F')^\times)^M) &= \ker(H^1(\bar{F}/F, \boldsymbol{\mu}_M) \to H^1(\bar{F}/F', \boldsymbol{\mu}_M)) \\ &= H^1(F(\boldsymbol{\mu}_M)/F, \boldsymbol{\mu}_M).\end{aligned}$$

Since $\mathrm{Gal}(F'/F)$ is cyclic,

$$\#(H^1(F(\boldsymbol{\mu}_M)/F, \boldsymbol{\mu}_M)) = \#(H^0(F(\boldsymbol{\mu}_M)/F, \boldsymbol{\mu}_M)) = \#(\boldsymbol{\mu}_M(F)) = 1.$$

Step IV. *Construction of λ satisfying* (i), (ii), *and* (iii).

Combining step III with the Kummer pairing in step II, we conclude

$$\mathrm{Gal}(F'(W^{1/M})/F') \approx \mathrm{Hom}(W, \boldsymbol{\mu}_M).$$

Fix a primitive M-th root of unity ζ_M and define $\iota: (\mathbf{Z}/M\mathbf{Z})[G] \to \boldsymbol{\mu}_M$ by $\iota(1_G) = \zeta_M$ and $\iota(g) = 1$ for $g \in G$, $g \neq 1_G$. Let $\gamma \in \mathrm{Gal}(F'(W^{1/M})/F')$ be the element corresponding to $\iota \circ \psi \in \mathrm{Hom}(W, \boldsymbol{\mu}_M)$. Then by definition of the Kummer pairing, $\iota \circ \psi(w) = \gamma(w^{1/M})/w^{1/M}$ for all $w \in W$.

Since $F'(W^{1/M}) \cap H = F$ we can choose $\delta \in \mathrm{Gal}(HF'(W^{1/M})/F)$ such that δ restricts to γ on $F'(W^{1/M})$ and to c on H. Let λ be any prime of F of degree 1, unramified in $HF'(W^{1/M})/\mathbf{Q}$, whose Frobenius in $\mathrm{Gal}(HF'(W^{1/M})/F)$ is the conjugacy class of δ. Since W is finite, the Chebotarev theorem guarantees the existence of unfinitely many such λ. We must verify that λ satisfies (i), (ii), and (iii). Let l be the rational prime below λ.

The identification of C with $\mathrm{Gal}(H/F)$ sends the class of λ to the Frobenius of λ, so (i) is immediate. Also, since δ is trivial on F', l splits completely in $\mathbf{Q}(\boldsymbol{\mu}_M)/\mathbf{Q}$ which proves (ii). The first assertion of (iii), that $[w]_l = 0$ for all $w \in W$, holds because λ is unramified in $F'(W^{1/M})/F$.

From the definition of φ_l, $\mathrm{ord}_\lambda(\varphi_l(w)) = 0$ if and only if w is an M-th power modulo λ. Also,

$$\begin{aligned} \mathrm{ord}_\lambda(\psi(w)\lambda) = 0 \quad &\Leftrightarrow \quad \iota \circ \psi(w) = 1 \quad \Leftrightarrow \quad \gamma(w^{1/M})/w^{1/M} = 1 \\ &\Leftrightarrow \quad w \text{ is an } M\text{-th power modulo } \lambda. \end{aligned}$$

Therefore there is a unit $u \in (\mathbf{Z}/M\mathbf{Z})^\times$ such that

$$\mathrm{ord}_\lambda(\varphi_l(w)) = u\, \mathrm{ord}_\lambda(\psi(w)\lambda) \qquad \text{for all} \quad w \in W.$$

It follows that the map $w \to \varphi_l(w) - u\psi(w)\lambda$ is a $\mathrm{Gal}(F/K)$-equivariant homomorphism into $\bigoplus_{\lambda'|l,\, \lambda' \neq \lambda} (\mathbf{Z}/M\mathbf{Z})\lambda'$, which has no nonzero $\mathrm{Gal}(F/K)$-stable submodules. This proves (iii).

§4. Example: The Ideal Class Group of $\mathbf{Q}(\boldsymbol{\mu}_p)^+$

In this section we give Kolyvagin's analysis of the ideal class group of $\mathbf{Q}(\boldsymbol{\mu}_p)^+$. Although not needed for the proof of the Mazur–Wiles theorem, this is the basic example of Kolyvagin's method.

We will apply the results of the previous sections with $F = \mathbf{Q}(\boldsymbol{\mu}_p)^+$, where p is an odd prime. Let C denote the p-part of the ideal class group of F, E the group of global units of F, and $\mathscr{E}$ the subgroup of cyclotomic units (see Chapter 7, §5). For every character χ of $G = \mathrm{Gal}(\mathbf{Q}(\boldsymbol{\mu}_p)^+/\mathbf{Q})$

define the χ-idempotent

$$e(\chi) = \frac{2}{p-1} \sum_{\gamma \in G} \chi^{-1}(\gamma)\gamma.$$

If Y is a $\mathbf{Z}_p[G]$-module, we write $Y(\chi) = e(\chi)Y$. We will be interested in $C(\chi)$ and $(E/\mathscr{E})(\chi)$, where $(E/\mathscr{E})(\chi)$ denotes the χ-component of the p-Sylow subgroup of the finite group $E/\mathscr{E}$.

Theorem 4.1. *For every character χ of G, $\#(C(\chi)) | \#[(E/\mathscr{E})(\chi)]$.*

Proof. We can assume $\chi \neq 1$, or else $C(\chi) = (E/\mathscr{E})(\chi) = 0$. Let

$$M = \#[(E/\mathscr{E})(\chi)]\#(C(\chi))p,$$

and let $\mathfrak{c}_1, \ldots, \mathfrak{c}_k$ be any collection of ideal classes which generate $C(\chi)$. We will choose inductively primes $\lambda_1, \ldots, \lambda_k$ of F such that the class of λ_i is $\mathfrak{c}_i$ and the rational prime l_i below λ_i satisfies $l_i \equiv 1 \pmod{M}$.

Suppose $1 \leqq i \leqq k$ and we have chosen $\lambda_1, \ldots, \lambda_{i-1}$ lying above the rational primes $l_1, \ldots, l_{i-1} \equiv 1 \pmod{M}$. We choose λ_i as follows. Write $r_i = \prod_{j<i} l_j$ (so $r_1 = 1$) and let W be the subgroup of $F^\times/(F^\times)^M$ generated by $e(\chi)\kappa_{r_i}$. Define t_i to be the largest divisor of M such that $e(\chi)\kappa_{r_i} \in (F^\times)^{t_i}/(F^\times)^M$, and define a map $\psi: W \to (\mathbf{Z}/M\mathbf{Z})[G]$ by $\psi(e(\chi)\kappa_{r_i}) = t_i e(\chi)$. Now apply Theorem 3.1 with $\mathfrak{c} = \mathfrak{c}_i$ and M, W, and ψ as above. Choose λ_i to be any prime of F satisfying Theorem 3.1 with this data, and let l_i be the rational prime below λ_i. Then $l_i \equiv 1 \pmod{M}$ and there is a $u_i \in (\mathbf{Z}/M\mathbf{Z})^\times$ such that $\varphi_{l_i}(e(\chi)\kappa_{r_i}) = u_i t_i e(\chi)\lambda_i$.

Let E' (resp. $\mathscr{E}'$) denote the image of E (resp. $\mathscr{E}$) in $F^\times/(F^\times)^M$. Then $E'(\chi)$ and $\mathscr{E}'(\chi)$ are cyclic, and $(E/\mathscr{E})(\chi) = E'(\chi)/\mathscr{E}'(\chi)$. Recall

$$\kappa_1 = (1 - \zeta_p)(1 - \zeta_p^{-1}),$$

so $e(\chi)\kappa_1$ generates $\mathscr{E}'(\chi)$. Thus $t_1 = \#[(E/\mathscr{E})(\chi)]$.

For each $i > 1$, by Proposition 2.4 the principal ideal $[e(\chi)\kappa_{r_i}] \in \mathscr{I}/M\mathscr{I}$ satisfies

$$\begin{aligned} [e(\chi)\kappa_{r_{i+1}}] &= \varphi_{l_i}(e(\chi)\kappa_{r_i}) + \sum_{j<i} [e(\chi)\kappa_{r_{i+1}}]_{l_j} \\ &= u_i t_i e(\chi)\lambda_i \quad \text{in } \mathscr{I}/(M\mathscr{I}, e(\chi)\lambda_1, \ldots, e(\chi)\lambda_{i-1}). \end{aligned} \tag{3}$$

Since $e(\chi)\kappa_{r_{i-1}} \in (F^\times)^{t_{i+1}}$ we see that $t_{i+1} | t_i$. In particular $t_{i+1} | t_1 = \#[(E/\mathscr{E})(\chi)]$, so $(M/t_{i+1})C(\chi) = 0$ and we can divide (3) by t_{i+1} and project into $C(\chi)$ to get

$$(t_i/t_{i+1})\mathfrak{c}_i = 0 \quad \text{in } C(\chi)/(\mathfrak{c}_1, \ldots, \mathfrak{c}_{i-1}).$$

Thus since $c_1, \ldots, c_k$ generate $C(\chi)$,

$$\#(C(\chi)) \Big| \prod_{i=1}^{k} (t_i/t_{i+1}) = t_1/t_{k+1} = \#[(E/\mathscr{E})(\chi)]/t_{k+1}.$$

Theorem 4.2 (Mazur–Wiles, Kolyvagin). *For every character χ of G,*

$$\#(C(\chi)) = \#[(E/\mathscr{E})(\chi)].$$

Proof. By the analytic class number formula (Theorem 5.1 of Chapter 3),

$$\prod_{\chi} \#(C(\chi)) = \#(C) = \#[(E/\mathscr{E}) \otimes \mathbf{Z}_p] = \prod_{\chi} \#[(E/\mathscr{E})(\chi)].$$

The theorem follows immediately from this and Theorem 4.1.

§5. The Main Conjecture

Fix a rational prime $p > 2$, and for every integer $n \geqq 0$ let

$$K_n = \mathbf{Q}(\boldsymbol{\mu}_{p^{n+1}}), \qquad K_\infty = \bigcup K_n.$$

Put

$$\Delta = \mathrm{Gal}(K_0/\mathbf{Q}) \approx (\mathbf{Z}/p\mathbf{Z})^\times \qquad \text{and} \qquad \Gamma = \mathrm{Gal}(K_\infty/K_0) \approx \mathbf{Z}_p;$$

then $\mathrm{Gal}(K_\infty/\mathbf{Q}) = \Delta \times \Gamma$. For $1 \leqq n < \infty$ write C_n for the p-part of the ideal class group of K_n, E_n for the group of global units of K_n, and $\mathscr{E}_n$ for the group of cyclotomic units of K_n. Write U_n for the group of local units of the completion of K_n above p which are congruent to 1 modulo the maximal ideal, and let $\bar{E}_n$ and V_n denote the closures of $E_n \cap U_n$ and $\mathscr{E}_n \cap U_n$, respectively, in U_n. We also define

$$C_\infty = \varprojlim C_n, \qquad E_\infty = \varprojlim \bar{E}_n, \qquad V_\infty = \varprojlim V_n, \qquad \text{and} \qquad U_\infty = \varprojlim U_n,$$

all inverse limits with respect to the norm maps. For $n \leqq \infty$ let Ω_n be the maximal abelian p-extension of K_n which is unramified outside of the prime above p, and write $X_n = \mathrm{Gal}(\Omega_n/K_n)$. Define the Iwasawa algebra

$$\Lambda = \mathbf{Z}_p[[\Gamma]] = \varprojlim \mathbf{Z}_p[\mathrm{Gal}(K_n/K_0)].$$

For every character χ of Δ define the χ-idempotent

$$e(\chi) = \frac{1}{p-1} \sum_{\delta \in \Delta} \chi^{-1}(\delta)\delta.$$

If Y is a $\mathbf{Z}_p[\Delta]$-module, we write $Y(\chi) = e(\chi)Y$. In particular, $C_\infty(\chi)$, $U_\infty(\chi)$, $E_\infty(\chi)$, $V_\infty(\chi)$, and $X_\infty(\chi)$ are all finitely generated Λ-modules, $C_\infty(\chi)$ is a torsion Λ-module, and if χ is an even character $X_\infty(\chi)$ and $U_\infty(\chi)/V_\infty(\chi)$ are torsion as well (§§5 and 6 of Chapter 5, §§2 and 5 of Chapter 7).

Recall that two Λ-modules are said to be quasi-isomorphic if there is a map between them with finite kernel and cokernel. By Theorem 3.1 of Chapter 5, every finitely generated torsion Λ-module Y is quasi-isomorphic to a module of the form $\bigoplus \Lambda/f_i\Lambda$ for some $f_i \in \Lambda$, and the characteristic ideal $(\prod f_i)\Lambda$ is a well-defined invariant of Y which we will denote by $\mathrm{char}(Y)$.

The following theorem is one of several equivalent forms of Iwasawa's "main conjecture", first proved by Mazur and Wiles in [M–W]. This theorem will be proved in §7. For a discussion of some of the other formulations see §8.

Theorem 5.1. *For all even characters χ of Δ,*

$$\mathrm{char}(C_\infty(\chi)) = \mathrm{char}(E_\infty(\chi)/V_\infty(\chi)).$$

§6. Tools from Iwasawa Theory

In using Proposition 2.4 and Theorem 3.1 to prove Theorem 5.1, we will need to know about the structure of C_n and $\bar{E}_n$ as $\mathbf{Z}_p[\mathrm{Gal}(K_n/\mathbf{Q})]$-modules. For any fixed n we know very little, but Theorem 3.1 of Chapter 5 (the classification theorem for Λ-modules) shows that C_∞ and E_∞ are well behaved as Λ-modules. In this section we relate C_n and $\bar{E}_n$ with C_∞ and E_∞. The major results of this section (Theorems 6.1 and 6.3) are due to Iwasawa [Iw 12], although our proof of Theorem 6.3 is different from his.

For every n let $\Gamma_n = \mathrm{Gal}(K_\infty/K_n)$ and let I_n denote the ideal of Λ generated by $\{\gamma - 1 : \gamma \in \Gamma_n\}$; if γ is a generator of Γ then $I_n = (\gamma^{p^n} - 1)\Lambda$. Write

$$\Lambda_n = \Lambda/I_n\Lambda \approx \mathbf{Z}_p[\mathrm{Gal}(K_n/K_0)].$$

If Y is a Λ-module we write

$$Y_{\Gamma_n} = Y/I_nY = Y \otimes \Lambda_n.$$

We need to study the natural maps

$$X_\infty(\chi)_{\Gamma_n} \to X_n(\chi), \qquad C_\infty(\chi)_{\Gamma_n} \to C_n(\chi), \qquad U_\infty(\chi)_{\Gamma_n} \to U_n(\chi),$$
$$E_\infty(\chi)_{\Gamma_n} \to \bar{E}_n(\chi), \qquad \text{and} \qquad V_\infty(\chi)_{\Gamma_n} \to V_n(\chi)$$

induced by the projection maps.

Theorem 6.1. *For every χ the natural map $C_\infty(\chi)_{\Gamma_n} \to C_n(\chi)$ is an isomorphism. If χ is even and $\chi \neq 1$ then the natural maps*

$$X_\infty(\chi)_{\Gamma_n} \to X_n(\chi), \qquad U_\infty(\chi)_{\Gamma_n} \to U_n(\chi), \qquad \text{and} \qquad V_\infty(\chi)_{\Gamma_n} \to V_n(\chi)$$

are isomorphisms.

Proof. For $C_\infty(\chi)$ this is Theorem 4.1 of Chapter 5. For $X_\infty(\chi)$ the proof is the same (but we now need $\chi \neq 1$); see §6 of Chapter 5. For $U_\infty(\chi)$ and $V_\infty(\chi)$ this is Theorems 2.2 and 5.1 of Chapter 7, respectively.

Lemma 6.2.

(i) *Suppose $0 \to W \to Y \to Z \to 0$ is an exact sequence of Λ-modules. For every n the kernel of the induced map $W_{\Gamma_n} \to Y_{\Gamma_n}$ is a quotient of Z^{Γ_n}.*
(ii) *If Z is a finitely generated Λ-module and Z_{Γ_n} is finite, then Z^{Γ_n} is finite.*

Proof. Fix a generator γ of Γ. Applying the snake lemma to the diagram

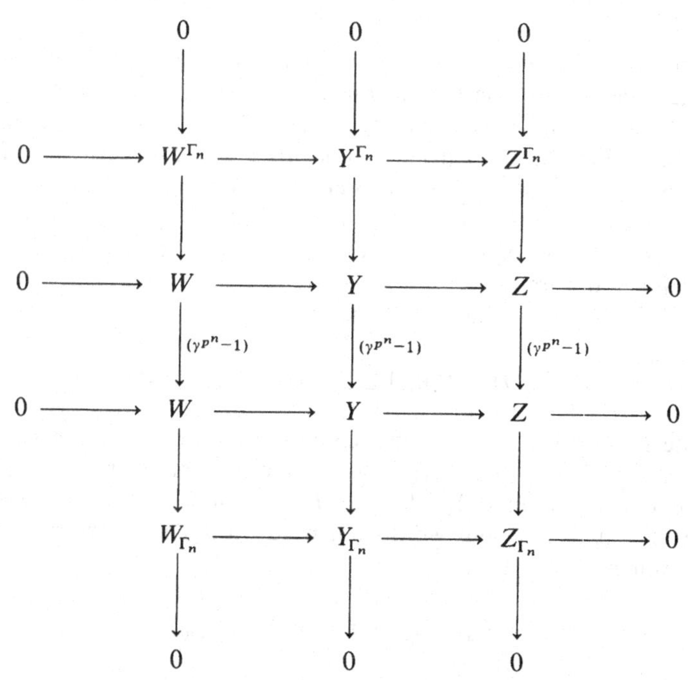

proves (i). For (ii), if Z_{Γ_n} is finite then the right-hand column of the diagram shows that Z/Z^{Γ_n} is quasi-isomorphic to Z. By Theorem 3.1 of Chapter 5 this is impossible unless Z^{Γ_n} is finite.

Theorem 6.3. *If χ is even and $\chi \neq 1$, then there is an ideal $\mathscr{A}$ of finite index in Λ such that for every n, $\mathscr{A}$ annihilates the kernel and cokernel of the natural map $E_\infty(\chi)_{\Gamma_n} \to \bar{E}_n(\chi)$. In particular the kernel and cokernel are finite with order bounded independently of n.*

Proof. Consider the two diagrams with exact rows

$$\begin{array}{ccccccccc} & & (U_\infty(\chi)/E_\infty(\chi))_{\Gamma_n} & \xrightarrow{\phi_1} & X_\infty(\chi)_{\Gamma_n} & \longrightarrow & C_\infty(\chi)_{\Gamma_n} & \longrightarrow & 0 \\ & & \downarrow{\scriptstyle \pi_{U/E}} & & \downarrow & & \downarrow & & \\ 0 & \longrightarrow & U_n(\chi)/\bar{E}_n(\chi) & \longrightarrow & X_n(\chi) & \longrightarrow & C_n(\chi) & \longrightarrow & 0 \end{array}$$

and

$$\begin{array}{ccccccccc} & & E_\infty(\chi)_{\Gamma_n} & \xrightarrow{\phi_2} & U_\infty(\chi)_{\Gamma_n} & \longrightarrow & (U_\infty(\chi)/E_\infty(\chi))_{\Gamma_n} & \longrightarrow & 0 \\ & & \downarrow{\scriptstyle \pi_E} & & \downarrow & & \downarrow{\scriptstyle \pi_{U/E}} & & \\ 0 & \longrightarrow & \bar{E}_n(\chi) & \longrightarrow & U_n(\chi) & \longrightarrow & U_n(\chi)/\bar{E}_n(\chi) & \longrightarrow & 0. \end{array}$$

The first one is induced by class field theory (Chapter 5, §5). Applying the snake lemma to the second diagram, since the map from $U_\infty(\chi)_{\Gamma_n}$ to $U_n(\chi)$ is an isomorphism by Theorem 6.1, we see that $\mathrm{coker}(\pi_E) \approx \ker(\pi_{U/E})$. Since the map from $X_\infty(\chi)_{\Gamma_n}$ to $X_n(\chi)$ is injective by Theorem 6.1, $\ker(\pi_{U/E}) = \ker(\phi_1)$. By Theorem 6.1, $C_\infty(\chi)_{\Gamma_n} \approx C_n(\chi)$ is finite, so Lemma 6.2 shows that $\ker(\phi_1)$ is a quotient of $C_\infty(\chi)_{\mathrm{finite}}$, the maximal finite Λ-submodule of $C_\infty(\chi)$. (Since C_∞ is finitely generated, $C_\infty(\chi)_{\mathrm{finite}}$ is well defined and finite.)

Similarly, $\ker(\pi_E) \approx \ker(\phi_2)$. Since

$$\#((U_\infty(\chi)/E_\infty(\chi))_{\Gamma_n}) \leq [U_n(\chi) : \bar{E}_n(\chi)]\,\#(\ker(\pi_{U/E}))$$

is finite for $\chi \neq 1$ (by Leopoldt's conjecture, Theorem 4.2 of Chapter 4), Lemma 6.2 shows that $\ker(\phi_2)$ is a quotient of $(U_\infty(\chi)/E_\infty(\chi))_{\mathrm{finite}}$, the maximal finite submodule of $U_\infty(\chi)/E_\infty(\chi)$. (In fact, one can show that $(U_\infty(\chi)/E_\infty(\chi))_{\mathrm{finite}} = 0$, but we do not need this.) Thus if we take $\mathscr{A}$ to be the annihilator in Λ of

$$C_\infty(\chi)_{\mathrm{finite}} \oplus (U_\infty(\chi)/E_\infty(\chi))_{\mathrm{finite}},$$

$\mathscr{A}$ satisfies the conditions of the theorem.

For every character χ of Δ fix a generator $h_\chi \in \Lambda$ of $\mathrm{char}(E_\infty(\chi)/V_\infty(\chi))$.

Corollary 6.4. *Suppose χ is even and $\chi \neq 1$. There is an ideal $\mathscr{A}$ of finite index in Λ such that for every $\eta \in \mathscr{A}$ and every n, there is a map $\theta_{n,\eta}: \bar{E}_n(\chi) \to \Lambda_n$ such that $\theta_{n,\eta}(V_n(\chi)) = \eta h_\chi \Lambda_n$.*

Proof. Since $U_\infty(\chi)$ is free of rank one over Λ (Theorem 2.1 of Chapter 7) and $0 \neq E_\infty(\chi) \subset U_\infty(\chi)$, $E_\infty(\chi)$ is torsion free and rank one. Therefore by the classification theorem (Chapter 5, Theorem 3.1) there is an injective homomorphism $\theta: E_\infty(\chi) \to \Lambda$ with finite cokernel.

We first claim that $\theta(V_\infty(\chi)) = h_\chi \Lambda$. Clearly θ induces a quasi-isomorphism $E_\infty(\chi)/V_\infty(\chi) \to \Lambda/\theta(V_\infty(\chi))$; since $V_\infty(\chi)$ is free of rank one over Λ (Theorem 5.1, Chapter 7),

$$\theta(V_\infty(\chi)) = \mathrm{char}(\Lambda/\theta(V_\infty(\chi))) = \mathrm{char}(E_\infty(\chi)/V_\infty(\chi)) = h_\chi \Lambda.$$

Let $\mathscr{A}$ be an ideal of Λ satisfying Theorem 6.3. Fix an n and let θ_n be the homomorphism from $E_\infty(\chi)_{\Gamma_n}$ to Λ_n induced by θ, and π_n the projection map from $E_\infty(\chi)_{\Gamma_n}$ to $\bar{E}_n(\chi)$. For any $\eta \in \mathscr{A}$ we can define $\theta_{n,\eta}: \bar{E}_n(\chi) \to \Lambda_n$ so that the following diagram commutes:

$$\begin{array}{ccc} E_\infty(\chi)_{\Gamma_n} & \xrightarrow{\theta_n} & \Lambda_n \\ \downarrow{\scriptstyle \pi_n} & & \downarrow{\scriptstyle \eta} \\ \bar{E}_n(\chi) & \xrightarrow[\theta_{n,\eta}]{} & \Lambda_n \end{array}$$

i.e., $\theta_{n,\eta}(u) = \theta_n(\pi_n^{-1}(\eta u))$. This is well defined because by Theorem 6.3, η annihilates $\mathrm{coker}(\pi_n)$, and $\ker(\pi_n)$ is finite so $\ker(\pi_n) \subset \ker(\theta_n)$. Then since $V_n(\chi) = \pi_n(V_\infty(\chi))$,

$$\theta_{n,\eta}(V_n(\chi)) = \eta\theta_n(V_\infty(\chi)) = \eta h_\chi \Lambda_n.$$

This concludes the proof of Corollary 6.4.

By the classification theorem, $C_\infty(\chi)$ is quasi-isomorphic to a module of the form

$$\bigoplus_{i=1}^{k} \Lambda/f_i\Lambda$$

with nonzero $f_i \in \Lambda$. In particular

$$\text{writing } f_\chi = \prod_{i=1}^{k} f_i, \qquad \text{we have} \qquad \mathrm{char}(C_\infty(\chi)) = f_\chi \Lambda.$$

Corollary 6.5. *Let $f_1, \ldots, f_k$ be as above. There is an ideal $\mathscr{B}$ of finite index in Λ and for every n there are classes $\mathfrak{c}_1, \ldots, \mathfrak{c}_k \in C_n(\chi)$ such* *that the annihilator* $\mathrm{Ann}(\mathfrak{c}_i) \subset \Lambda_n$ *of* $\mathfrak{c}_i$ *in* $C_n(\chi)/(\Lambda_n\mathfrak{c}_1 + \cdots + \Lambda_n\mathfrak{c}_{i-1})$ *satisfies* $\mathscr{B}\,\mathrm{Ann}(\mathfrak{c}_i) \subset f_i\Lambda_n$.

Proof. On torsion Λ-modules, the quasi-isomorphism relation is reflexive, so there is an exact sequence

$$0 \to \bigoplus_{i=1}^{k} \Lambda/f_i\Lambda \to C_\infty(\chi) \to Z \to 0$$

with a finite Λ-module Z. By Theorem 6.1 and Lemma 6.2, tensoring with $\Lambda_n = \Lambda/I_n\Lambda$ yields

$$Z^{\Gamma_n} \to \bigoplus_{i=1}^{k} \Lambda_n/f_i\Lambda_n \to C_n(\chi) \to Z_{\Gamma_n} \to 0.$$

Let $\mathscr{B}$ be the annihilator of the finite module Z and choose $\mathfrak{c}_i$ to be the image of 1 in the i-th summand $\Lambda_n/f_i\Lambda_n$ under the map above. The corollary now follows.

Lemma 6.6. *Let* χ *be an even character of* Δ, *and let*

$$f_\chi\Lambda = \mathrm{char}(C_\infty(\chi)) \qquad \textit{and} \qquad h_\chi\Lambda = \mathrm{char}(E_\infty(\chi)/V_\infty(\chi))$$

as above.

(i) *For every* n, $\Lambda_n/f_\chi\Lambda_n$ *and* $\Lambda_n/h_\chi\Lambda_n$ *are finite.*
(ii) *There is a positive constant* c *such that for all* n,

$$c^{-1} \leqq \frac{\#(C_n(\chi))}{\#(\Lambda_n/f_\chi\Lambda_n)} \leqq c, \qquad c^{-1} \leqq \frac{\#(\bar{E}_n(\chi)/V_n(\chi))}{\#(\Lambda_n/h_\chi\Lambda_n)} \leqq c.$$

(iii) *If* $\chi = 1$ *then* f_χ *and* h_χ *are units in* Λ.

Proof. First suppose $\chi \neq 1$. From a quasi-isomorphism

$$C_\infty(\chi) \to \bigoplus_{i=1}^{k} \Lambda/f_i\Lambda$$

we get, for every n, a map

$$C_n(\chi) \approx C_\infty(\chi)_{\Gamma_n} \to \bigoplus_{i=1}^{k} \Lambda_n/f_i\Lambda_n$$

with kernel and cokernel finite and bounded independently of n. In particular each $\Lambda_n/f_i\Lambda_n$ is finite. It follows easily that $\Lambda_n/f_\chi\Lambda_n$ is finite for every n, and by Theorem 1.2 of Chapter 5,

$$\#(\Lambda_n/f_\chi\Lambda_n)/\#\left(\bigoplus_{i=1}^{k} \Lambda_n/f_i\Lambda_n\right)$$

is bounded above and below independently of n (in fact, one can show that

$$\#(\Lambda_n/f_\chi\Lambda_n) = \#\left(\bigoplus_{i=1}^{k}\Lambda_n/f_i\Lambda_n\right)\Big).$$

This proves the first part of (i) and (ii). Similarly we have maps

$$(E_\infty(\chi)/V_\infty(\chi))_{\Gamma_n} \to \Lambda_n/h_\chi\Lambda_n,$$
$$(E_\infty(\chi)/V_\infty(\chi))_{\Gamma_n} \to \bar{E}_n(\chi)/V_n(\chi)$$

with kernel and cokernel finite and bounded independently of n (using Theorems 6.1 and 6.3 for the second map). This gives the second part of (i) and (ii), for $\chi \neq 1$.

Now suppose $\chi = 1$, and write $\mathbf{Q}_n = K_n^\Delta$. Then $C_n(\chi) = C_n^\Delta$ is the p-part of the ideal class group of $\mathbf{Q}_n$ for every n, and $K_0^\Delta = \mathbf{Q}$. Therefore $C_n(\chi) = 0$ for every n by Theorem 4.3 of Chapter 5, and f_χ is a unit.

Also when $\chi = 1$, $\bar{E}_n(\chi) = \bar{E}_n^\Delta$ and $V_n(\chi) = V_n^\Delta$. The analytic class number formula for the field $\mathbf{Q}_n$ (the analogue of Theorem 5.1 of Chapter 3, with the same proof), together with Leopoldt's conjecture (Theorem 4.2 of Chapter 4), shows that $[\bar{E}_n(\chi): V_n(\chi)] = \#(C_n(\chi)) = 1$ for every n. Therefore h_χ is a unit and the inequalities of (ii) hold with $\chi = 1$, $c = 1$.

§7. Proof of Theorem 5.1

We now come to the main result. For this section fix n and write $C = C_n$, $E = E_n$, and $V = V_n$. We will apply the results of §§2 and 3 with $F = K_n^+$. Note that by Theorems 4.2 and 4.3 of Chapter 3, if χ is even we can identify $C_n(\chi)$ with the χ-component of the ideal class group of F. If l is a rational prime splitting completely in F and $w \in F^\times$, recall that $(w)_l \in \mathscr{I}_l$ is the portion of the principal ideal (w) which is supported on the primes above l and $[w]_l \in \mathscr{I}_l/M\mathscr{I}_l$ its projection. If λ is a prime of F above l then $\mathscr{I}_l(\chi) = e(\chi)(\mathscr{I}_l \otimes \mathbf{Z}_p)$ is free of rank one over Λ_n, generated by $\lambda(\chi) = e(\chi)\lambda$, and we define $v_\lambda = v_{\lambda,\chi}: F^\times \to \Lambda_n$ by $v_\lambda(w)\lambda(\chi) = e(\chi)(w)_l$. We will write $\bar{v}_\lambda$ for the corresponding map from $F^\times/(F^\times)^M$ to $\Lambda_n/M\Lambda_n$ satisfying $\bar{v}_\lambda(w)\lambda(\chi) = e(\chi)[w]_l$.

Recall that for $r \in \mathscr{S}_M$ we have $\kappa_r \in F^\times/(F^\times)^M$ as defined in §2.

Lemma 7.1. *Suppose $r \in \mathscr{S}_M$, $l|r$, and λ is a prime of F above l. Let B be the subgroup of the ideal class group C generated by the primes of F dividing r/l. Write $\mathfrak{c} \in C(\chi)$ for the class of $e(\chi)\lambda$ and write W for the Λ_n-submodule of $F^\times/(F^\times)^M$ generated by $e(\chi)\kappa_r$. If η, $f \in \Lambda_n$ are such that the annihilator $\operatorname{Ann}(\mathfrak{c}) \subset \Lambda_n$ of $\mathfrak{c}$ in $C(\chi)/B(\chi)$ satisfies $\eta \operatorname{Ann}(\mathfrak{c}) \subset f\Lambda_n$, $\Lambda_n/f\Lambda_n$ is finite, and*

$$M \geq \#(C(\chi))\#((\mathscr{I}_l(\chi)/M\mathscr{I}_l(\chi))/\Lambda_n[e(\chi)\kappa_r]_l),$$

then there is a Galois-equivariant map $\psi: W \to \Lambda_n/M\Lambda_n$ *such that*

$$f\psi(e(\chi)\kappa_r) = \eta\bar{v}_\lambda(\kappa_r).$$

Proof. Let β be any lift of $e(\chi)\kappa_r$ to $F^\times$. Then

$$e(\chi)(\beta) = v_\lambda(\beta)\lambda(\chi) + \sum_{q \neq l} e(\chi)(\beta)_q.$$

By Proposition 2.4(i), $(\beta)_q \in M\mathscr{I}_q$ if $q \nmid r$. Since M annihilates $C(\chi)$ we conclude that $v_\lambda(\beta)\lambda(\chi)$ projects to 0 in $C(\chi)/B(\chi)$ and therefore $\eta v_\lambda(\beta) \in f\Lambda_n$. Write $\delta = \eta v_\lambda(\beta)/f$; division by f is uniquely defined because $\Lambda_n/f\Lambda_n$ is finite.

Define $\psi: W \to \Lambda_n/M\Lambda_n$ by $\psi(\rho e(\chi)\kappa_r) = \rho\delta$ for all $\rho \in \mathbf{Z}[\mathrm{Gal}(K_n/K_0)]$. This map has the desired property but we need to show that it is well defined. Suppose $\rho e(\chi)\kappa_r = 0$, that is, $\rho\beta = x^M$ with $x \in F^\times$. Then in particular $\rho[e(\chi)\kappa_r]_l = 0$. Write $h = \#(C(\chi))$. By our assumption on M, $(M/h)(\mathscr{I}_l(\chi)/M\mathscr{I}_l(\chi)) \subset \Lambda_n[e(\chi)\kappa_r]_l$, so we must have $\rho \in h\Lambda_n$. Then

$$\begin{aligned} e(\chi)(x) &= \sum_q e(\chi)(x)_q \\ &= M^{-1}e(\chi)(\rho\beta)_l + \sum_{q|(r/l)} e(\chi)(x)_q + \sum_{q \nmid r} he(\chi)(\rho/h)(M^{-1}(\beta)_q) \\ &\equiv M^{-1}e(\chi)(\rho\beta)_l \quad \left(\mathrm{mod} \bigoplus_{q|(r/l)} \mathscr{I}_q(\chi), h\mathscr{I}(\chi)\right). \end{aligned}$$

Since h annihilates $C(\chi)$ we conclude that $M^{-1}e(\chi)(\rho\beta)_l$ projects to 0 in $C(\chi)/B(\chi)$. Thus $M^{-1}v_\lambda(\rho\beta)\mathfrak{c} = 0$ in $C(\chi)/B(\chi)$, so $\rho\delta f = \eta v_\lambda(\rho\beta) \in Mf\Lambda_n$ and $\psi(\rho e(\chi)\kappa_r) = \rho\delta \in M\Lambda_n$. This concludes the proof of the lemma.

Recall $\mathrm{char}(E_\infty(\chi)/V_\infty(\chi)) = h_\chi\Lambda$ and $\mathrm{char}(C_\infty(\chi)) = f_\chi\Lambda$, where $f_\chi = \prod_{i=1}^k f_i$.

Theorem 7.2. *For every even character* χ *of* Δ, $\mathrm{char}(C_\infty(\chi))$ *divides* $\mathrm{char}(E_\infty(\chi)/V_\infty(\chi))$.

Proof. If $\chi = 1$, this is true because both characteristic ideals are trivial by Lemma 6.6. Thus we may assume $\chi \neq 1$.

Recall that κ_1 is represented by $\xi = (\zeta_{p^n} - 1)(\zeta_{p^n}^{-1} - 1) \in F^\times$, and $\xi(\chi) = \xi^{e(\chi)}$ generates $V_n(\chi)$. Let $\mathfrak{c}_1, \ldots, \mathfrak{c}_k \in C(\chi)$ be as in Corollary 6.5. Also choose one more class $\mathfrak{c}_{k+1}$ which can be any element of $C(\chi)$ at all (for example, $\mathfrak{c}_{k+1} = 0$). Fix an ideal $\mathscr{C}$ of finite index in Λ satisfying both Corollaries 6.4 and 6.5, and let $\eta \in \mathscr{C}$ be any element such that $\Lambda_m/\eta\Lambda_m$ is finite for all m (i.e. for all m, η is prime to $\gamma^{p^m} - 1$, where γ is a generator of Γ). Let $\theta = \theta_{n,\eta}: \bar{E}(\chi) \to \Lambda_n$ be the map given by Corollary 6.4 with this choice of η; without loss of generality we can normalize θ so that $\theta(\xi(\chi)) = \eta h_\chi$.

Let h be any integer such that $p^h \geq \#(\Lambda_n/\eta\Lambda_n)$ and $p^h \geq \#(\Lambda_n/h_\chi\Lambda_n)$, which is finite by Lemma 6.6. Fix $M = \#(C(\chi))p^{n+(k+1)h}$.

We will use Theorem 3.1 inductively to choose primes λ_i of F lying above l_i of $\mathbf{Q}$ for $1 \leq i \leq k+1$ satisfying:

$$(4) \qquad \lambda_i \in \mathfrak{c}_i, \qquad l_i \equiv 1 \pmod{M},$$

$$(5) \quad \bar{v}_{\lambda_1}(\kappa_{l_1}) = u_1\eta h_\chi, \qquad f_{i-1}\bar{v}_{\lambda_i}(\kappa_{r_i}) = u_i\eta\bar{v}_{\lambda_{i-1}}(\kappa_{r_{i-1}}) \qquad \text{for } 2 \leq i \leq k+1,$$

where $r_i = \prod_{j \leq i} l_j$ and $u_i \in (\mathbf{Z}/M\mathbf{Z})^\times$.

For the first step take $\mathfrak{c} = \mathfrak{c}_1$, $W = (E/E^M)(\chi)$, and

$$\psi\colon W \to \bar{E}(\chi)/\bar{E}(\chi)^M \xrightarrow{\theta} \Lambda_n/M\Lambda_n \xrightarrow{e(\chi)} e(\chi)(\mathbf{Z}/M\mathbf{Z})[\mathrm{Gal}(F/\mathbf{Q})].$$

Let λ_1 be any prime satisfying Theorem 3.1 with this data, and l_1 the rational prime below λ_1. Then (i) and (ii) of Theorem 3.1 imply (4). By Theorem 3.1(iii) and Proposition 2.4(ii), for some $u_1 \in (\mathbf{Z}/M\mathbf{Z})^\times$,

$$\begin{aligned}\bar{v}_{\lambda_1}(\kappa_{l_1})\lambda_1(\chi) = e(\chi)[\kappa_{l_1}]_{l_1} &= e(\chi)\varphi_{l_1}(\kappa_1) = u_1\psi(\kappa_1)\lambda_1(\chi)\\ &= u_1\theta(\xi(\chi))\lambda_1(\chi) = u_1\eta h_\chi\lambda_1(\chi).\end{aligned}$$

Since $(\mathcal{I}_{l_1}/M\mathcal{I}_{l_1})(\chi)$ is free over $\Lambda_n/M\Lambda_n$, generated by $\lambda_1(\chi)$, this proves (5) for $i = 1$.

Now suppose $2 \leq i \leq k+1$ and we have chosen $\lambda_1, \ldots, \lambda_{i-1}$ satisfying (4) and (5). We will define λ_i. Let $r_{i-1} = \prod_{j<i} l_j$. By (5), $\bar{v}_{\lambda_{i-1}}(\kappa_{r_{i-1}})$ divides $\eta^{i-1}h_\chi$, so

$$\#[(\mathcal{I}_{l_{i-1}}/M\mathcal{I}_{l_{i-1}})/\Lambda_n[\kappa_{r_{i-1}}]_{l_{i-1}}] \leq \#(\Lambda_n/\eta^{i-1}h_\chi\Lambda_n) \leq p^{ih}.$$

Let W_i be the Λ_n-submodule of $F^\times/(F^\times)^M$ generated by $e(\chi)\kappa_{r_{i-1}}$. Using Corollary 6.5 and Lemma 6.6, Lemma 7.1 (with $r = r_{i-1}$, $l = l_{i-1}$) yields a map $\psi_i\colon W_i \to \Lambda_n/M\Lambda_n$ such that

$$f_{i-1}\psi_i(e(\chi)\kappa_{r_{i-1}}) = \eta\bar{v}_{\lambda_{i-1}}(\kappa_{r_{i-1}}).$$

Now choose λ_i satisfying the conclusions of Theorem 3.1 with $\mathfrak{c} = \mathfrak{c}_i$, $W = W_i$, $\psi = e(\chi)\psi_i$, and M as above. Then (i) and (ii) of Theorem 3.1 imply (4) for i. By Proposition 2.4(ii) and Theorem 3.1(iii) there is a $u_i \in (\mathbf{Z}/M\mathbf{Z})^\times$ so that

$$\begin{aligned}f_{i-1}\bar{v}_{\lambda_i}(\kappa_{r_i})\lambda_i(\chi) = f_{i-1}e(\chi)[\kappa_{r_i}]_{l_i} &= f_{i-1}\varphi_{l_i}(e(\chi)\kappa_{r_{i-1}})\\ &= f_{i-1}u_i\psi_i(e(\chi)\kappa_{r_{i-1}})\lambda_i(\chi) = u_i\eta\bar{v}_{\lambda_{i-1}}(\kappa_{r_{i-1}})\lambda_i(\chi).\end{aligned}$$

This proves (5) for i.

Continue this induction process $k+1$ steps. Combining all of the relations (5) gives

$$\eta^{k+1}h_\chi = u\left(\prod_{i=1}^{k} f_i\right)\bar{\nu}_{\lambda_{k+1}}(\kappa_{r_{k+1}}) \quad \text{in } \Lambda_n/M\Lambda_n$$

for some $u \in (\mathbf{Z}/M\mathbf{Z})^\times$. Thus $f_\chi = \prod_{i=1}^{k} f_i$ divides $\eta^{k+1}h_\chi$ in $\Lambda_n/p^n\Lambda_n$. This holds for every n, so $f_\chi | \eta^{k+1}h_\chi$ in Λ.

To conclude the proof we need to remove the extra factor of η^{k+1}. Recall that $\mathscr{C}$ is an ideal of finite index in Λ and η is any element of $\mathscr{C}$ such that $\Lambda_n/\eta\Lambda_n$ is finite for every n. Thus we can choose η to be a power of p, and then use the Ferrero–Washington theorem (Theorem 2.3 of Chapter 10) which says that f_χ is not divisible by p. However, one can also work without using the Ferrero–Washington result, by verifying that it is always possible to make two choices of η which are relatively prime. Since Λ is a unique factorization domain, it follows that f_χ divides h_χ and the proof is complete.

Still writing $f_\chi = \text{char}(C_\infty(\chi))$ and $h_\chi = \text{char}(E_\infty(\chi)/V_\infty(\chi))$, let

$$f = \prod_\chi f_\chi \qquad \text{and} \qquad h = \prod_\chi h_\chi.$$

We will show $f\Lambda = h\Lambda$, and it will then follow from Theorem 7.2 that $f_\chi\Lambda = h_\chi\Lambda$ for every χ. The proof will use the analytic class number formula. If a_n, b_n are two sequences of positive integers we will write $a_n \sim b_n$ to mean that a_n/b_n is bounded above and below independently of n.

Lemma 7.3. *Suppose $g_1, g_2 \in \Lambda$, $g_1 | g_2$, and $\#(\Lambda/g_1\Lambda)_{\Gamma_n} \sim \#(\Lambda/g_2\Lambda)_{\Gamma_n}$. Then $g_1\Lambda = g_2\Lambda$.*

Proof. This is immediate from Theorem 1.2 of Chapter 5.

Proof of Theorem 5.1. By Theorem 1.2 of Chapter 5 and Lemma 6.6,

$$\#(\Lambda/f\Lambda)_{\Gamma_n} \sim \prod_\chi \#(\Lambda/f_\chi\Lambda)_{\Gamma_n} \sim \prod_\chi \#(C_n(\chi)) = \#(C_n),$$

$$\#(\Lambda/h\Lambda)_{\Gamma_n} \sim \prod_\chi \#(\Lambda/h_\chi\Lambda)_{\Gamma_n} \sim \prod_\chi [\bar{E}_n(\chi) : V_n(\chi)] = [\bar{E}_n : V_n].$$

But the analytic class number formula (Chapter 3, Theorem 5.1) together with Theorem 4.2 of Chapter 4 shows that $\#(C_n) = [\bar{E}_n : V_n]$. Therefore $(\Lambda/f\Lambda)_{\Gamma_n} \sim (\Lambda/h\Lambda)_{\Gamma_n}$. By Theorem 7.2, $f|h$ so by Lemma 7.3, $f\Lambda = h\Lambda$. Again using Theorem 7.2 we conclude that $f_\chi\Lambda = h_\chi\Lambda$ for all χ, i.e.

$$\text{char}(C_\infty(\chi)) = \text{char}(E_\infty(\chi)/V_\infty(\chi)).$$

§8. Other Formulations and Consequences of the Main Conjecture

Write $\omega_\infty: \Gamma \to 1 + p\mathbf{Z}_p$ for the character giving the action of Γ on $\boldsymbol{\mu}_{p^\infty}$ and ω for the Teichmuller character giving the action of Δ on $\boldsymbol{\mu}_p$. For every nontrivial even character χ of Δ let $g_\chi \in \Lambda$ be the p-adic L-function appearing in Theorem 5.2 of Chapter 7. Then g_χ is characterized by the values

$$\omega_\infty^k(g_\chi) = L_p(1-k, \chi) = -(1 - \chi\omega^{-k}(p)p^{k-1})\frac{1}{k}\mathbf{B}_{k,\chi\omega^{-k}} \tag{6}$$

for positive integers k, where we view $\chi\omega^{-k}$ as a primitive Dirichlet character. The following theorem is another form of Iwasawa's main conjecture; it is weaker than the consequence of Vandiver's conjecture mentioned at the end of Chapter 7.

Theorem 8.1. *For all nontrivial even characters χ of Δ,*

$$\mathrm{char}(X_\infty(\chi)) = g_\chi\Lambda.$$

Proof. By class field theory (see §5 of Chapter 5, especially Theorem 5.1), there is an exact sequence

$$0 \to U_\infty(\chi)/E_\infty(\chi) \to X_\infty(\chi) \to C_\infty(\chi) \to 0.$$

Also we clearly have

$$0 \to E_\infty(\chi)/V_\infty(\chi) \to U_\infty(\chi)/V_\infty(\chi) \to U_\infty(\chi)/E_\infty(\chi) \to 0.$$

Since the characteristic ideal is multiplicative in exact sequences,

$$\mathrm{char}(X_\infty(\chi)) = \mathrm{char}(U_\infty(\chi)/V_\infty(\chi)) = g_\chi\Lambda$$

by Theorem 5.1 of this Appendix and Theorem 5.2 of Chapter 7.

We now turn our attention to odd characters χ. Define $A_\infty = \varinjlim C_n$. If χ is any odd character of Δ, then $\chi^{-1}\omega$ is even and using the Kummer duality of Chapter 6 we can relate $A_\infty(\chi)$ with $X_\infty(\chi^{-1}\omega)$.

Theorem 8.2. *For every odd character χ of Δ, Kummer theory gives a nondegenerate pairing*

$$A_\infty(\chi) \times X_\infty(\chi^{-1}\omega) \to \boldsymbol{\mu}_{p^\infty}.$$

Proof. This is immediate from Theorems 2.2 and 2.3 of Chapter 6. The only thing to check is that with Ω_E defined as in §2 of Chapter 6,

$\mathrm{Gal}(\Omega_E/K_\infty)(\chi^{-1}\omega) = 0$. This can be seen easily by Kummer theory since $\chi^{-1}\omega$ is an even character. (Caution on notation: the C_∞ of Chapter 6 is what we are calling A_∞ in this Appendix.)

Define a "twisting" map $\iota: \Lambda \to \Lambda$ by $\iota(\gamma) = \omega_\infty(\gamma)\gamma$ for $\gamma \in \Gamma$.

Corollary 8.3. *For every odd character $\chi \neq \omega$ of Δ,*

$$\mathrm{char}(\mathrm{Hom}(A_\infty(\chi), \mathbf{Q}_p/\mathbf{Z}_p)) = \iota(g_{\chi^{-1}\omega})\Lambda.$$

Proof. Exercise, from Theorems 8.1 and 8.2.

Remark. By the theorem of Iwasawa [Iw 12] mentioned in Chapter 6, §2, $C_\infty(\chi)$ is quasi-isomorphic to $\mathrm{Hom}(A_\infty(\chi), \mathbf{Q}_p/\mathbf{Z}_p)$, so one can conclude from Corollary 8.3 the following result:

Theorem 8.4. *For every odd character $\chi \neq \omega$ of Δ,*

$$\mathrm{char}(C_\infty(\chi)) = \iota(g_{\chi^{-1}\omega})\Lambda.$$

This was the form of Iwasawa's main conjecture proved by Mazur and Wiles.

Remark. When $\chi = \omega$, $C_\infty(\chi) = 0$ by Corollary 2, §3 of Chapter 1.

As a final consequence of the main conjecture we wish to determine the orders of the groups $C_n(\chi)$ for odd characters χ. This is the analogue of Theorem 4.2 which gave the orders of $C_0(\chi)$ for even χ. Recall $\Gamma_n = \mathrm{Gal}(K_\infty/K_n)$.

Proposition 8.5. *For every odd character $\chi \neq \omega$ of Δ, $A_\infty(\chi)^{\Gamma_n} = C_n(\chi)$.*

Proof. Since χ is odd and $\chi \neq \omega$, $\bar{E}_n(\chi) = \{1\}$ (see Theorem 4.1 of Chapter 3). It follows by Theorem 4.3(i), Chapter 6, that the maps $C_n(\chi) \to C_m(\chi)$ are injective for all $m \geq n$, and so $C_n(\chi) \subset A_\infty(\chi)^{\Gamma_n}$. If γ denotes any generator of Γ_n we have an exact sequence

$$0 \longrightarrow C_m(\chi)^{\Gamma_n} \longrightarrow C_m(\chi) \xrightarrow{\gamma-1} C_m(\chi) \longrightarrow C_m(\chi)/(\gamma - 1)C_m(\chi) \longrightarrow 0.$$

By Theorem 4.1(iii) of Chapter 5, $C_m(\chi)/(\gamma - 1)C_m(\chi) \approx C_n(\chi)$ and we conclude that $\#(C_m(\chi)^{\Gamma_n}) = \#(C_n(\chi))$ for all $m \geq n$. Thus $\#(A_\infty(\chi)^{\Gamma_n}) = \#(C_n(\chi))$ and the proposition follows.

Lemma 8.6. *Suppose Y is a finitely-generated torsion Λ-module with no nonzero finite Λ-submodules, $\gamma \in \Gamma$, $a \in 1 + p\mathbf{Z}_p$ and $Y/(\gamma - a)Y$ is finite.*

Then

$$\#(Y/(\gamma - a)Y) = \#(\Lambda/(\mathrm{char}(Y), (\gamma - a)\Lambda)).$$

Proof. Fix a quasi-isomorphism $Y \to \bigoplus \Lambda/f_i\Lambda$ with finite cokernel Z. Since Y has no finite submodules the kernel must be trivial, and $\mathrm{char}(Y) = \prod f_i\Lambda$. We have an exact diagram

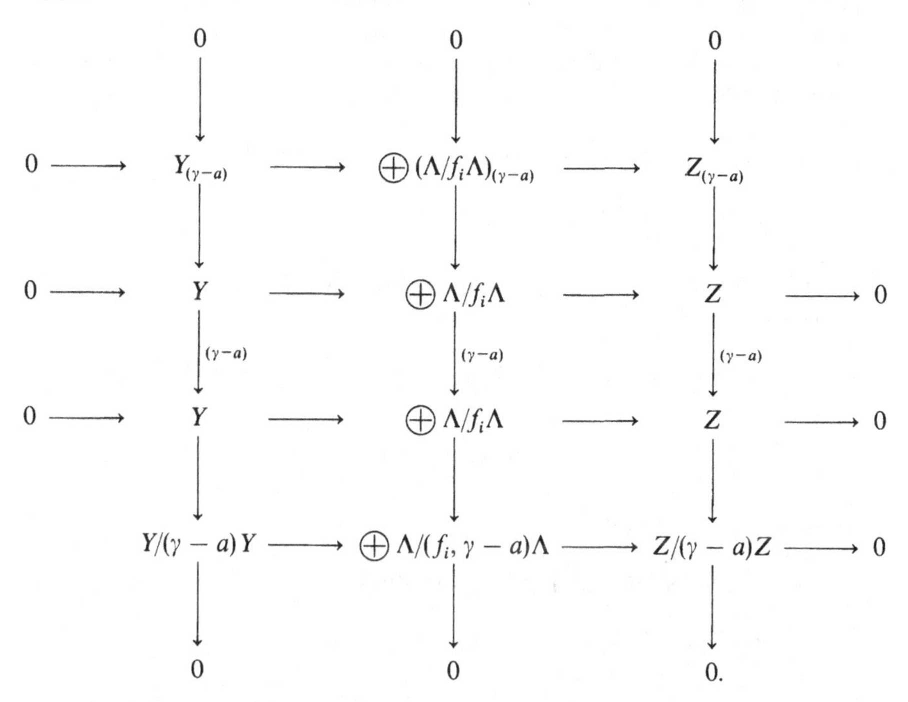

Here $Y_{(\gamma-a)}$ denotes the kernel of $(\gamma - a)$ on Y and similarly for $(\Lambda/f_i\Lambda)_{(\gamma-a)}$, $Z_{(\gamma-a)}$. We see

$$\mathrm{rank}_{\mathbf{Z}_p}\left(\bigoplus (\Lambda/f_i\Lambda)_{(\gamma-a)}\right) = \mathrm{rank}_{\mathbf{Z}_p}(Y_{(\gamma-a)}) = \mathrm{rank}_{\mathbf{Z}_p}(Y/(\gamma - a)Y) = 0$$

so each f_i is prime to $\gamma - a$ and $\bigoplus (\Lambda/f_i\Lambda)_{(\gamma-a)} = 0$. Also $\#(Z_{(\gamma-a)}) = \#(Z/(\gamma - a)Z)$ since Z is finite, so by the snake lemma

$$\#(Y/(\gamma - a)Y) = \#\left(\bigoplus \Lambda/(f_i, \gamma - a)\Lambda\right).$$

It is left as an exercise to show that

$$\#\left(\bigoplus \Lambda/(f_i, \gamma - a)\Lambda\right) = \#\left(\Lambda/\left(\prod f_i, \gamma - a\right)\Lambda\right).$$

(Hint: If $f, g \in \Lambda$ and f is prime to $\gamma - a$, then multiplication by f gives an isomorphism $\Lambda/(g, \gamma - a)\Lambda \approx (f, \gamma - a)\Lambda/(fg, \gamma - a)\Lambda$.)

The next result was proved by Iwasawa ([Iw 12], Theorem 18).

Lemma 8.7. *For every even character χ of Δ, $X_\infty(\chi)$ has no nonzero finite Λ-submodules.*

Proof. By Theorem 8.2 it will suffice to show that $A_\infty(\chi^{-1}\omega)$ has no proper Λ-submodules of finite index. Suppose $A \subset A_\infty(\chi^{-1}\omega)$ is stable under Γ and $\#(A_\infty(\chi^{-1}\omega)/A) = p^k$ is finite. Then for some sufficiently large N, $\mathrm{Gal}(K_\infty/K_N)$ acts trivially on $A_\infty(\chi^{-1}\omega)/A$. If $m \geq N$, since the norm map $\mathbf{N}_{K_{m+k}/K_m} : C_{m+k} \to C_m$ is surjective (Chapter 5, §4),

$$C_m(\chi^{-1}\omega) = \mathbf{N}_{K_{m+k}/K_m} C_{m+k}(\chi^{-1}\omega) \subset A.$$

Thus $A_\infty(\chi^{-1}\omega) \subset A$, which proves the lemma.

The following is a strengthening of Stickelberger's theorem, Theorem 3.1 of Chapter 1.

Theorem 8.8 (Mazur–Wiles, Kolyvagin). *For every odd character $\chi \neq \omega$ of Δ,*

$$\#(C_0(\chi)) = p^{m(\chi)},$$

where $m(\chi) = \mathrm{ord}_p(\mathbf{B}_{1,\chi^{-1}})$.

Proof. By Theorem 8.2 and Proposition 8.5,

$$C_0(\chi) = \mathrm{Hom}(X_\infty(\chi^{-1}\omega), \boldsymbol{\mu}_{p^\infty})^\Gamma = \mathrm{Hom}(X_\infty(\chi^{-1}\omega)/(\gamma - \omega_\infty(\gamma))X_\infty(\chi^{-1}\omega), \boldsymbol{\mu}_{p^\infty}),$$

where γ is a generator of Γ. Thus by Lemmas 8.6 and 8.7, Theorem 8.1, and (6),

$$\begin{aligned}
\#(C_0(\chi)) &= \#(X_\infty(\chi^{-1}\omega)/(\gamma - \omega_\infty(\gamma))X_\infty(\chi^{-1}\omega)) \\
&= \#(\Lambda/(g_{\chi^{-1}\omega}, \gamma - \omega_\infty(\gamma))\Lambda) \\
&= \#(\mathbf{Z}_p/\omega_\infty(g_{\chi^{-1}\omega})\mathbf{Z}_p) \\
&= \#(\mathbf{Z}_p/\mathbf{B}_{1,\chi^{-1}}\mathbf{Z}_p),
\end{aligned}$$

which is the desired equality.

Remark. Write $\mathscr{R}_n$ for the Stickelberger ideal in $\mathbf{Z}_p[\mathrm{Gal}(K_n/\mathbf{Q})]$ as defined in Chapter 2. Theorem 8.8 can be viewed as stating that for odd characters $\chi \neq \omega$,

$$\#(C_0(\chi)) = [\mathbf{Z}_p[\Delta](\chi) : \mathscr{R}_0(\chi)].$$

Without much extra difficulty one can use Theorems 8.1 and 8.2 and Proposition 8.5 above to show that

$$\#(C_n(\chi)) = [\mathbf{Z}_p[\mathrm{Gal}(K_n/\mathbf{Q})](\chi) : \mathscr{R}_n(\chi)].$$

Further, using results from Chapters 2 and 4 one can prove for all odd $\chi \neq \omega$ another result of Iwasawa

$$\varprojlim (\mathbf{Z}_p[\mathrm{Gal}(K_n/\mathbf{Q})]/\mathscr{R}_n)(\chi) \approx \Lambda/\imath(g_{\chi^{-1}\omega})\Lambda.$$

Together with Theorem 8.4 this gives one more equivalent version of the main conjecture:

Theorem 8.9. *For all odd characters $\chi \neq \omega$ of Δ,*

$$\mathrm{char}(C_\infty(\chi)) = \mathrm{char}\left(\Lambda/\varprojlim \mathscr{R}_n(\chi)\right).$$

Bibliography

[Am] Y. AMICE, Les nombres p-adiques, Presse Universitaire de France, 1975

[A–F] Y. AMICE and J. FRESNEL, Fonctions zeta p-adiques des corps de nombres abéliens reels, *Acta Arith.* **XX** (1972) pp. 353–384

[A–H] E. ARTIN and H. HASSE, Die beiden Erganzungssätze zum Reziprozitäts gesetz der l^n-ten Potenzreste im Korper der l^n-ten Einheitswurzeln, *Abh. Math. Sem.* Hamburg 6 (1928) pp. 146–162

[Ba] D. BARSKY, Transformation de Cauchy p-adique et algèbre d'Iwasawa, *Math. Ann.* **232** (1978) pp. 255–266

[Ba] H. BASS, Generators and relations for cyclotomic units, *Nagoya Math. J.* **27** (1966) pp. 401–407

[Bo] M. BOYARSKY, p-adic gamma functions and Dwork cohomology, *Trans. AMS*, **257**, No. 2 (1980) pp. 359–369

[Br] A. BRUMER, On the units of algebraic number fields, *Mathematika*, **14** (1967) pp. 121–124

[Ca] L. CARLITZ, A generalization of Maillet's determinant and a bound for the first factor of the class number, *Proc. AMS* **12** (1961) pp. 256–261

[Ca–O] L. CARLITZ and F. R. OLSON, Maillet's determinant, *Proc. AMS* **6** (1955) pp. 265–269

[Co 1] J. COATES, On K_2 and some classical conjectures in algebraic number theory, *Ann. of Math.* **95** (1972) pp. 99–116

[Co 2] J. COATES, K-theory and Iwasawa's analogue of the Jacobian, Algebraic K-theory II, Springer *Lecture Notes* **342** (1973) pp. 502–520

[Co 3] J. COATES, p-adic L-functions and Iwasawa's theory, Durham conference on algebraic number theory and class field theory, 1976

[Co 4] J. COATES, Fonctions zeta partielles d'un corps de nombres totalement réel, Seminaire Delange–Pisot–Poitou, 1974–75

[C–L] J. COATES and S. LICHTENBAUM, On l-adic zeta functions, *Ann. of Math.* **98** (1973) pp. 498–550

[C–S 1] J. COATES and W. SINNOTT, On p-adic L-functions over real quadratic fields, *Invent. Math.* **25** (1974) pp. 253–279

[C–S 2] J. COATES and W. SINNOTT, An analogue of Stickelberger's theorem for higher K-groups, *Invent. Math.* **24** (1974) pp. 149–161

[C–S 3] J. COATES and W. SINNOTT, Integrality properties of the values of partial zeta functions, *Proc. London Math. Soc.* **1977** pp. 365–384

[C–W 1] J. COATES and A. WILES, Explicit Reciprocity laws, Proceedings, Conference at Caen, *Soc. Math. France Astrisque* **41–42** (1977) pp. 7–17

[C–W 2] J. COATES and A. WILES, On the conjecture of Birch and Swinnerton-Dyer, *Invent. Math.* **39** (1977) pp. 223–251

[C–W 3] J. COATES and A. WILES, Kummer's criterion for Hurwitz numbers, Kyoto Conference on Algebra Number Theory, 1977

[C–W 4] J. COATES and A. WILES, On the conjecture of Birch–Swinnerton-Dyer II, to appear

[Col] R. COLEMAN, Division values in local fields, *Invent. Math.* **53** (1979) pp. 91–116

[D–H] H. DAVENPORT and H. HASSE, Die Nullstellen der Kongruenz-zetafunktionen in gewissen zyklischen Fällen, *J. reine angew. Math.* **172** (1935) pp. 151–182

[Di 1] J. DIAMOND, The p-adic log gamma function and p-adic Euler constant, *Trans. Am. Math. Soc.* **233** (1977) pp. 321–337

[Di 2] J. DIAMOND, The p-adic gamma measures, *Proc. AMS* **75** (1979) p. 211–217.

[Di 3] J. DIAMOND, On the values of p-adic L-functions at positive integers, *Acta Aithm XXXV* (1979) pp. 223–237

[Dw 1] B. DWORK, On the zeta function of a hypersurface, *Publ. Math. IHES* **12** (1962) pp. 5–68

[Dw 2] B. DWORK, A deformation theory for the zeta function of a hypersurface, *Proc. Int. Cong. Math.* Stockholm, 1962

[Dw 3] B. DWORK, On the zeta function of a hypersurface II, *Ann. of Math.* **80** (1964) pp. 227–299

[Dw 4] B. DWORK, p-adic cycles, *Publ. Math. IHES* **37** (1969) pp. 28–115

[Dw 5] B. DWORK, Bessel functions as p-adic functions of the argument, *Duke Math. J.* **41** No. 4 (1974) pp. 711–738

[Dw–Ro] B. DWORK and P. ROBBA, On ordinary linear p-adic differential equations, *Trans. AMS* Vol. **231** (1977) pp. 1–46

[En 1] V. ENNOLA, On relations between cyclotomic units, *J. Number Theory* **4** (1972) pp. 236–247

[En 2] V. ENNOLA, Some particular relations between cyclotomic units, *Ann. Univ. Turkuensis* **147** (1971)

[Fe 1] B. FERRERO, Iwasawa invariants of abelian number fields, *Math. Ann.* **234** (1978) pp. 9–24

[Fe 2] B. FERRERO, An explicit bound for Iwasawa's λ-invariant, *Acta Arithm.* **XXXIII** (1977) pp. 405–408

[Fe–Gr] B. FERRERO and R. GREENBERG, On the behavior of p-adic L-functions at $s = 0$, *Invent. Math.* **50** (1978–79) pp. 91–102

[Fe-W] B. FERRERO and L. WASHINGTON, The Iwasawa invariant μ_p vanishes for abelian number fields, *Ann. Math.* **109** (1979) pp. 377-395

[Fre] J. FRESNEL, Nombres de Bernoulli et fonctions L p-adiques, *Ann. Inst. Fourier* **17** (1967) pp. 281-333

[Fro] A. FROHLICH, Formal groups, Springer *Lectures Notes in Mathematics* **74**, 1968

[Gi 1] R. GILLARD, Unités cyclotomiques, unités semi-locales et $\mathbf{Z}_l$-extensions, *Ann. Inst. Fourier* **29** (1979) pp. 49-79

[Gi 2] R. GILLARD, Extensions abéliennes et répartition modulo 1, *Journées arithmétiques de Marseille*, Asterisque, 1979.

[Gr 1] R. GREENBERG, A generalization of Kummer's criterion, *Invent. Math.* **21** (1973) pp. 247-254

[Gr 2] R. GREENBERG, On a certain l-adic representation, *Invent. Math.* **21** (1973) pp. 117-124

[Gr 3] R. GREENBERG, The Iwasawa invariants of Γ-extensions of a fixed number field, *Amer. J. Math.* **95** (1973) pp. 204-214

[Gr 4] R. GREENBERG, On the Iwasawa invariants of totally real number fields, *Amer. J. Math.* **93** (1976) pp. 263-284

[Gr 5] R. GREENBERG, On p-adic L-functions and cyclotomic fields, *Nagoya Math. J.* **56** (1974) pp. 61-77

[Gr 6] R. GREENBERG, On p-adic L-functions and cyclotomic fields II, *Nagoya Math. J.* **67** (1977) pp. 139-158

[Gr 7] R. GREENBERG, A note on K_2 and the theory of Z_p-extensions, *Am. J. Math.* **100** (1978) pp. 1235-1245

[Gr] B. GROSS, p-adic L-series at $s = 0$, *J. Fac. Sci. Univ. Tokyo* (Sec IA) **28** (1982) pp. 979-994

[Gr-Ko] B. GROSS and N. KOBLITZ, Gauss sums and the p-adic gamma function, *Ann. of Math.* **109** (1979) pp. 569-581

[Ha 1] H. HASSE, Uber die Klassenzahl abelschen Zahlkorper, Akademie Verlag, Berlin, 1952

[Ha 2] H. HASSE, Bericht...., Teil II, Reziprozitätsgesetz, Reprinted, Physica Verlag, Würzburg, Wien, 1965

[Ha 3] H. HASSE, Theorie der relativ zyklischen algebraischen Funktionenkörper insbesonderen bei endlichen Konstantenkörper, *J. reine angew. Math.* **172** (1935) pp. 37-54

[Ho] T. HONDA, On the formal structure of the Jacobian variety of the Fermat curve over a p-adic integer ring, Symposia Mathematica Vol. **XI**, Academic Press, NY, 1973, pp. 271-284

[Hu-V] L. K. HUA and H. S. VANDIVER, *Proc. Nat. Acad. Sci. USA* **34** (1948) pp. 258-263

[Iw 1] K. IWASAWA, On Γ-extensions of algebraic number fields, *Bull. Amer. Math. Soc.* **65** (1959) pp. 183-226

[Iw 2] K. IWASAWA, A note on the group of units of an algebraic number field, *J. Math. pures et app.* **35** (1956) pp. 189-192

[Iw 3] K. IWASAWA, Sheaves for algebraic number fields, *Ann. of Math.* **69** (1959) pp. 408-413

[Iw 4] K. IWASAWA, On some properties of Γ-finite modules, *Ann. of Math.* **70** (1959) pp. 291–312

[Iw 5] K. IWASAWA, On the theory of cyclotomic fields, *Ann. of Math.* **70** (1959) pp. 530–561

[Iw 6] K. IWASAWA, On some invariants of cyclotomic fields, *Amer. J. Math.* **80** (1958) pp. 773–783

[Iw 7] K. IWASAWA, A class number formula for cyclotomic fields, *Ann. of Math.* **76** (1962) pp. 171–179

[Iw 8] K. IWASAWA, On some modules in the theory of cyclotomic fields, *J. Math. Soc. Japan* Vol. 16, No. 1 (1964) pp. 42–82

[Iw 9] K. IWASAWA, Some results in the theory of cyclotomic fields, *Symposia in Pure Math.* Vol. VIII, AMS 1965, pp. 66–69

[Iw 10] K. IWASAWA, On explicit formulas for the norm residue symbol, *J. Math. Soc. Japan* Vol. 20, Nos. 1–2 (1968) pp. 151–165

[Iw 11] K. IWASAWA, On p-adic L-functions, *Ann. of Math.* **89** (1969) pp. 198–205

[Iw 12] K. IWASAWA, On Z_l-extensions of algebraic number fields, *Ann. of Math* **98** (1973) pp. 246–326

[Iw 13] K. IWASAWA, A note of cyclotomic fields, *Invent. Math.* **36** (1976) pp. 115–123

[Iw 14] K. IWASAWA, Lectures on p-adic L-functions, *Ann. of Math. Studies* No. 74

[Iw 15] K. IWASAWA, On the μ-invariants of Z_l-extensions, Conference on number theory, algebraic geometry, and commutative algebra in honor of Y. Akizuki, Tokyo (1973) pp. 1–11

[Iw 16] K. IWASAWA, Some remarks on Hecke characters, *Algebraic Number Theory*, Kyoto International Symposium, 1976, Japan Society for Promotion of Science, Tokyo 1977, pp. 99–108

[Iw 17] K. IWASAWA, A note of Jacobi sums, *Symposia Mathematica XV*, (1975) pp. 447–459

[Iw 18] K. IWASAWA, Analogies between number fields and function fields, see *Math. Reviews* Vol. **41**, No. 172

[Jo] W. JOHNSON, Irregular primes and cyclotomic invariants, *Math. Comp.* **29** (1975) pp. 113–120

[Ka 1] N. KATZ, Formal groups and p-adic interpolation, *Journées Arithmétiques de Caen, Astérisque* **41–42** (1976) pp. 55–65

[Ka 2] N. KATZ, On the differential equations satisfied by period matrices, *Publ. Math. IHES* **35** (1968) pp. 71–106

[Ka 3] N. KATZ, Une formule de congruence pour la fonction zeta, SGA 7 II, Expose 22 p. 401, Springer *Lecture Notes* **340**

[Ka 4] N. KATZ, Another look at p-adic L-functions for totally real fields, *Math. Ann.* (1981) pp. 43–83

[Ka 5] N. KATZ, Divisibilities, congruences, and Cartier duality, *J. Fac. Sci. Univ. Tokyo* (Sec IA) **28** (1982) pp. 667–678

[Ko 1] N. KOBLITZ, Interpretation of the p-adic log gamma function and Euler constants using the Bernoulli measures, *Trans. AMS* (1978) pp. 261–269.

[Ko 2] N. KOBLITZ, A new proof of certain formulas for p-adic L-functions, *Duke J.* (1979) pp. 455–468

[Kol] V.A. KOLYVAGIN, Euler systems, to appear (translated by Neal Koblitz)

[Ku 1] D. KUBERT, A system of free generators for the universal even ordinary $\mathbf{Z}_{(2)}$ distribution on $\mathbf{Q}^{2k}/\mathbf{Z}^{2k}$, *Math. Ann.* **224** (1976) pp. 21–31

[Ku 2] D. KUBERT, The Universal ordinary distribution, *Bull. Soc. Math. France* **107** (1979) pp. 179–202

[KL 1] D. KUBERT and S. LANG, Units in the modular function field, I, Diophantine applications, *Math. Ann.* **218** (1975) pp. 67–96

[KL 2] D. KUBERT and S. LANG, Idem II, A full set of units, pp. 175–189

[KL 3] D. KUBERT and S. LANG, Idem III, Distribution relations, pp. 273–285

[KL 4] D. KUBERT and S. LANG, Idem IV, The Siegel functions are generators, *Math. Ann.* **227** (1977) pp. 223–242

[KL 5] D. KUBERT and S. LANG, Distributions on toroidal groups, *Math. Zeit.* (1976) pp. 33–51

[KL 6] D. KUBERT and S. LANG, The p-primary component of the cuspidal divisor class group on the modular curve $X(p)$, *Math. Ann.* **234** (1978) pp. 25–44

[KL 7] D. KUBERT and S. LANG, The index of Stickelberger ideals of order 2 and cuspidal class numbers, *Math. Ann.* **237** (1978) pp. 213–232

[KL 8] D. KUBERT and S. LANG, Stickelberger ideals, *Math. Ann.* **237** (1978) pp. 203–212

[KL 9] D. KUBERT and S. LANG, Iwasawa theory in the modular tower, *Math. Ann.* **237** (1978) pp. 97–104

[KL 10] D. KUBERT and S. LANG, Modular units inside cyclotomic units, *Bull. Soc. Math. France* **107** (1979) pp. 161–178

[KL 11] D. KUBERT and S. LANG, Cartan–Bernoulli numbers as values of L-series, *Math. Ann.* **240** (1979) pp. 21–26

[KL 12] D. KUBERT and S. LANG, *Modular Units*, Springer-Verlag, 1981

[Ku–L] T. KUBOTA and H. LEOPOLDT, Eine p-adische Theorie der Zetawerte, *J. reine angew. Math.* **214/215** (1964) pp. 328–339

[L 1] S. LANG, *Algebraic Number Theory*, Addison-Wesley, 1970

[L 2] S. LANG, *Elliptic functions*, Addison-Wesley, 1973

[L 3] S. LANG, *Introduction to modular forms*, Springer-Verlag, 1976

[L 4] S. LANG, *Elliptic curves: Diophantine analysis*, Springer-Verlag, 1978

[L 5] S. LANG, Units and class groups in number theory and algebraic geometry, *Bull. AMS* **6**, No. 3 (1982) pp. 253–316

[Leh] D. H. LEHMER, *Applications of digital computers*, in *Automation and Pure Mathematics*, Ginn, Boston (1963) pp. 219–231

[Le 1] H. W. LEOPOLDT, Zur Geschlechtertheorie in abelschen Zahlkörpern, *Math. Nach.* **9**, 6 (1953) pp. 351–363

[Le 2] H. W. LEOPOLDT, Uber Einheitengruppe und Klassenzahl reeller Zahlkörper, Abh. Deutschen Akad. Wiss. Berlin, Akademie Verlag, Verlag, Berlin, 1954

[Le 3] H. W. LEOPOLDT, Über ein Fundamentalproblem der Theorie der Einheiten algebraischer Zahlkörper, Sitzungsbericht Bayerischen Akademie Wiss. (1956), pp. 41–48

[Le 4] H. W. LEOPOLDT, Eine Verallgemeinerung der Bernoullischen Zahlen, *Abh. Math. Sem. Hamburg* (1958) pp. 131–140

[Le 5] H. W. LEOPOLDT, Zur Struktur der l-Klassengruppe galoisscher Zahlkörper, *J. reine angew. Math.* (1958) pp. 165–174

[Le 6] H. W. LEOPOLDT, Uber Klassenzahlprimteiler reeller abelscher Zahlkörper, *Abh. Math. Sem. Hamburg* (1959) pp. 36–47

[Le 7] H. W. LEOPOLDT, Uber die Hauptordnung der ganzen Elemente eines abelschen Zahlkörpers, *J. reine angew. Math.* **201** (1959) pp. 113–118

[Le 8] H. W. LEOPOLDT, Uber Fermatquotienten von Kreiseinheiten und Klassenzahlformeln modulo p, *Rend. Circ. Mat. Palermo* (1960) pp. 1–12

[Le 9] H. W. LEOPOLDT, Zur approximation des p-adischen Logarithmus, *Abh. Math. Sem. Hamburg* **25** (1961) pp. 77–81

[Le 10] H. W. LEOPOLDT, Zur Arithmetik in abelschen Zahlkörpern, *J. reine angew. Math.* **209** (1962) pp. 54–71

[Le 11] H. W. LEOPOLDT, Eine p-adische Theorie der Zetawerte II, *J. reine angew Math.* **274–275** (1975) pp. 224–239

[Li 1] S. LICHTENBAUM, Values of zeta functions, étale cohomology, and algebraic K-theory, Algebraic K-Theory II, Springer-Verlag *Lecture Notes in Mathematics* **342** (1973) pp. 489–501

[Li 2] S. LICHTENBAUM, Values of zeta and L-functions at zero, *Soc. Math. France Asterisque* **24–25** (1975)

[Li 3] S. LICHTENBAUM, On p-adic L-functions associated to elliptic curves, to appear

[Lu] J. LUBIN, One parameter formal Lie groups over p-adic integer rings, *Ann. of Math.* **80** (1964) pp. 464–484

[L–T] J. LUBIN and J. TATE, Formal complex multiplication in local fields, *Ann. of Math.* **8** (1965) pp. 380–387

[Man] J. MANIN, Cyclotomic fields and modular curves, *Russian Math. Surveys* Vol. 26, No. 6, Nov-Dec 1971, pp. 7–78

[Mas 1] J. MASLEY, On Euclidean rings of integers in cyclotomic fields, *J. reine angew. Math.* **272** (1975) pp. 45–48

[Mas 2] J. MASLEY, Solution of the class number two problem for cyclotomic fields, *Invent. Math.* **28** (1975) pp. 243–244

[M–M] J. MASLEY and H. MONTGOMERY, Cyclotomic fields with unique factorization, *J. reine angew. Math.* **286** (1976) pp. 248–256

[Maz] B. MAZUR, Analyse p-adique, Bourbaki report, 1972

[M–SwD] B. MAZUR and H. SWINNERTON-DYER, Arithmetic of Weil curves, *Invent. Math.* **18** (1972) pp. 183–266

[M–W] B. MAZUR and A. WILES, Class fields of abelian extensions of **Q**, *Invent. Math.* **76** (1984) pp. 179–330

[Me 1] T. METSÄNKYLÄ, Class numbers and μ-invariants of cyclotomic fields, *Proc. Am. Math. Soc* **43** No. 2 (1974) pp. 299–300

[Me 2] T. METSÄNKYLÄ, On the growth of the first factor of the cyclotomic class number, *Ann. Univ. Turku* Ser A1 **155** (1972)

[Mi] J. MILNOR, Introduction to algebraic K-theory, *Ann. of Math. Studies* **72** (1971)

[Mi] H. MITCHELL, The generalized Jacobi-Kummer function, *Transactions AMS* (1916) pp. 165-177

[Mo] Y. MORITA, A p-adic analogue of the Γ-function, *J. Fac. Sci. Tokyo* Section 1A Vol. **22** (1975) pp. 255-266

[No.] A. P. NOVIKOV, Sur le nombre de classes des extensions abeliennes d'un corps quadratique imaginaire, *Izv. Akad. Nauk SSSR* **31** (1967) pp. 717-726

[Oe] J. OESTERLE, Bourbaki report on Ferrero-Washington, Bourbaki Seminar, February 1979

[Pol] F. POLLACZEK, Uber die irregulären Kreiskorper der l-ten ond l^2-ten Einheitswürzeln, *Math. Zeit.* **21** (1924) pp. 1-38

[Qu 1] D. QUILLEN, Finite generation of the groups K_l of rings of algebraic integers, Algebraic K-Theory 1, Springer *Lecture Notes* **341** (1973) pp. 179-198

[Qu 2] D. QUILLEN, Higher algebraic K-theory I, in Algebraic K-theory I, Springer *Lecture Notes in Mathematics* **341** (1973) pp. 85-147

[Ra] K. RAMACHANDRA, On the units of cyclotomic fields, *Acta Arith.* **12** (1966) pp. 165-173

[Ri] K. RIBET, A modular construction of unramified p-extensions of $\mathbf{Q}(\mu_p)$, *Invent. Math.* **34** (1976) pp. 151-162

[Ro] G. ROBERT, Nombres de Hurwitz et unités élliptiques, *Ann. scient. Ec. Norm. Sup. 4e série* t. 11 (1978) pp. 297-389

[Ru] K. RUBIN, Global units and ideal class groups, *Invent. Math.* **89** (1987) pp. 511-526

[S-T] A. SCHOLZ and O. TAUSSKY, Die Hauptideale der kubischen Klassenkörper imaginär-quadratischer Zahlkörper, *J. reine angew. Math.* **171** (1934) pp. 19-41

[Se 1] J.-P. SERRE, Classes des corps cyclotomiques, d'après Iwasawa, Seminaire Bourbaki, 1958

[Se 2] J.-P. SERRE, Formes modulaires et fonctions zeta p-adiques, Modular functions in one variable III, Springer *Lecture Notes* **350** (1973)

[Se 3] J.-P. SERRE, Endomorphismes complètement continus des espaces de Banach p-adiques, *Pub. Math.* IHES **12** (1962) pp. 69-85

[Se 4] J.-P. SERRE, Sur le résidu de la fonction zeta p-adique d'un corps de nombres, *C. R. Acad. Sci. France* t. **287** (1978) pp. 183-188

[Sh] T. SHINTANI, On evaluation of zeta functions of totally real algebraic number fields at non-positive integers, *J. Fac. Sci. Univ. Tokyo* IA Vol. 23, No. 2 (1976) pp. 393-417

[Si 1] C. L. SIEGEL, Uber die Fourierschen Koeffizienten von Modulformen, *Göttingen Nachrichten* **3** (1970) pp. 15-56

[Si 2] C. L. SIEGEL, Zu zwei Bemerkungen Kummers, *Nachr. Akad. Wiss. Gottingen* **6** (1964) pp. 51-57

[Sin 1] W. SINNOTT, On the Stickelberger ideal and the circular units. *Annals of Mathematics* **198** (1978) pp. 107-134

[Sin 2] W. SINNOTT, On the μ-invariant of the Γ-transform of a rational function, *Invent. Math.* **75** (1984) pp. 273-282

[St] H. STARK, L-functions at $s = 1$, III: Totally real fields and Hilbert's twelfth problem, *Adv. Math.* Vol. 22, No. 1 (1976) pp. 64–84

[Ta 1] J. TATE, Letter to Iwasawa on a relation between K_2 and Galois cohomology, in Algebraic K-Theory II, Springer *Lecture Notes* **342** (1973) pp. 524–527

[Ta 2] J. TATE, Relations between K_2 and Galois cohomology, *Invent. Math.* **36** (1976) pp. 257–274

[Ta 3] J. TATE, Symbols in arithmetic, *Actes Congrès Intern. Math.* 1970, Tome 1, pp. 201–211

[Th] F. THAINE, On the ideal class groups of real abelian number fields, *Ann. of Math.* **128** (1988) pp. 1–18

[Va 1] H. S. VANDIVER, Fermat's last theorem and the second factor in the cyclotomic class number, *Bull. AMS* **40** (1934) pp. 118–126

[Va 2] H. S. VANDIVER, Fermat's last theorem, *Am. Math. Monthly* **53** (1946) pp. 555–576

[Wag] S. WAGSTAFF, The irregular primes to 125,000, *Math. Comp.* **32**, (1978) pp. 583–591

[Wa 1] L. WASHINGTON, Class numbers of elliptic function fields and the distribution of prime numbers, *Acta Arith.* **XXVII** (1975) pp. 111–114

[Wa 2] L. WASHINGTON, Class numbers and $\mathbf{Z}_p$-extensions, *Math. Ann.* **214** (1975) pp. 177–193

[Wa 3] L. WASHINGTON, A note on p-adic L-functions, *J. Number Theory* **8** Vol. 2 (1976) pp. 245–250

[Wa 4] L. WASHINGTON, On Fermat's last theorem, *J. reine angew. Math.* **289** (1977) pp. 115–117

[Wa 5] L. WASHINGTON, Units of irregular cyclotomic fields, *Ill. J. Math.* (1979) pp. 635–647

[Wa 6] L. WASHINGTON, The class number of the field of 5^nth roots of unity, *Proc. AMS*, **61** 2 (1976) pp. 205–208

[Wa 7] L. WASHINGTON, The calculation of $L_p(1, \chi)$, *J. Number Theory*, **9** (1977) pp. 175–178

[Wa 8] L. WASHINGTON, Euler factors for p-adic L-functions, *Mathematika* **25** (1978) pp. 68–75

[Wa 9] L. WASHINGTON, The non-p part of the class number in a cyclotomic $\mathbf{Z}_p$ extension, *Inv. Math.* **49** (1979), pp. 87–97

[Wa 11] L. WASHINGTON, Kummer's calculation of $L_p(1, \chi)$, *J. reine angew. Math.* **305** (1979) pp. 1–8

[Wa 12] L. WASHINGTON, The derivative of p-adic L-functions, *Acta Arithm.* **40** (1981–82) pp. 109–115

[We 1] A. WEIL, Number of solutions of equations in finite fields, *Bull. AMS* **55** (1949) pp. 497–508

[We 2] A. WEIL, Jacobi sums as Grossencharaktere, *Trans. AMS* **73** (1952) pp. 487–495

[We 3] A. WEIL, Sommes de Jacobi et caracteres de Hecke, *Gött. Nach.* (1974) pp. 1–14

[We 4] A. WEIL, On some exponential sums, *Proc. Nat. Acad. Sci. USA* **34**, No. 5 (1948) pp. 204–207

[Wi] A. Wiles, Higher explicit reciprocity laws, *Ann. of Math.* 107 (1978) pp. 235–254

[Ya 1] K. Yamamoto, The gap group of multiplicative relationships of Gaussian sums, *Symposia Mathematica* No. 15, (1975) pp. 427–440

[Ya 2] K. Yamamoto, On a conjecture of Hasse concerning multiplicative relations of Gaussian sums, *J. Combin. Theory* **1** (1966) pp. 476–489

Index

P

Q

R

S

T

U

V

W

Z

Graduate Texts in Mathematics

continued from page ii

48 SACHS/WU. General Relativity for Mathematicians.
49 GRUENBERG/WEIR. Linear Geometry. 2nd ed.
50 EDWARDS. Fermat's Last Theorem.
51 KLINGENBERG. A Course in Differential Geometry.
52 HARTSHORNE. Algebraic Geometry.
53 MANIN. A Course in Mathematical Logic.
54 GRAVER/WATKINS. Combinatorics with Emphasis on the Theory of Graphs.
55 BROWN/PEARCY. Introduction to Operator Theory I: Elements of Functional Analysis.
56 MASSEY. Algebraic Topology: An Introduction.
57 CROWELL/FOX. Introduction to Knot Theory.
58 KOBLITZ. p-adic Numbers, p-adic Analysis, and Zeta-Functions. 2nd ed.
59 LANG. Cyclotomic Fields.
60 ARNOLD. Mathematical Methods in Classical Mechanics. 2nd ed.
61 WHITEHEAD. Elements of Homotopy Theory.
62 KARGAPOLOV/MERZIJAKOV. Fundamentals of the Theory of Groups.
63 BOLLABÁS. Graph Theory.
64 EDWARDS. Fourier Series. Vol. I. 2nd ed.
65 WELLS. Differential Analysis on Complex Manifolds. 2nd ed.
66 WATERHOUSE. Introduction to Affine Group Schemes.
67 SERRE. Local Fields.
68 WEIDMANN. Linear Operators in Hilbert Spaces.
69 LANG. Cyclotomic Fields II.
70 MASSEY. Singular Homology Theory.
71 FARKAS/KRA. Riemann Surfaces.
72 STILLWELL. Classical Topology and Combinatorial Group Theory.
73 HUNGERFORD. Algebra.
74 DAVENPORT. Multiplicative Number Theory. 2nd ed.
75 HOCHSCHILD. Basic Theory of Algebraic Groups and Lie Algebras.
76 IITAKA. Algebraic Geometry.
77 HECKE. Lectures on the Theory of Algebraic Numbers.
78 BURRIS/SANKAPPANAVAR. A Course in Universal Algebra.
79 WALTERS. An Introduction to Ergodic Theory.
80 ROBINSON. A Course in the Theory of Groups.
81 FORSTER. Lectures on Riemann Surfaces.
82 BOTT/TU. Differential Forms in Algebraic Topology.
83 WASHINGTON. Introduction to Cyclotomic Fields.
84 IRELAND/ROSEN. A Classical Introduction to Modern Number Theory.
85 EDWARDS. Fourier Series: Vol. II. 2nd ed.
86 VAN LINT. Introduction to Coding Theory.
87 BROWN. Cohomology of Groups.
88 PIERCE. Associative Algebras.
89 LANG. Introduction to Algrebraic and Abelian Functions. 2nd ed.
90 BRØNDSTED. An Introduction to Convex Polytopes.
91 BEARDON. On the Geometry of Discrete Groups.
92 DIESTEL. Sequences and Series in Banach Spaces.

Printed in the United States
By Bookmasters